CAMBRIDGE LIBRARY COLLECTION

Books of enduring scholarly value

Life Sciences

Until the nineteenth century, the various subjects now known as the life sciences were regarded either as arcane studies which had little impact on ordinary daily life, or as a genteel hobby for the leisured classes. The increasing academic rigour and systematisation brought to the study of botany, zoology and other disciplines, and their adoption in university curricula, are reflected in the books reissued in this series.

Memoir and Correspondence of the Late Sir James Edward Smith, M.D.

Originally published in 1832, this two-volume account of the life of Sir James Edward Smith (1759–1828) was posthumously compiled by his wife, Pleasance (1773–1877). Smith trained originally as a doctor, but his independent wealth enabled him to pursue botany. Hugely influenced by the work of Linnaeus, he benefited greatly from the purchase of the latter's library and herbarium in 1783, upon the advice of his friend, Sir Joseph Banks. He was highly regarded throughout Europe as a botanist, and in 1788 founded the Linnean Society. He published various botanical works, of which the most important was *The English Flora* (1824–8), and assisted in the publication of many more. His wife recounts his 'religious, social and scientific character' as well as his achievements, and Volume 2 includes correspondence from Alexander von Humboldt, and concludes with an appendix in which short papers by Smith present a variety of topics.

Memoir and Correspondence of the Late Sir James Edward Smith, M.D.

VOLUME 2

EDITED BY PLEASANCE SMITH

CAMBRIDGE UNIVERSITY PRESS

Cambridge, New York, Melbourne, Madrid, Cape Town,
Singapore, São Paolo, Delhi, Tokyo, Mexico City

Published in the United States of America by Cambridge University Press, New York

www.cambridge.org
Information on this title: www.cambridge.org/9781108037082

This edition first published 1832
This digitally printed version 2011

ISBN 978-1-108-03708-2 Paperback

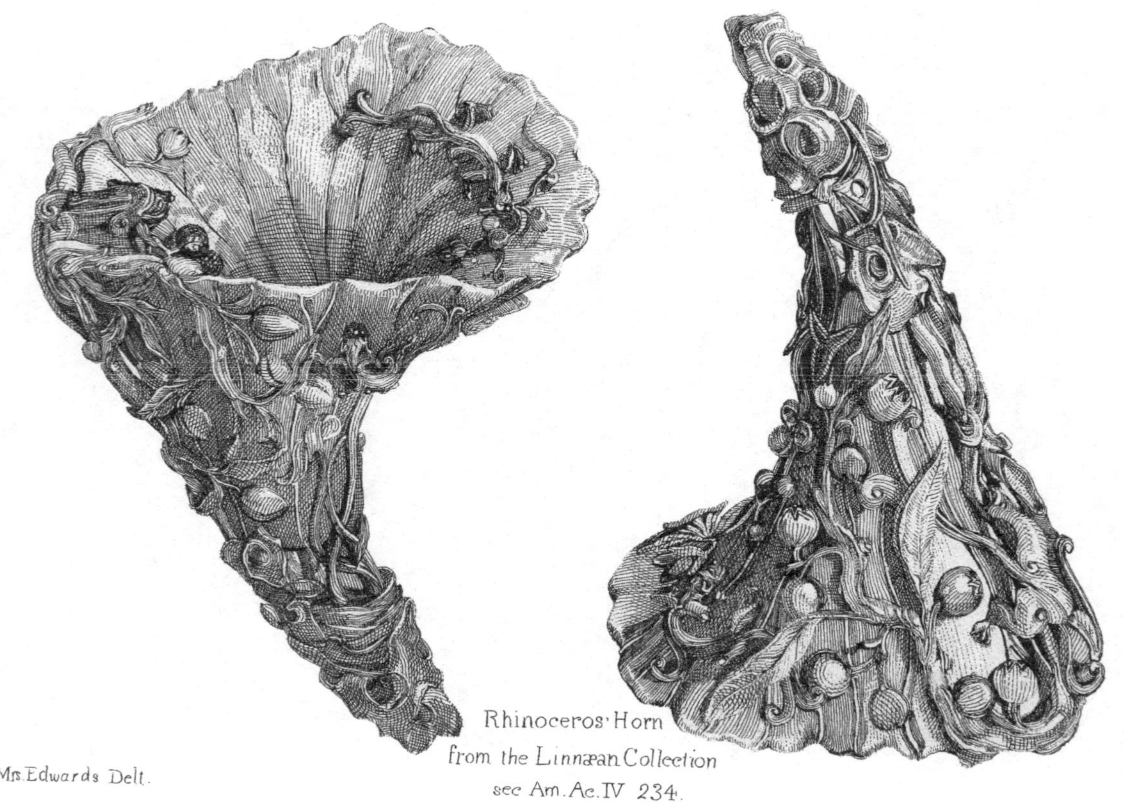

Rhinoceros' Horn
from the Linnæan Collection
see Am. Ac. IV 234.

Mrs. Edwards Delt.

MEMOIR

AND

CORRESPONDENCE

OF

THE LATE

SIR JAMES EDWARD SMITH, M.D.

FELLOW OF THE ROYAL SOCIETY OF LONDON;
MEMBER OF THE ACADEMIES OF
STOCKHOLM, UPSAL, TURIN, LISBON, PHILADELPHIA, NEW YORK, ETC. ETC.
THE IMPERIAL ACAD. NATURÆ CURIOSORUM,
AND
THE ROYAL ACADEMY OF SCIENCES AT PARIS;
HONORARY MEMBER OF THE HORTICULTURAL SOCIETY OF LONDON;
AND
PRESIDENT OF THE LINNÆAN SOCIETY.

EDITED
BY LADY SMITH.

"How delightful and how consolatory it is, among the disappointments and anxieties of life, to observe Science, like Virtue, retaining its relish to the last!"
Sketch of a Tour on the Continent, vol. ii. p. 60.

IN TWO VOLUMES.
VOL. II.

LONDON:
PRINTED FOR
LONGMAN, REES, ORME, BROWN, GREEN, AND LONGMAN,
PATERNOSTER ROW.

1832.

CONTENTS.

VOL. II.

CHAPTER XI.

CHAPTER XII.

CHAPTER XIII.

CHAPTER XIV.

MEMOIR

AND

CORRESPONDENCE

OF

SIR JAMES EDWARD SMITH.

CHAPTER VII.

Correspondence of Edmund Davall, Esq.,—Sir James Edward Smith,—and the Marchioness of Rockingham;—and two Letters from Professor Afzelius.

THE late Mr. Davall, of Orbe, was one whose fondness for natural science led him to cultivate an acquaintance with the subject of these pages, which soon settled into a warm personal affection on either side, and remained unimpaired through their lives.

An Englishman by birth, he was destined by circumstances to reside in Switzerland: but although he lived in a beautiful country, surrounded by objects most pleasing to him, yet he seems to suffer the pangs of an exile whenever he writes to his friend. The yearnings of desire to be among those who assimilate in pursuits, in intellectual and moral taste, cannot be more forcibly expressed than in the observation,—that even the grand scenes of nature, most precious in the estimation of pure and enthusiastic minds,—even these grow flat, stale and unprofitable, without the presence of a congenial friend to share in the enjoyments which they impart.

A sensibility which became morbid, affected his tender spirits; and he experienced towards England the true *maladie du pays* which the inhabitants of his adopted country feel when long absent from their native soil.

Mr. Davall had projected, and indeed made some progress in, a work on the plants of Switzerland, which he never completed. Ill-health, and a too anxious care for its being faultless, retarded, and finally stopped the publication of his accurate and ingenious labours.

A love of botany was the ruling passion of his mind, and was indeed but an effect of the adoration he paid to the beauty of creation in all its forms. Actuated by a pure love of nature, he was free from the restless passions of ambition or fame:—but the passions of others too often troubled his repose; and when he discovered the degrading traits of suspicion and reserve, where he reposed confidence, his ingenuous spirit was vexed and grieved more than it ought to have been, and he was led to charge himself with misanthropy, when all he felt was wounded love.

A little sketch of his history, which he had written to his friend sometime about the year 1795, will at once exhibit the sensitive disposition of this excellent man.

"Orbe, Canton de Berne, en Suisse.

"I never yet remained so long without writing to you, my dearest and best friend: you will not be surprised at this, when I shall have explained the cause of my silence. My health has been very unsettled, as it too generally is.—You know how ardent is my love of plants. Among the various contrarieties I have been obliged to bear, the want of sufficient room for my books and herbarium, &c. drove me from a very small study which I formerly occupied, and to which I was confined, because a contiguous room, which would have suited me, was appropriated to a much more important use,—the admittance of some card parties, when another room more usually frequented could not suffice. I was therefore obliged, during the latter part of the reign of these accursed cards, to take refuge in a less limited room on the ground-floor.—Not to be too prolix, and to come to the point, I discovered last summer, that in this position, from a certain degree of damp, though not very great, my herbarium has been in a great measure spoilt; a great number of specimens are become mouldy, and among these some of the most precious.

"You will pardon me, surely, if I unload in some degree my heart, by opening myself somewhat further to you, who give me such kind proofs of your friendship.

"It was just at the epoch of the death of my father that I began to have some vague notion of botany; that I perceived it was possible to acquire some knowledge of that enchanting study, by means of solitary application. I had from my cradle a latent germen, which unfortunately,—not having had the happiness of reckoning among my early acquaintance any one properly given to natural history,—had waited for direction till that moment. Having always been fond of gardening, I had bought Miller's Calendar, and it was there I first saw a sketch of the Linnæan system. I was suddenly inspired; and at this critical period resolved to pursue a study which I was certain would be productive of better happiness than any plan that might lead to pecuniary advantage.

"With my father I lost my home in my native country; for my mother, after this separation, was earnestly desirous of finishing her days in the land which gave her birth. Thus I became from the nature of my circumstances exiled as it were to this country.

"On my return here from my last visit to England, —which is surely the most memorable and the happiest period of my whole life, as it procured me not only your acquaintance but also your inestimable friendship, which I prefer to every other blessing that Providence could grant me,—my ardour became greater than ever; and for that very reason the perpetual card parties in the house I inhabit, the never-ending histories of Spadille and Manille of my good old aunts, became more irksome to me than before. I discontinued totally their *sociétés*, which I found quite intolerable; and the more so, as almost without a single exception the society of this little place is composed of that too numerous herd, who are constantly desirous of killing time they know not how to employ! and who express their esteem for persons of a studious disposition by—a shrug.

"Thus recluse, and little thinking how soon every one of these good women would be obliged to lay aside her cards (the last died literally with cards in her hands), I too often felt the want of a present friend.

"Having mentioned these several matters, which I could not resist communicating to you, I shall forget them for a while to turn to other subjects; for I have two letters from you!!—unanswered! and moreover I must talk a little botany, to convince you that no circumstances whatever can divert me totally from that delightful source of happiness, to which I shall have recourse till my last breath.

"I have something to tell you which will give you pleasure, affording a new proof how much certain good discoveries in botany are casual, and the effect of some happy moment.

"I went one day in August to my mountain Suchet. In going up the fir woods at the side, I learnt by the waving of the trees and the uncommon murmur, where I am accustomed to delightful "*horror ac silentium*", that a strong wind had arisen. When I got to the ridge with the intention of gaining the summit, I found the gale so strong, full against me from the N.E., that after tying my handkerchief over my hat and under my chin, every effort to get on was useless; when a leg was lifted, I was almost blown back;—you know by experience what is the power of wind in such positions: I therefore declined it.

The mountain to the N.W. is very abrupt; yet wishing to get to a somewhat less boisterous situation, I looked out for the first break in the rocks where it might be at all practicable to get down; and this I executed, not indeed without some difficulty; though I had still more to make myself followed by a boy, who carried my large vasculum, "*parcequ'il n'y avoit point de* sentier, *et qu'il n'avoit pas envie de casser le cou.*" Here I proceeded slowly, sometimes getting down a steep narrow pass, securing my progress by laying hold of branches of *Cratægus Chamæmespilus!* sometimes finding a narrow green place over the rocks, hence sliding myself down a crevice just sufficient to receive the body, to get at another shelf, &c. In one of the crevices, for the first time in the Jura I had the happiness of seeing *Cyathea montana,* in small quantity and so much advanced that I brought home but one specimen; as the only one I had left was very poor, having given my best to Dickson on my arrival in England, and one, not very good, to you. So now I have it in my reach, and, please God, you shall be properly supplied with characteristic specimens as soon as I can. It seems so rare, that I keep it to myself, and shall not mention it till we are supplied. In the name of Heaven reflect seriously, my dearest friend, that this mountain is so well within my reach! that I go quietly and easily to the summit, up one side and down the other, and get home by 7 or 8 p.m. Think you have a home, a good stock of books, &c. within its reach, and tell me whether this cannot once tempt you?—You might work at least at English Botany, and make many notes useful even for you here as in Marlborough-street. Moreover, by sleeping out one night I would make you mow *Saxifraga Hirculus*, gather *Carex leporina*, Linn., &c. When I think how possible, how easy the execution of all this might be, I am almost out of my senses for joy on thinking, hoping the time *will* come,—as much dejected with the fear of my passing to another world, before you can contrive to come here.

"*Orchis abortiva* and *coriophora* are within a moderate walk of my house!! But I have also the Alps daily before my eyes, and I know what might be done there. I wait for you, and then!!!—I WILL return. Had I but the certainty of seeing this part of my plan realized, I should bear with patience the circumstances which render impossible a visit to my native land."

It was to Mr. Davall that Sir James was indebted for an introduction to the Marchioness of Rockingham, who continued her friendly attentions and correspondence till her death in 1804. At her residence at Hillingdon he enjoyed many delightful visits, and heard many anecdotes of her illustrious husband.

The memory of this great statesman was cherished by her with a fondness that delighted in speaking of him to those she esteemed, and in connecting some reminiscence of him with every pursuit of her life.

The following letter from her, though a little

anticipated, as its date will show, shall be inserted first. It was written only two years before her death; but the share the noble Marquis has in it will excuse its being out of place.

The Marchioness of Rockingham to J. E. Smith.

Hillingdon House, May 21, 1802.

What shall I say to my good friend Dr. Smith, who is always showing me attentions which I am very unworthy of? But it is only *seeming;* for I am in reality always much pleased and gratified by the remembrance of those I esteem, though my untoward health will not often allow me to express it; and I am so little conscious of any neglect of the mind, that without hesitation I will beg that, when you come again to London, you will steal a day from your business to let me see you at this place; for it is a long time since I had that satisfaction. I had mislaid your letter, and have been hunting in my drawers all day; I have at last found it, and inexpressibly ashamed I am to see the date; and not only so, but to see that that very letter of November is complaining of my silence. I can only repeat the same, —that my conscience stands clear of ever forgetting a friend, or intentionally neglecting one: but I am free to acknowledge that I may give way too much to those incapacities that so frequently come upon me, even in the slender exertion of writing a letter. Poor Mr. Spragg would have said, *Oh! a nervous affection;* but my correspondents (I think) have greater reason to say, a nervous *dis*affection.—But what is to be said now! for thus far I had wrote, before this day brought me another very pleasing mild letter from Dr. Smith, who might justly reprove me sharply:—but I will still trust my cause to himself, who knows how *journaliere* my health has long been. You had congratulated me upon the blessing of peace; and I thought when I received your first letter, how soon I would answer it in the same sentiment, and say a great deal, the loss of which you have no cause to lament. I believe nobody's mind composes more epistles than mine, for it is always full of thoughts that I wish to utter to my particular friends; but they generally have the good luck to escape them. As you are coming to London, might I hope that you will let me see you before you leave it again?

I do not know in what manner you celebrate your Linnæan anniversary; nor did I know the day of the month till your letter mentioned it. How much it would gratify me, if a certain name could be without impropriety mentioned at that meeting! He might truly be called a rising botanist of great promise; for his mind had the happiness of taking a most comprehensive view of a subject at one glance, which seemed to require ages of investigation. He was a most attached disciple of Linnæus: his collection of plants and books was valuable and elegant, as far as it had gone; and the *24th of May* was *his birthday*. I do think that my friend Dr. Smith and (perhaps) Sir Joseph Banks would find a moment at their feast to drink a small bumper to the memory of this true brother in their science; as great a friend and encourager of science, and himself as near perfection in every line of life and character as human nature ever admitted of.

You will excuse this sally of zeal upon my observing that the 24*th of May* happened to be your *anniversary*, and will not treat it with any attention that may be inconvenient. It was an impulse of the heart that I could not restrain.

My dear Sir, I have many thanks to offer (too long postponed) for your kind present of excellent dried fruit and biscuits; the latter are, I perceive, recovering their peculiar taste and good quality—which I will now venture to say were quite gone. Those cruel times of real or pretended scarcity ruined every sort of eatables where flour was concerned; neither do I think that bread or any other flour compositions have yet been in, nor perhaps ever will return to, the same perfection as before.

The sermon you were so good as to send me is excellent; it must (as it were) have compelled a large collection;—in its address to the congregation*, I think it puts me in mind of something in Saurin. I venture to pay in your own coin, by sending you a sermon preached before many dignitaries of the Church at a visitation, by a great friend of mine, whom you have seen at Hillingdon, upon

* A Sermon preached at the Octagon Chapel, Norwich, August 30, 1801, for the Benefit of the Norfolk and Norwich Hospital, by the Rev. Pendlebury Houghton,—from the text: "I was sick, and ye visited me." Matth. ch. xxv. ver. 36.

a subject I think I may call the parent of yours, as piety and sound doctrine must of course produce charity and benevolence.

I have insensibly wrote two sheets of paper instead of one, and neither of them is very legible; for my poor eyes are far from recovered of that uneasy complaint in the lids:—reading or writing always occasions an unpleasant sensation to the eyes and across the forehead.

Dear Sir,

Your very sincere and obliged humble Servant,

M. ROCKINGHAM.

This letter and the preceding passage from one of Mr. Davall's later epistles will serve as introductions to the following correspondence, beginning sixteen years earlier than that of the Marchioness's here inserted.

The Marchioness of Rockingham to J. E. Smith.

Hillingdon House, Dec. 10, 1788.

The *Portlandia* flowers that I had the pleasure of sending to Dr. Smith yesterday were, I hope, moderately worthy of a place in an herbarium; but the first was so little so, that I was quite unwilling to send it. The evening coach yesterday brought me your packet, and I am quite at a loss how to express my thankfulness for the extreme obliging attention you have paid to my botanical library, in completing the works of Linnæus by so scarce a volume, which I should have despaired of obtaining; but I am ashamed that you should have taken the trouble of imitating the binding of my books. One out of the actual Linnæan library, though ever so old and musty, would have been sufficiently valuable; and it is greatly enhanced by your politeness in sparing it me. If any thing should happen to the remaining duplicate, I must beg that you will without scruple recall it back to its place. I am much obliged to you for the ingenious little tract you have favoured me with, and for the melon seeds. I shall be very happy to receive the little plant, whenever it is proper to be sent; and if you have any green-house or stove where any of mine would be acceptable, they will be extremely at your service.

My coming to town is always very precarious; but if I should come within a week or ten days, I shall be very glad to see your valuable collections, though with the most ignorant eye they ever were viewed with.

I beg you will give my sincere compliments to Mr. Davall, and assure him I am much concerned to find he has lost a valuable beloved parent; he does not mention how soon he shall be obliged to leave England. I shall be very glad if I have the opportunity of offering him my good wishes before his departure.

I am, Sir,

Your very obliged humble Servant,

M. Rockingham.

Mr. Davall to J. E. Smith.

Orbe, May 5, 1789:
Canton de Berne en Suisse.

My dear Sir,

For a considerable time past I had formed the project of sitting down to write to you. A journey to Berne, another to Lausanne, the illness and death of one of my aunts with its consequent employments, together with the care of my garden, had occasioned me to defer writing till I was most agreeably awakened by your kind letter.

If I may conclude from the inexpressible satisfaction and delight I experience at the sight of a letter from you,—from the portion of my thoughts which you occupy, from the anxiety concerning your health,—you are of all other beings the one who interests me most; and were botany quite out of the question, I think I could say as much from the high esteem your conversation and character have created in me.

While I think of it, let me tell you, that of late we have had flights of *Ampelis garrulus;* and that the other day a friend sent me the *Oriolus Galbula*, shot among twelve or fifteen. I examined an insect yesterday, which I make *Hister quadrimaculatus.* Is this rare?—I have a specimen, at this moment in press for you, of *Ribes alpinum*, &c. &c.

One of the most lovely plants of this country is coming into flower in my little garden, the *Astragalus monspessulanus.*

N.B. I have seen an old book in German, the title "Feuillée, Description of the Medicinal Plants of Chili and Peru, 1709: 2nd edit. Nurnberg, 1756, by George Leonard Huth." At vol. ii. p. 89. tab. 47. is a plant named *Thilco*; anne *Fuchsia coccinea?*

Sowerby is an admirable man, as he unites great excellence in engraving to considerable botanical intelligence.

The Collector of the Convent of the Great St. Bernard dined with us not long since, when I promised to go and see him in July, chiefly (on my honour) for you, my good friend, in hopes of your *Lichen cucullatus?* a small bit of which I think is in my Reynier's herbarium as *une variété developpée de* nivalis.

May Heaven preserve you for those who love and respect you, and for the improvement of botany!

Your devoted Friend,

E. Davall.

J. E. Smith to Mr. Davall.

My dear Sir, London, July 2, 1789.

I have so much to thank you for, and so many things to say to you, that I know not where to begin.—First, above all accept my thanks for your very friendly expressions, and be assured, as our excellent friend Dr. Goodenough says, "that there is no love lost between us." Now indeed I feel the value of botany, when it procures me such friends! Now for your letter of May 5th. *Hister quadrimaculatus* is excessively rare indeed; I have it, but pray take all you can.

The *Thilco* of Feuillée is quoted in Kew. Cat. for *Fuchsia coccinea*. I am not quite sure that it is the same species.

How you profane the cards of your good aunts to write botany upon them! What vile cards they are! No. 15 is true *Schœnus ferrugineus*, admirable! Send! send! send! "The horseleech (saith the Scripture) hath three daughters, crying Give! give! give!" I am your horseleech.

I am going to lose François, for a time at least; his mother and brother are dead, and his friends want him; he longs to see home, but promises to return. I shall certainly take him again, for I part from him with great regret. He sets out for Milan next week.

I go to Matlock this summer. At Kew the beautiful *Fuchsia* is multiplied without end, and Lee has plenty of it; it will soon be very common.

I wish I could come with Mandrot to see you: I live in hope, and am

Yours,

J. E. Smith.

The Marchioness of Rockingham to J. E. Smith.

Hillingdon, Sept. 19, 1789.

I was in hopes that Dr. Smith would have reached Wentworth before the great bustle of preparation began, that he might have had a quiet view of that charming place, and the many things worthy of inspection there; instead of which, I perceived from your letter that your arrival was in the midst of getting every thing in order for the Royal Guest, and that your politeness hurried you so quick away, that you seemed not to have even thought of the plants,—which was a pity, for your road to Wakefield being under the terrace wall, at the end of which are the conservatory and green-houses, you might have stepped in there without being in the way at all (though perhaps Henderson the gardener might be in a bustle like all the rest). I regret it the more, because I should so much have liked your account of the plants, which were all duplicates of the collection at Wimbledon (which I have here); and the finest plants were always sent to Wentworth: however, I hope whenever you pass that road again you will call and see all.

Without disfiguring the *Aralia* I could not send a specimen fine enough for a drawing; therefore I have had Sowerby here, and have also set him to make a drawing of a size he never did before, which is both extravagant and perhaps ridiculous; but really the plant has flowered in such magnitude, and is so very beautiful an object, that I could not resist having it done the size of life. I send you in the box one of the eight flowers, which are all at this moment in full perfection, two rising rather higher at the top, and the other six round them in the form of an umbrella. It is what hitherto has been called the *Pancratium amboinensis*, but I never had one flower before in this immense dimension (though very fine); the fragrance of it is prodigious, exactly like the *Cactus grandiflora*. The circumference of the umbrella which the *Pancratium* makes is a yard and three quarters; and the bulb twelve inches and a half. The leaves are not yet full out*.

My *Portlandia* and *Catesbæa* have been flowering a second time. Mr. Dryander plucked off the first flower upon the *Martinea*, which I perceive is likely to get a new name. I am sorry there is not another to send to you now, but there will be plenty by your return.

I am, &c.

M. Rockingham.

Mr. Davall to J. E. Smith.

Orbe, Oct. 27, 1789.

I must begin as you have done, my best and dearest friend:—I have so much to thank you for, and so many things to say to you, that I know not where to begin. Your friendly and most interesting letter of July the 2nd, which accompanied the parcel, by Mandrot's strange management did not arrive here till Friday evening, 23rd of October. He had brought me the fruits you were so kind as to send for Wyttenbach, the diplomas, &c., but had sent the more interesting packet the roundabout road with his merchandise. I delivered Van Berchem's diploma to him as I passed, at his house, on my way to the Great St. Bernard. This person gave me a complete farce on the occasion, which would occupy too much space in this sheet: thanked me, as the occasion of his being received! and, such is the ridiculous vanity of the people of

* Probably a variety of *P. caribbæum.*—J. E. S.

this Pays de Vaud, the father having apparently misconceived the account I gave him of the Institution, boasted at dinner, in presence of fourteen persons, that there were only four honorary members, and that his son had been chosen as one! They suffered no little *avanie* when I corrected the mistake, observing at the same time, that distinguished knowledge and rank united were essentially requisite.

I can hardly follow any method in the multiplicity of things I have to say to you. I sincerely thank Dr. Goodenough as well as yourself for the assurance that there is no love lost between you and me :—nothing can make me so perfectly happy as that assurance, nor anything more afflict me than if the contrary were the case. I am very well aware that to be on good terms with you is enviable, that I am highly interested in cultivating your good-will, that from your peculiar advantages and knowledge, the common run of men should court you; but, in the name of God, I can assure you that, were botany quite out of the question, I know not a man on earth whose friendship I so much desire, nor whose sentiments harmonize so perfectly with mine, as yours. I love you from my heart, and know you well enough to be certain of a return.

I have just received a cabinet on the plan of yours.

Bravo! I like to hear of your superb works with coloured plates,—and to be dedicated to the Marchioness of Rockingham: it were almost necessary, *entre nous*, to persuade her *Aralia* to bear some new kind of flower. But to be serious,—has she been in town? Pray when you see her present my compliments. Nothing can be more unpleasant than the difficulty of speedy and safe communication between this place and England. I have a tolerable drawing, I may say a good one, of *Micropus erectus*, but cannot find a conveyance to England. I should have more frequent opportunities were I at Jamaica.

You give me such longings in telling me the New Holland seeds rise at Lee's, that I am melancholy when I feel the impossibility of beholding them. The peasants here have a proverb, "*Où la chèvre est attachée il faut qu'elle broute:*" so I must *broute* my neighbouring *Pyrolas*, and not think of New Holland fruits,—*Proteas*, *Rousseas!* *Thouinias!* Yet let me not depart this life without being able to show a *Smithia* in my little herbarium. La Chenal writes to nobody.

I have been looking out for your *Lichen cucullatus?* and I believe have got it; yet I sought long, and began to despair before I found the fructification.

François ought to pass here on his return to England. I shall be very glad to see him; but it would affect me to see him without his master, and that must, and as Corporal Trim or some one says, by G— *shall come*. I see my phrase is obscure, but it will not be so to you.

God bless you! Write if you can, were it but a few lines, to give me new existence. I must work at my garden while the weather permits.

Yours,

E. Davall.

J. E. Smith to Mr. Davall.

My dear Friend, London, Nov. 17, 1789.

I was really very near writing to you to know what was become of you, when your most welcome letter arrived. Many thanks for your kind expressions of regard, which I hope to be ever worthy of, inasmuch as similar feelings for you may deserve them. But how is it that you tell *me* not one word of your marriage? Could you think I should not be interested in an event so important to your happiness? Who is the lady? May she be worthy of you, and, if possible, no enemy to Flora!

Since you went away I am become very intimate with your friend Forster, whom indeed I had long known as a botanist, and now esteem highly as a most amiable and sensible man. Your introducing me to Lady Rockingham has been peculiarly fortunate, and I rejoice to owe such an advantage to you. She is a worthy woman; and her sentiments accord with mine in many things. She treats me with peculiar affability and attention, and sometimes consults me medically as well as botanically.

She has given me her six volumes of *Hortus Malabaricus*, a very useful and magnificent present. I go to Hillingdon on Wednesday, to present young Jacquin to her.

I now begin to answer your letter.—Thank you for your care of the diplomas, and Van Berchemian anecdotes thereunto belonging!—how expressive! The Linnæan Society goes on admirably. We have fixed on some papers for printing: Dryander has given us a most excellent one on *Begonias*. When will you send us something out of the abundance of your discoveries? I could wish to see your name in the first volume.

We had a very comfortable day at Lady Rockingham's, and came back late on Thursday. She is to give Jacquin a fine drawing of her *Aralia*, (which is really an *Hedera*,) by Sowerby.

The blunder of making your *Gentiana* a variety of *campestris* is Linnæus's, in Mant. 2. All Murray's are *manual* blunders; there is no *idea* in the case. I will write to François your kind invitation: his return is at present uncertain. I always flatter myself with the hopes of seeing you and Switzerland again.

Dickson has brought many things from Scotland. We have got seeds from the East Indies of *Smithia*, and when they grow you shall have some.

Bring out your work, that I may publish a *Davallia*. Adieu, my dear friend: do not forget

Your ever affectionate

J. E. SMITH.

Mr. Davall to J. E. Smith.

Orbe, Nov. 24, 1789.

I hope, my good friend, that you have long ere this received the first part of this letter, sent to the post on the 27th of October. I have an *Arenaria*; —shall send you several of my best specimens, as it is within my reach on the nearest mountain, Suchet, mentioned often by Haller, ex. gr. under *Stachys alpina*, *Festuca decumbens*, &c.

I have *Sonchus canadensis* on the same mountain, which I call *my* mountain: a *Ribes* which I think *petræum*, Jacq.;—but you shall become acquainted personally with my good friend Suchet, where I have also found *Hypericum dubium*, Leers, in profusion. I could send you a thousand specimens, and will send.

I fear that among *Banksias*, *Fuchsias*, *Proteas*, and such heavenly plants, the poor *Potentillas* will hardly find a place; yet you will not neglect even an *Adoxa*, if you could attend to the *Chenopodium* which I wrote to you about. This is certainly an handsome *Potentilla*. I shall say nothing of the *Arenarias*, otherwise than that such a plant of Haller is such a plant of Linn. Herb. All my discriminating notes are for my private information: the rest is yours by every reason, and by our agreement:—moreover, my dear friend, no one is equal to the task but you. Have you received the *Arenarias* of Piedmont? &c. I long to know how you go on in that charming genus.

God bless you for your charming and precious specimen of *Diapensia lapponica*, gathered by Linnæus. A piece of the Cross is less precious to a bigot Catholic: it shall soon be framed and glazed. *Aristolochia Pistolochia* gave me great pleasure. Lichens will be treasures; and Grasses,—Oh! how I long to understand them well;—the *Agrostis* tribe, &c.

My paper is full, and I have not done; but I cannot say on paper all I would say; I must not take up too much of your time, nor write too often, although my heart will, I fear, oblige me to recommence ere long. What will you say, when I tell you I am married!—but it is to a person of the best merit. She has so often heard me speak to her of you, she has remarked the extreme pleasure your letters give me, the readiness with which I sit down to write to you rather than to any other person, that she longs to see you as my best friend. She desires her compliments, and invites you here as cordially as myself.

I wish to send you some few trifles. I wish to offer you some Chamouny honey. Pray did you receive the chamois horns? I do not ask for thanks, for I shall always be indebted to you.

Believe me, with the most affectionate and grateful remembrance,

Your devoted Friend,

E. DAVALL.

From the same.

My dear Friend, Orbe, Jan. 25—29, 1790.

I hope you received the continuation of my letter of the 27th of October, dated 24th of November. On the 7th of December I had the happiness of receiving your delightful letter of November 17—24. It is quite impossible to express the effect of your letters on me;—yet there is a passage in your last that thoroughly dejected me, where (as you well might say) after expressing your friendship, and thanking me for mine, you add " *but* how is it that you tell *me* not one word of your marriage?" This *but* has been to me as an accusation to the most conscience-wounded criminal. I shall all my life regret that you learnt this event from any other than myself; yet I might have assured you that the person was no enemy to Flora. I was so dissatisfied with my position, in a house with two good old aunts (for one died since my return here), who have no other ideas than those of that herd of silly beings called *gens du monde,* eternal players of quadrille, in a place where I find no reasonable creature to converse with, that I yielded to the idea of uniting myself to a young lady whose conversation I had always found more agreeable than any other here, and whose way of thinking agrees so well with mine. She is a year older than myself; only daughter of one of the best families in this country. Her name was De Cottens. Her eldest brother, who is about twenty-three, and not of age here till twenty-five, is C. Lieut. in the Swiss Guards at the Hague:—the other, Ensign and Aide Major in a Swiss regiment in the King of Sardinia's service. But as the world goes, I would rather have a wife of her disposition and way of thinking, with little fortune, than a worldly-minded girl with a great one. My wife is so much attached to you from my speaking of you to her as my best friend, and the person of all others I most esteem and love, that she insists on learning to dry plants to send to you, and often says, ... "*Presses le devenir; faites lui mes complimens en l'assurant que nous ferons tout ce qui est possible pour rendre ce séjour agréable; nous en aurons si grand soin que je suis sure qu'il se trouvera bien avec nous.*" I answer, that must and shall be some time or other; and if he will not come, I shall never die contented.

You make me very happy in what you say of Lady Rockingham. I had wished she might have the good thought of placing so judiciously her duplicate *Hort. Mal.* Pray, present my compliments when you see her. I have so many things to say that I hardly know where to begin, and shall be obliged to scrawl a second sheet.

If young Jacquin is still in England, pray give him my compliments, and tell him I expect he will not forget me in his journey through this part of the world. What a wretched figure his father has given of *Arenaria liniflora* in his second volume of *Collectanea!*

I can send you a most charming specimen in fine preservation of one of the *rarissimæ* plants of Switzerland, gathered by me this last season on Great St. Bernard, *Sisymbrium tanacetifolium;* but I ought not to have asked you whether you want it. —Pray tell me frankly as you should, and I know you will, whether I may copy in the course of my work, your text concerning *Sonchus alpinus* and *Stellaria dichotoma.* If you do not wish it, you will give me a new proof of friendship in saying No. —Although I must now look to close œconomy, I long most irresistibly to ornament my study with the plates of *Limodorum Tankervillia,* and *Strelitzia,* in colours ; and if Lee would permit Sowerby to copy Miss Lee's drawing of *Protea mellifera,* and you direct the choice of some fourth rare and *spe-*

ciosa planta of which Sowerby might make me a good drawing, all four in exact colouring, I would live six months on bread and water to pay for them. What with you, and the English gardens, I have some cruel moments of *maladie du pays*. Your letters and these drawings will be a temporary cure. I could not exist under the idea of not returning to England, but I certainly cannot during my aunt's life.

Jacq. *Fl. Austr.* gives as a true distinction of *Senecio sarracenicus* and *nemorensis*, 8 *radii* in *sarracen.* and 5 in *nemorens.* Now all the plants I have seen in Switzerland answer to Jacquin's *nemorens.* which, according to Haller, is *non satis certa civis;* but Haller's authority in the citation of Linnæus is to me of little force. Do you find Jacquin's observation good in the specimens of Herb. Linn.? I have made no note on the radii. I am more and more embarrassed with *Hieracium alpinum*, and have several to communicate to you:—some *scapo nudo H. alp.* Allion. *non* Linn. Herb.;—another with a narrow leaf on the upper part of the stalk, somewhat like Lightfoot's figure; and I almost suspect that the *pili* take the rufous colour in the dried specimens from age, as I have seen gradations in the Herbariums. As to the colour, I hardly dare say anything of this plant, and would wish to leave it to you, and will communicate what I have of specimens. I do not mean to say that the plant of Allioni and many others be the *alpinum:* I think it is not;—but the other is much less frequent, which I have gathered on the heights of Great St. Bernard, and which I have from the Grimsel. I am the more disposed to think the *pili* may take the rufous colour from age, as the pubescence of the *calyces* in *Trifolium incarnatum* becomes quite rufous in dried specimens, and as I observed no such colour in the recent plant which I saw in La Chenal's garden. There is a *Tussilago* in your Linn. Herb. marked *paradoxa?* J. E. S. *conf. T. spuria Retz. Obs. Bot. Fasc.* I. p. 29. Tab. 2. Forgive me: you know my intentions. The *Chenopodia* are terrible. I now find that what Haller refers to, *rubrum* Linn. (which I have not yet found here) is *murale.* This is not uncommon, and I did not see *rubrum* even in Haller's habitation, the *pays d'aigle,* which is the seat of *Chenopodia.*

How I long to see your new coloured work! You tell me I am good, because I have been in search of your *Lichen cucullatus.* Remember, my good friend, that all the little I can do to render you service, is to me the most agreeable duty, and the greatest pleasure I can know. I am dissatisfied in not having in my power the hundredth part of what I could wish to do. It is curious that Wyttenbach should have given you *Stellaria cerastoides* for *Cerastium alpinum.* He sent it to me by the name of *Arenaria multicaulis.* He should stick to his minerals. I have not yet *Cerastium alpinum,* however strange it may appear.

Your inestimable present of *Diapensia lapponica* is framed and glazed, with your label, as your handwriting is to me a treasure. I have had two cases made for my Herbarium, as yours are with a top in pyramid, and Wedgwood's medallion of Linnæusin the middle. But alas! my divisions are not full, although they contain many excellent things. I almost become ambitious, and would wish to have in my herbarium other than Swiss plants, although they must and shall be my principal object. I could wish to have plants that neither Wyttenbach, Haller, or La Chenal can produce. —What will you think of me? I have thrown off some of my moderate and humble ideas, and may say as you tell me, "You may thank yourself." If I no longer think myself a cipher, may it not be, that having left every path that might have led to an amelioration of fortune, I wish to have an adequate collection? adequate recompence? When I think of some men, among whom I may in conscience, without an atom of flattery, reckon J. E. Smith, I hide my head;—when I think of some others, I draw aside the veil.

Adieu, my dear friend. Remember your most affectionate and devoted

E. Davall.

J. E. Smith to Mr. Davall.

My good Friend, London, March 9, 1790.

Your two delightful long letters were most acceptable, and have not been answered yet for that very reason, for I have not been able to sit down to answer them as they deserve, so have deferred it. Nevertheless, I must thank you for them now, and answer them fully when I can. I am now in the midst of my zoological and botanical lectures, which I am obliged to compose every day before I deliver them. I have a private pupil besides, who comes to my house, (Lady Hume,) and I am printing my second fasciculus of *Icones ineditæ;* writing the letter-press for *Icones pictæ* (of which, by the by, Lady Rockingham means to beg your acceptance of a copy from her). Besides this, I have two or three other works in hand. Judge then whether I am busy or not! I indeed owe you a grudge for suspecting I should not rejoice with all my heart at your marriage, and quite forget botany itself in so interesting an occurrence. I admire the principles upon which you have made your choice; they are certainly the only ones upon which a wise man can be happy. How I long to know the lady! and how much I think myself obliged to her for her kind message! Tell her, a plant of her drying shall meet with its due reverence; I shall even be rejoiced to hear she adopts as an amusement your favourite pursuit; for certainly a similarity of tastes, especially the more elegant ones, endears a companion vastly to us. But I shall no less esteem your lady, if I hear that the cares of a family (on which I congratulate you on having a prospect) should take her off even from botany.

You inquire about my old faithful servant Francois. He is come back, but was robbed on his way. The Archduke of Milan has, however, been very kind to him, and even sent him a present. I hope one day to have him as a botanical servant, as he is above being a mere footman. Perhaps the time will come, he may travel with me,—but these are visions. This summer will probably determine many things respecting me. In short, I rely on the blessing of Providence which has hitherto been most indulgent to me; and as long as I wish for what is only really best, I cannot be materially unhappy. Excuse this grave unbotanical letter. Remember me most respectfully to your lady. I will write you a longer letter when I can. I am remarkably well this fine winter,—no snow at all!

Yours,

J. E. Smith.

Mr. Davall to J. E. Smith.

My good Friend, Orbe, March 23, 1790.

I last night received your most friendly letter of the 9th, which gave me equal pleasure and equal pain. Yet even the pain, from the degree in which I felt it, and now do feel it, was to me a delight, as a fresh and most convincing proof of my warm and sincere attachment to you, and of your proportionate return. Oh now I do indeed conceive most forcibly, how much you might be dissatisfied with my temporary silence. You might well, as you say, owe me a grudge; but I can swear by all that is sacred and dear to me, that, as it is the first, it shall be the last. I had adopted the idea that I was not doomed to enjoy that greatest object of my ambition, a warm and bosom friend; and I dare add, that at the same time I felt within me all that was proper for maintaining such a desirable connexion, and you were above every other the person whose attachment I had most desired. My wife, who offers you her most friendly compliments, desires you with me, never to forget that you have now a home at Orbe. We join in entreating you not to refuse us a favour we have to ask of you; which is, in case our child comes safe into the world, that you will be so good as to consent to be its godfather. Nothing can make us more happy; and you may safely consent in a moral point of view, as we shall ourselves take care of its principles.

Adieu, my best friend, and believe me for my life

Your devoted and affectionate

E. DAVALL.

P. S. *Smithia sensitiva* is framed and glazed in my study, and I never enter without bowing to *Diapensia Lapponica!!* These two are together in the sacred corner between the bookcase which contains Linnæus's works, and my (little as yet) Herbarium.

Heaven preserve your health; and, as the ghost of Hamlet says, "Farewell,—remember me."

J. E. Smith to Mr. Davall.

My dear Friend, London, April 25, 1790,

Bring out your book as soon as possible, and by all means call it *Illustrationes Hallerianæ,* or some such title. I received your box by Lord Sinclair:

thanks for *L. cucullatus*. I have only one point to answer in your last of the 23rd of March.

I feel deeply the proof of your affection in wishing me to be godfather to your child; but you shall judge whether I ought, or not. I must be quite open with you in this as in all things. I could not conscientiously promise it. I have taken much pains to settle my faith; and, thank God, it is settled so as to make me very happy. I am no enthusiast: I look up to one God, and delight in referring all my hopes and wishes to him. I consider the doctrine and example of Christ as the greatest blessing God has given us, and that his character is the most perfect and lovely we ever knew, except that of God himself. This is my religion, and I hope it is not unsound. I have found great good from it; and if not all the good I ought, I feel it is my own fault. But to the point in question. I think godfathers and godmothers an unnecessary form, and even worse, as a religious form that means nothing cannot be innocent. If by it is meant only that I am to be attached nearer to the child than any one except its parents, that I may presume more to advise and study its happiness,—very well; but *that* I shall do without any form or ceremony. I had rather the ceremony, as far as respects me, were dispensed with. But if you and Mrs. Davall still persist in your desire, and think it will in any manner attach me closer to you, I submit.

Pray give my best respects and wishes to her. May she soon be a happy mother! and may you ever be happy in each other! My next favourite hope is to come and see you one day.—I long to see Switzerland again, and *you* I must see; I like to indulge cheerful hope. Thank God I have enjoyed much, and hope to enjoy more;—if not, His will be done, and I shall not be the worse for having hoped. Adieu!

Yours ever,

J. E. Smith.

Mr. Davall to J. E. Smith.

My dear Friend, Orbe, May 31, 1790.

My time has been so taken up, what with the care of my garden and excursions to collect some of the plants I wished to send you, that I find myself close to Mandrot's departure.—Now for your last invaluable letter: but I must first say, that I told you in my letter of the 27th April, that I should have a good opportunity of sending specimens. It was by a most worthy man, and a very good friend of mine, who was on his departure for England, Mr. Schutz, the Queen's equerry; he dined with me on a Thursday, was to set off in ten days, was taken suddenly ill four days after I saw him, and died. I regret him most sincerely.

Nothing can be more perfectly open, sincere and friendly, than your answer to my request of being godfather to my child, if it comes well into this world. I assure you, my good friend, although I have been educated in the Church of England, I consider the office of godfathers and godmothers as unnecessary; as you will judge, that if I had not thought that parents ought to attend to the care of their children's principles, I should not have asked a person—for my misfortune far too distant from me, to fill this ceremony. I wished that my child might have you for his friend, and I am certain that it will be so. I may safely say that the reasoning you make use of is in such perfect concord with good sense, that if I were obliged to make a formal declaration of my ideas, they would agree very well with yours; and I would readily infuse them into my child. My devotion is mostly private, though I frequent the church here, and I believe I may any one that is Protestant, without any great impropriety.

I at no time feel myself so deeply penetrated with veneration for the Divinity as in the contemplation of his works, and hope that the tears of delight and adoration which often escape me in my herborizations, and in examining the wonderful structure and providence so striking in every object, are as acceptable to Him, and as fit to secure me His blessing, as my prayers; and indeed I experience now, with gratitude, more happiness than I could ever have hoped.

May Heaven preserve you, my best friend, for botany and me! You are and ever will be the first object of my prayer. My wife and aunts desire their compliments.

Your devoted and most affectionate,

E. Davall.

J. E. Smith to Mr. Davall.

My dear Friend, London, July 11, 1790.

Nothing could be more unlucky than my being out of town when Mr. Mandrot called. Accept my most hearty thanks for your letter; you and I can never differ much, I believe, even in opinion. I should not have obtruded religious matters on you had I not been obliged, but now am glad I did so. We can never go far wrong with such principles as yours. May the God whom we sincerely endeavour to adore as we ought, bring your child safe into the world, and preserve its mother!

July 14.—I have now got your inestimable packet, my dear friend, and know not how to find words to thank you for it. The views* delight me above all; I cannot cease from looking at them. How kind and attentive you are to what may give me pleasure! Your harvest on Mount Suchet was very rich.

Now, my dear friend, let me exhort you not to lose any time, but bring out your work directly. You must not wait to make it anything like *complete*. There is no end of that.

I will now confess to you a fault I have been guilty of, as I ought not "to dissemble or cloke before you". I distributed the seeds you sent among Fairbairn, Dickson, Lee, Curtis, and Goodenough; from some carelessness I gave none to Aiton. Will you pardon me for this, and send anything you please to make my peace with him;—I shall mind

* of Switzerland; published by Aberli.

no expense, and will deliver them directly. Pray be not angry, and write soon to your ever affectionate friend,

J. E. Smith.

J. E. Smith to his Mother.

Honoured Madam, August 30, 1790.

Yesterday Dr. Younge and I set off for Wentworth House, where we were received with all kindness and attention. Lord Fitzwilliam was to have been at York races, but was not well enough to go; so we dined alone, and afterwards went to the new mausoleum erected for the Marquis of Rockingham, which is very elegant and grand: saw the gardens, &c. We were attended by the chaplain and steward, two old servants of Lord and Lady Rockingham's, who thought they could not do enough for me as her friend. Indeed it was pleasing to see how many old servants and workmen gathered about me, making inquiries concerning her, and sending their duty to their old mistress. Lord Fitzwilliam being better in the afternoon, we drank tea with him; and little Lord Milton, who is not five years old, brought me all the botanical books he could find, and told me the Linnæan names of some plants I had just brought out of the garden; he was very loth to leave me to go to-bed. We went to prayers in the evening in the chapel.—Next day after breakfast saw the house at our leisure.

Lord and Lady Harewood are at Scarborough,

so I shall not have an opportunity of going to see them.

I am in great haste, as a messenger waits for my letter. Adieu!

J. E. Smith.

J. E. Smith to Mrs. Davall.

Madame, Londres, ce 28 Sept. 1790.

Quoique je ne sçaurois que regretter l'accident qui m'a procuré l'honneur de recevoir une lettre si flatteuse et si intéressante de votre part, permettez moi, puisque le mal s'est passé, de me féliciter d'avoir mérité de quelque manière que ce soit les sentiments que vous avez pour moi. C'est avec joie que je vous félicite aussi du rétablissement (parfait j'ose espérer à present) de votre estimable et digne époux, dont l'amitié m'est si chère. Je l'ai conjuré de ne pas exposer sa santé à de nouveaux dangers, mais c'est à vous de le faire obeir à nos souhaits. J'espère qu'il s'occupera un peu de sa jeune fille par préférence; je craindrois presque si vous lui présentiez plusieurs de *cette* espèce de jolies fleurs, que celles dont il s'est occupé avec tant d'empressement jusqu'ici ne soient negligées. J'en serai un peu jaloux.

Adieu, Madame: continuez de faire le bonheur de mon ami et de tous ceux qui vous connoissent, et n'oubliez pas que vous avez des amis sincères ici qui vous souhaitent toute la felicité que vous méritez si bien.

Je suis toujours, Madame,

Votre très humble et très obéissant Serviteur,

J. E. Smith.

Mr. Davall to J. E. Smith.

My very dear Friend, Orbe, Feb. 11, 1791.

How can I ever acknowledge your goodness, and I may say your preferable attention and favour, to me, in giving up so much of your precious time to me and to the settling of my doubts; in truth, will it not be to *you* to whom students will be obliged, if my work contains any thing good? I am, God knows, as grateful as any man possessed of a heart uncorrupted and capable of the warmest attachment can be;—but this is nearly all I have to offer you. May I never hope the blessing of Providence, if ever you discover a friend more cordial and sincere! I hardly think but of you, and, together with my best female friend and my little girl, I exist but for and through you.

The weather has been mild of late; the snow has melted; we have had rain, and what they properly call here "*un hiver pourri*": at a time when I ought to work assiduously at the execution of my delayed plan, I have had *longings* and *longings* like those of a pregnant woman, to ramble after *Lichens*. I have been but once; hereafter I must ramble. Having observed t'other day a charming Lichen on a rock, which I could not well detach, as the weather was dry, I returned yesterday under favour of a little rain, which procured me a good scolding from my good old aunts, and some little reprimand from my wife; but I was soon *consolé*, as I believe my Lichen may be worth your attention. There is one specimen remarkably fine; *this* naturally is destined for J. E. S.

I am almost sure my plant is *Psora testacea* (*Lichen testaceus*) Hoffman, *Plantæ Lichenosæ.* Tell me whether you have it, and whether it be not good, that I may not have been brought to task *for nothing.* Obs. I do not find this plant in Haller;—if so, it will be another addition to *Fl. Helv.* It seems to have affinity to your *gypsaceus*, but mine has fine red *tubercula.* N.B. One time or other, when you are at Sir J. Banks's without much to do, read Saussure, *Voyages dans les Alpes,* vol. ii. p. 465. chap. xlv. Glacier de la Valsorey, which is very good. You will then *see* for a moment the spot where I gathered *Gentiana tenella.*

I crossed the glacier, and looked into the *Gouille à Vassu.*

To return to what I was talking of:—You see, my dearest friend, that, from the non-execution of my plan, I appear irresolute,—slow,—and I know not what; my disordered health is very much against me, not less so the indispensable care of my garden, and the *absolute impossibility* of resisting the delightful verdure that surrounds me during the good season, and which I have here under every form of wood, marsh, pasture, corn-fields, mountains;—and how are they to be resisted? Let me hereafter do all the little I can, by excursions and the aid of my garden, to be of use to you. Let me be as it were a part, an unworthy part, of yourself; and let every step, every remark be subservient to your use. This is my ambition and my warmest desire as long as

I exist in this world. If I am called on to bid adieu to my wife, to plants, and to you, let my books and my plants be admitted to your use. Preserve them in memory of a true and cordial friend, who dying with this assurance will be happier even in Paradise.

English Botany pleases me much. The motto is admirable, and the more so as original. Adieu, my good friend! May Heaven preserve you, and continue your health!

Your very affectionate,

E. Davall.

Mr. Davall to J. E. Smith.

Orbe, April 30, 1791.

Having learnt, my very dear friend, that a voiturier of my neighbourhood sets off on Monday, and hoping him more honest than his brethren, I eagerly embrace the opportunity of writing a few lines in haste, which accompany a little barrel of Chamouny honey, that I wish you to taste. The bees who have made it feed chiefly on *Pinus Larix*.

The *Anemone sylvestris*, which Curtis's Magazine says often becomes troublesome in gardens, is, alas! one of the greatest *desiderata* in mine.—Pray let it be *Potamogeton* ob*tusum*. If you knew me sufficiently well, my good friend, you would have directed it so without even consulting me, who wish for nothing so much as to be guided in all matters by you, and whose highest ambition is (however unworthy perhaps) to exist as your and the Linnæan Society's delegate in this country. All my views tend thither; hence, after my own private and desolate enjoyment, every excursion—each solitary climbing of the mountains—the source of new vigour, when sitting to recover my breath at the foot of some black fir in dismal woods, where the woodpecker is the only being who salutes me. Had I never known you, I should have wanted the energy which has brought my garden to what it already is, and what more I hope it will be. To divert my thoughts I have taken a turn in that part of my garden which I arrogantly name the *Alps*. *Iberis rotundifol.* F.; *Viola biflora* F.; *Laserpitium simplex* coming to F; *Ranunculus Thora* in bud; *Ranunculus rutæfolius* and *parnassifol.* with leaves in true Alpine vigour, not garden luxuriance; *Turritis cærulea*—all coming to flower; *Hieracium alpinum*, H.L.; *Hierac. montanum*, formerly *Hypochæris pontana; Phaca alpina, frigida* and *australis*, in tolerable condition; *Phellandrium Mutellina* coming out of the ground. What say you to this? Is it not worthy of you? Yet nothing pleases me so much as *Thlaspi alpestre*, (which is marked *J. E. S. Matlock*,) and *Astragalus leucophæus*, Smith,—both in flower, and both from you.

My little girl has two teeth.—May Heaven preserve and give you every blessing!

E. DAVALL.

J. E. Smith to Mr. Davall.

London, July 1791.

I thank you for your honey, which is an emblem and a proof of your friendship. I shall "eat it in remembrance of you." The honest voiturier charged nothing.

No. 50. *Polypodium montanum*, Allioni. THANK YOU. I must have the two words printed, as they are what you have decreed; if I trust to my pen, it will unavoidably transgress.

29. This I found at Vaucluse in wet places. 'Tis quite new, and so thinks Dickson Suppose we call it *fastigiatus*, if there be no such name already; it is so like *Fucus fastigiatus* in miniature.

42. *chrysoleucus*, J. E. S. Act. Linn. vol. i. 82. t. 4. f. 5. *imbricatus, foliolis lobatis obtusis: supra pallide sulphureis; subtus atro-viridibus, scutellis aureis*, from Mount Cenis. It certainly is not umbilicated, though your specimen looks a little so. But we must advert to the habit and *scutellæ*.

43. *encaustus*, J. E. S. ibid. 82. t. 4. f. 6. *imbricatus, foliolis linearibus dichotomis: supra albis nitidis; subtus nigris opacis, scutellis badiis*. Montanvert near Chamonix.

These, my dear friend, I look upon as two of the most original and interesting of my discoveries, being *quite* unknown and undescribed. Add to these No. 29. and *L. cucullatus*, and reflect how much better *you* and *I* search for plants, than anybody else: but let us not be too vain. Comparison with you may be in danger of making me so;

for, without a compliment, you are more accurate than almost anybody I ever knew. But what shall I say to your most precious black and gold Lichen in the round box! It never blest my eyes before. How like *geographicus*, and yet how distinct! It must come among the *imbricati*. What think you of *chrysomelanos* for a name? But I would not dictate, as it is yours alone!

I have inquired about you from Madame Combe, who is so good as to indulge me with talking about you. She draws me a charming picture, which must be true, of your happiness. Your little girl is "*jolie comme un cœur, mais petite ; et les tantes si bonnes! et si heureuses! et toute la famille si heureuse!*" *Tant mieux*, thought I; cards then and botany are come to a treaty of peace, of which I guess the *chère épouse* is the connecting link; and if so, it is best of all. Mademoiselle Combe says you have played at quadrille this last winter: even *I* play at whist. I hope we shall be rewarded!

How I long to see the first sheet of your work! Your gold and black Lichen Hudson and Dickson are positive is *geographicus*. They say it is altered by growing on *earth*, not *stones*. May be so; it is very like it.

As I send so few playthings to papa, I must send one for the *jolie petite ;* so pray teach her my name when she can speak, that she may learn to sound the *th* early. The two tinkling bells may serve to promote exercise and trying to walk; and I send two, that there may be one in store when the other is broken, which I presume may soon happen.

Nymphæa Nelumbo has been in flower at Bulstrode lately (the first time in Europe). It is a new genus, and is to be called *Portlandia,* as I must make the old one a *Catesbæa.*

Adieu, my kind and excellent friend!

Believe me ever yours most truly,

J. E. SMITH.

Mr. Davall to J. E. Smith.

Orbe, July 22, 1791.

Flaction delivered to me your parcel containing your very charming *Spicilegium Botanicum;* and just now your letter, and parcel of plants with true *Arenaria saxatilis!* and *Acrostichum Marantæ!* But, my best friend, all these precious and most excellent gifts have found me and the incomparable (I may well say) partner of my cares in the most cruel and heartfelt affliction. We had just then been separated from our beloved child, to whom we were dotingly attached. She had enjoyed better health than the greater number of us experience at so tender an age; had successively (perhaps too precipitately) cut six teeth, without any convulsions. She was to all appearance so well, so strong, that we began to hope she might pass safely over the critical period of teething. It is true, we were frequently alarmed, perhaps from a sort of superstitious terror, at what was surely unnatural intelligence in a child who had not completed the eleventh month. She was to all appearance perfectly well on the third of this month,—in good spirits,—slept well for nine hours on that night. Early on the morning of the fourth she was seized with a heaviness, looked pale, and could not keep up her head. We called in our physician, and sent an express eight miles off for another who is much esteemed here in the treatment of young children. They administered some remedies,—applied blisters to her legs: she appeared somewhat better for an hour, looked up to her mother and me when we spoke to her. In the night she grew worse, was seized with convulsions; and about nine o'clock the next morning the dear little angel expired. No mother, I believe, has ever shown more assiduous care from the moment of her child's birth. So constant has her attention been, day and night, that her own health is much affected by the fatigue attending the tender care of her beloved offspring. Her affliction is so great, so deep,—and not expressed, as less serious grief, by tears,—that I am now anxious for her. I have already taken her, endeavouring to divert her thoughts, to spend a week at a friend's at some distance; and in a few days we undertake a journey wholly for the same purpose. I shall take her to Basle, where she has never been, and we shall stay a few days with La Chenal.

You will readily pardon me, my dearest and very best friend, if I have entered too minutely on a subject which has dejected me beyond all expression. My fondness for my child was excessive,—by many it would be called unmanly; but I have always been too susceptible of attachments. So I feared I had been towards you, my worthy friend, during my last visit to my native land. After having seen you repeatedly,—after having perceived, which I presently did, that you were the man who by character, conformity of mind, and every other reason, was formed as the resting-place of my best friendship,—had I not experienced on my return here such marks of reciprocal good-will, as are greater far than I had any reason or hope to expect, I should now be, as it were, annihilated: I should be the most surly misanthrope on earth.

I had known many disappointments, and was too much disposed to feel them. Botany, which now and ever is the chief "balm of my hurt mind," would have been abandoned: I had been an useless burthen on the earth. Now, through you, I may be not an useless member of society. Is it fit, and dare I recall to your memory here, that "Full many a flower is born to blush unseen; and *waste* its sweetness on the desert air"?

In the circumstances which now oppress me, I have often thought it would be a satisfaction to me if you had seen my child. If she had smiled on you, as she did daily on us;—if she had stretched forth her little arm to you, you would then at least be better able to conceive our affliction, which, in my calmer moments, I readily allow to be perfect egotism, as she is surely more happy than she could have been with us. My wife and myself were extremely pleased (though with a mixture of cruel emotion) at all the kind attentions you have shown to our beloved child. We shall keep the playthings, when they come, with the greatest care, and shall frequently visit them with sentiments of the kindest nature for you, and resignation to the will of Providence.

My wife joins me in kindest regards.

Yours ever most affectionately,

E. DAVALL.

J. E. Smith to Mr. Davall.

London, Sept. 3, 1791.

I should not have deferred a moment, my very dear friend, answering your last letter, had I been in town when it came, and had I been sure of my answer meeting you at home, which this probably will. I cannot express how much I feel for your great and sudden affliction. As much as one who is not a father can enter into your affliction, I do; for you will do me the justice to think I am not deficient in the feelings of a friend. It was surely a very strange and sudden illness that deprived you of your dear little girl; and from your account it was independent of teething, and rather some very violent fever, for which there could be no help from the beginning. I am anxious to hear how Mrs. Davall bears her distress. It must go hard with her. But you are happy in each other, and have nothing to reproach yourselves with. "The Lord gave, and 'tis he that hath taken away;" and he will,

I trust, send you many blessings to supply the place of that you have lost. I cannot help sending you a few lines from memory, part of an epitaph on four children, who died of the plague about 150 years ago. They are engraved on a monument in a field somewhere in England. The style is not perfect, but the sentiments very affecting:—

> "Good and Great God, to thee we do resign
> Our four sweet sons, for they were chiefly thine;
> And, Lord, we were not worthy of the name
> To be the sons of faithful Abraham,
> If we had not been willing for thy sake
> To yield our all,—as he his son Isa-ac!"

I regret much that I may have innocently added to your affliction by the toys I sent, and yet I think they will (after a time) not have that effect.

Cannot you visit England once more? I have many reasons to wish you could. My connections are now much increased, and I could procure you more amusement than when you were here before. We would visit the good Marchioness of Rockingham, with whom I am now quite intimate; and I assure you she is a most excellent character. How often do I wish for you when I am there!

I have also made an acquaintance I like very much in Sir Abraham and Lady Hume, a charming family indeed. I have just been ten days at their house in Hertfordshire. I spent two days at Bulstrode lately, and was very kindly treated. All these places, and others, we will one day visit together; but we must not confine ourselves to grandees. I have one family with whom I must make you acquainted, though alas! embarrassed, and in some respects unhappy, but of most amiable, accomplished, elegant and feeling minds.—Come then, and let us be happy together in pleasures which Heaven itself must contemplate with complacence, and whose very disappointments are more delightful than what nine tenths of the world so fondly pursue.—I have almost filled my sheet without a word of botany!

Adieu! God bless and comfort you both, says your ever affectionate

J. E. SMITH.

The Marchioness of Rockingham to J. E. Smith.

Hillingdon House, March 10, 1792.

I was extremely glad to receive Dr. Smith's letter, but I must lament the not seeing him and Mr. Afzelius at Hillingdon, before the latter departs for Sierra Leone.

There is something in the countenance and manner of that little man that shows a goodness which interests very much one's good wishes; and I sincerely hope he will find both advantage and satisfaction in his botanical researches in this new settlement. I have heard Governor Wentworth mention it, and shall ask him on Sunday the particulars of the botanizing arrangements there. May I beg the favour of you to pay my debts to Mr. Afzelius for those works of Thunberg's. Might I also ask you to present ten guineas to Mr. Afzelius to purchase some trifle of botanical use, as a remembrance of Hillingdon.

How can I have been so long before I have congratulated upon all the honours and glories that are surrounding Dr. Smith! My letter ought to have begun with a proper homage upon the occasion: perhaps even this note may have to follow him into my royal neighbourhood; but if not, I hope in Marlborough-street he may condescend to eat a couple of pines which accompany my note.

I hope the travellers have not yet left you. I think you are very good to give up your François, but I think he should be a little sorry to leave you.

Your sincere, &c.
M. Rockingham.

Mr. Davall to J. E. Smith.

Orbe, Oct. 12, 1792.

For several days past, my very dear friend, I have been intending to write to you for my comfort and my mind's ease. I have waited in hopes of telling you I have received the delightful books I owe to your friendship;—they at last arrived the evening of the day before yesterday, having been obliged to take a great *detour* in Germany to avoid the armies, &c. &c.

Receive my best thanks for your continuation of the lovely and most interesting *Spicilegium*, which during the whole of yesterday was "balm to my hurt mind", as also your inimitable, and *ne plus ultra, Icones pictæ.*

I must at present, to my great regret, lay aside botany during the rest of this letter, and I shall certainly derive strength from discharging my mind with the person of all others I most love, and in whom I have the greatest confidence. If I experience great uneasiness and some degree of melancholy from what I am going to relate, I no doubt must attribute it in a great measure to the sad state of my health ever since my last serious illness.

I have often returned my warmest thanks to Providence for the state in which I am placed,—able to pursue the study which, with your friendship and the attachment of my wife, are the three great and only objects which make me value my existence. I hoped ever to enjoy these blessings in peace and security; yet God only knows whether I shall be in that state when you receive this. You probably know that our restless neighbours are in possession of the greater part of Savoy. The Genevans, seeing them at their gates and fearing the same lot, have called on their allies the Swiss for succour; near 2000 Swiss, or rather 2140, (1500 of Berne and 640 of Zuric,) have entered Geneva. The French resident remonstrated against this admission, and has left Geneva, leaving a letter, in which he says "the admission of the Swiss cannot be considered by the French but as a coalition with the other powers against Liberty!" The end of this letter is a direct declaration of war. There are a great number of French close to Geneva, with artillery for a siege, &c. &c.; and it is expected they will shortly begin firing on that place. The Swiss march in every direction towards the frontiers of our French neighbours, and are admirably well disposed to receive them. If really they do attack this country, there will certainly be bloody work, unless the French show their heels, as they sometimes do, at a first onset. I am close to the frontiers (at about seven miles). Now to speak openly, all my uneasiness at this moment arises from my having a wife to whom I am much attached: were it not for this, and seeing my books, my herbarium, my hearth, and every thing at stake, I should not hesitate one moment to make one among the number, shoulder a musket if necessary, and be cut to pieces (if Heaven would have it so), with the hope of cutting down a wretch or two in the fields, where I am accustomed to look for plants and think of you; rather than see my books and plants burnt, and be quietly carved to pieces in my study. As things are, come what come may, I shall not stir from hence. These French in fact, as noxious insects, are only to be feared from their immense numbers. Nothing can be more delightful than the disposition of the common people here. We have had (not being on the high road) only two *bataillons* who have passed here, among them many men who have left their home without having done sowing, yet no appearance of discontent. Those of the German part, i. e. true Swiss, have a grave and manly appearance, are perfectly well armed, their artillery of the best sort, and all in the most excellent order. From what I can conjecture we have, or shall have shortly, between 30,000 and 40,000 men stirring. I had yesterday a young man with me who has escaped from the massacre of the Swiss Guards at Paris, in a manner almost miraculous.

He is quite eager to come to blows; and it is chiefly that sad affair which animates every individual.—God bless you. Write as soon as you can.

E. DAVALL.

J. E. Smith to Mr. Davall.

My very dear Friend, London, Nov. 9, 1792.

I write upon gilt paper for joy! If I had a harlequin sheet, such as Sterne puts in his Tristram Shandy, I would send it you to express my great joy at your escape. I hope you will have no further alarms: I trust that the cut-throat party in France will be got under, and that they will settle into that manly rational liberty at which they professed to aim. I should not despair of them, if they had but some principle of religion and less debauchery; on the contrary, their characters in these respects make me fear that nothing good can come out of them: but let us leave politics and politicians for our more virtuous study. I am come to town for two days from Windsor. I return this day. I am delighted with my pupils: we all sit together at a round table. I lecture from my notes, which the Queen takes home from Frogmore to Windsor to make extracts from; and my audience occasionally ask questions and make remarks very much to the purpose, and a conversation of half an hour or more follows the lecture. Nothing can be more polite and pleasing. I shall be thus engaged till next month. I have good news of Afzelius and François. Yours, in haste,

J. E. SMITH.

Mr. Afzelius to J. E. Smith.

Dear Sir, Freetown, Sierra Leone, July 2, 1792.

I am happy to acquaint you with our safe arrival to this place the 6th of May last, and with our perfect state of health since that time, except that Borone has been a little ill with a fever, from which he is now entirely recovered. But I am sorry to say that the inconveniences we here meet with, seeing the colony in confusion and divided into parties, and not having a separate house or room where we can dry our plants, preserve our things from the inclemency of the rainy season, and write down our observations, have hitherto prevented us from collecting any plants. This is certainly a cruel situation, particularly when we are surrounded by so many curious and unknown productions; but we must wait for a more prosperous time. Meanwhile I am vexed that we cannot have the pleasure of sending you, with the Sierra Leone packet, Captain Phipps, who now returns to London, anything but some bulbs and seeds, which I have only numbered; and I beg you will keep these numbers till another time, when I can be able to give them their proper names. I leave it entirely to yourself to distribute them as you please: I only wish that no description or figure of them may be published before I can come back myself. Amongst the bulbs are even those of the beautiful *Hæmanthus multiflorus* and the charming *Amaryllis ornata*. Borone behaves exceedingly well, and I am very happy to have got so amiable a young man as my companion in my travels.

I wish you could be good enough to send me Gærtner *de Seminibus*, Schreber's edition of *Genera Plantarum*, and Gmelin's edition of the System, only the Vegetable Kingdom. Though I like the territory of Sierra Leone very much, the climate being fine, the spot where the town is built healthy, the soil good, the woods abounding in game, the water with fine fish, and the whole country with the most wonderful plants, of which I have already seen about the half part of them figured in Sir Joseph Banks's book; nevertheless I should already have been tired with this place, from the inconveniences I have suffered, had not Mr. Clarkson shown both me and Borone so much favour and so many civilities, and promised us to have very soon a house for ourselves. This worthy gentleman is now almost the only man who supports the colony, and endeavours to put all things in order again: but it will be a long time before he will gain his laudable intentions; and I think it impossible without a speedy assistance from home. I have begun a long letter on our wretched situation and its causes, but I have not now time to finish it. You shall have it with the next vessel; and then I hope I shall have the pleasure of writing to you with greater satisfaction.

Present my respects to Lady Rockingham. I hope I need not apologize for my bad English, but am assured that you, as a friend, will always excuse my errors.

I am always, Sir, your sincere Friend,

AD. AFZELIUS.

Mr. Davall to J. E. Smith.

My very dear Friend, Orbe, Jan. 11, 1793.

I have had the very great pleasure of receiving your two last kind letters, which have done me more good than any article of the *materia medica;* I was, if possible, more impatient than I commonly am to have some news of you. I have been prevented writing to you by an unexpected and very sudden event. In rising from table one day after dinner, the eldest of my aunts, aged 84 years and 3 months, in all appearance in better health than myself (the son of a sister 20 years younger than her), fell after having half crossed the room, exactly as if struck by lightning, and from that moment never uttered one word. Her last conversation was as gay as you can imagine; and the evening before she had played at quadrille as usual. By-the-by, such is their eternal habit of cards, that we have had an example this autumn of one man dying, his cards in his hands. Great God! what a death! Let me rather break my neck down some precipice (if such be thy will) endeavouring to obtain some new proof of thy greatness, or in the desire of rendering service to a beloved friend!

The death of my good old aunt, (for a good woman she was,) and the sad moments that necessarily follow such events; the miserable state of my wife's health, now near her confinement, and who had been affected very seriously in seeing her fall; some disorder in my own health, &c.—have carried me on till now.

I am very happy, my dearest friend, to find by your letter you are so gay, as thence I judge well of your health. I who, as you know, live merely as a hermit, have lately found myself at table with Lady Spenser, the Duchess of Devonshire, and Lady Duncannon, giving them the names of some plants they had collected: they winter at Pisa. I mention this merely to give you notice, that you may not be surprised. I told them I had a friend in England infinitely better able to assist them than myself. By-the-by, they found me wonderfully clever, because they had seen nothing but some blundering obsolete *amateurs*, and were surprised just as I am when I go to England, when I constantly say to myself on arriving, "How well these people speak English!" were a sailor or a fishwoman the first persons I meet.

I shall not touch on politics, as I understand nothing at all of those matters. I should be very happy in seeing the establishment of a *virtuous* liberty, and a state of things free from the abominable effects of intrigue and corruption. But men will never be perfect; and I know not whether it were not better to remain in a state accompanied with some inconveniences, if not quite intolerable, as they were in France, than run the risk of being still worse in endeavouring to be better. As for the French, I know not whether they are, as a nation, capable of doing such great things: they always run into extremes, and overshoot reason. Perhaps I am narrow-minded; but I love them not, nor can I ever think them capable of solidity: I know there are exceptions. They will all be heroes, and give reason, liberty and atheism to the world. With respect to the execution of their wonderful perfections, I shall only assure you that many of the peasants in Savoy wish them, as I do, in another planet. They dare not, under pain of death, bring us over any sort of provisions as formerly; and the liberty they have acquired, is that of being obliged to give their hogs, corn and chestnuts to these French brethren for paper instead of money of better value, which they receive here, paying as much for taxes as before, under new names, and sending men to be shot at who would have cultivated the country. Miserable as they were, many are still more so now.

Totus tuus,

E. DAVALL.

The Marchioness of Rockingham to J. E. Smith.

Hillingdon House, March 3, 1793.

I am ashamed that I have not much sooner thanked Dr. Smith for the seeds from Sumatra and Sierra Leone. These seeds have been some time sown, and yet Dr. Smith is but now reaping my acknowledgements: so strange is human nature, that often, even against inclination, it neglects and delays the most pleasing occupation of expressing gratitude for favours and attentions that the mind is all the while sensible of. I little thought, five years ago, that I could have felt so much concern for the death of Mr. Aiton; but I had not seen him then, and only looked upon him as the Kew gardener:

but the single quarter of an hour that he was with me here occasioned an instantaneous conversion. I was quite charmed with the plainness of his manners, without a grain of that pomposity one might have expected; but on the contrary, quite pleasant and communicative in his profession: in short, he took my fancy so much that I cannot help feeling infinite regret that so great and good a man in his line should now be no more.

My *Strelitzia* is advancing into flower, and an odd *Amaryllis*, that nobody knows any thing about, I hope is doing the same; and much I shall wish for a visit from Dr. Smith when they are ready to receive company.

I am, with much sincerity,

Your obedient Servant,

M. ROCKINGHAM.

Mr. Davall to J. E. Smith.

Orbe, Aug. 16, 1793.

My dearest and best Friend,

All I can do for you, all I can send you, and ten times more, did my health, faculties and circumstances permit, I can never, never consider otherwise than as my indispensable *duty;*—a duty I owe you, as in my mind holding the reins of botany, as a small return for the great friendship you showed me in England, even at a time when I was no otherwise known to you than as a man fond of plants, and who had undertaken a long and expensive journey in search of the *Linnæa.*

I have often been surprised that with all you do, and all you have to do, you can so regularly, so amply elucidate every object I send. My work (if my health will at last permit me to get through it) will derive every merit it may have from your generous information. I have moreover the ambition to possess in time a rich herbarium, and one more correct than any who have plants from the gardens of Göttingen, Tubingen, Kursberchtolfgaxen, &c. &c. By-the-by, what a terrible number of *oides* have fallen on the fine Austral plants!

I am quite delighted with English Botany. What is above all precious to me, are the maritime plants; they are quite new to me.

Among all the excellent things in English Botany, nothing intrigues me more than *Galium pusillum*. The Swiss *Galia* are by no means clear.

The *Orchideæ* delight me, especially *Ophrys Loeselii* and *Malaxis paludosa!* as strangers to me.

E. Davall.

J. E. Smith to Mr. Davall.

My dear Friend, London, Sept. 24, 1793.

Where in the name of wonder is *Hortus Kursberchtolfgaxensis**?

Afzelius and François are safe returned, thank

* "I laughed heartily," says Mr. Davall in reply, "at your astonishment concerning *Kursberchtolfgaxen*. It is a mere imaginary specimen of German soft names. If you know some little French poems of M. de Florian, you will find this pretty word in his *Poule de Caux*."

God! and have brought many fine things, though not many specimens, because the Sierra Leone climate is very damp and *insectiferous*. Their fruit and capsules are the most wonderful. I recollect the Swiss *Melampyrum cristatum* struck me at Geneva as *uncommonly* beautiful. Can ours be different? There I found it in a grove. Ours is a corn plant?!

You say *Galium pusillum intrigues* you: if you talk thus, Mrs. Davall will become jealous.

Adieu! God bless you.

J. E. Smith.

Mr. Afzelius to J. E. Smith.

Freetown, Sierra Leone,
Nov. 19, 1794.

My dear Friend,

I am sorry to be a messenger of bad news; but an event that has taken place so openly cannot be concealed; it must be known sooner or later; and as I have myself unfortunately been a party concerned in the common calamity, I hasten to inform you of this melancholy affair.

The 28th of September last the French arrived here with a force far superior to what we had to oppose them, and they made themselves therefore soon masters of the place. They landed and plundered all the houses and the store. They carried on board their own vessels what they wanted themselves, and the rest they either broke to pieces, or plunged into the sea, or consecrated to the flames. They killed all the live stock in the colony. They burnt the public buildings and all the houses belonging to white people, as well as the shipping, except two vessels, which they took with them; and one of these was the long-expected Harpy, our last resource, which unfortunately happened to arrive some days before the enemy departed.

After they had done us all the mischief they wished, after having burnt Bance Island and the vessels in that port, and after having taken about ten prizes in the Sierra Leone river, they sailed away the 13th of October last to the leeward, and left us in the most wretched situation, without provision to live on, without any rag to put on our backs, and without comfortable habitations. But we were happy that they did not murder us every one, and that they left the huts of the settlers untouched. (There were only two persons killed, and three or four wounded, which happened the first day during the attack.) We suffered very much the first week; but after our friends in the neighbourhood had sent us some supply, and our settlers,—who have farms in the country, or rather in the woods, where the French did not dare to penetrate,—had brought us theirs, we have been pretty well off for provision, and there is at present not the least danger of starving. But we are in want of so many other necessaries and conveniences of life, that we are in real distress, till any vessel arrives from Europe and brings us some relief. I, for my part, am particularly unhappy, as being deprived of my activity, and not knowing what to do; for I have lost all my tools and implements for writing and making collections: I should otherwise soon make up some of my losses, as descriptions, dry plants, insects, shells, animals both stuffed and in spirits, &c. &c., the whole of which amounted to larger collections than I ever had at once in this country before; but the loss of some of my manuscripts, and particularly my journal, which I had kept from my arrival here to the present time, is irreparable; and if I should value all my losses, I could not make them less than 1500*l.* or 1600*l.*; though manuscripts, particularly when they are intended for publication, are invaluable, and may be worth double the sum you first thought of. I am in very little hopes that the Directors will repay me anything; and if they do, it will be a paltry sum I know. I must therefore rely upon my own exertions and the assistance of my friends, till I may be able to acknowledge their kindness. The present time is the worst, and before any vessel can arrive, which I hope will supply me again with all necessary articles; in the mean time I must be patient, and do as well as I can.

When you see the Marchioness of Rockingham, I wish you would tell her something of my distress, if you think proper. After having told you that I am now reduced quite to a beggar, I do not need to apologize for my scrawl and bad paper: I am glad to have any thing to write with and upon.

I am yours, affectionately,

Ad. Afzelius.

P.S. Just this moment, when I was going to seal up the letter, arrives a chartered ship, the Achilles of Liverpool, Captain Hogg, after seven weeks' passage, and brings the agreeable intelligence that both the Ocean and the Amy were safe arrived; and thus I hope you got long ago what was intended for you. The Achilles brings only salt and crockery ware; instead of which I wish her cargo had been provision, cloth, linen, furniture, &c. &c.

J. E. Smith to Mr. Davall.

Norwich, March 16, 1795.

It is so long since I heard from you or wrote to you, my excellent friend, that I can refrain no longer.

I have nothing botanical to communicate, nor indeed anything pleasant, except (as I hope) expressions of my unceasing regard may be acceptable. I have lately had the misfortune to lose my most indulgent and excellent father. There never was a more honest, sensible, or judicious man, nor a more excellent parent*.

I have another melancholy piece of news, which has made a very deep impression on my mind,

* The annexed lines are expressive of Sir James's feelings of respect and attachment upon the loss here recorded.

Dear tender Father! if thy honour'd shade
Still hovers o'er thy children, bless our tears!
May every pang, by which our hearts now own thee,
Impress thy virtues and thy precepts there!
Be thou our guardian angel, while we strive,
Though not with equal steps, to follow thee;
And may thy spirit welcome us to peace.

In St. Peter's Church, Norwich, where his parents and others though now my thoughts are taken up otherwise. Poor François is no more! He had a good journey with Dr. Sibthorp as far as Athens, and they were much satisfied with each other *.

I shall dedicate a genus to him, as a martyr to botany, and try to do justice to his merits in the New Holland Botany very soon. I know you will lament his unfortunate fate; it is consolatory to hear how much everybody esteemed him. It is a severe loss to me, for he was truly good and amiable, as well as intelligent. I think he had more acuteness in finding out specific differences of plants than anybody I ever knew.

Poor Afzelius has lost all his MSS. and collections by the descent of the French savages at Sierra Leone; but happily he is alive and well. His friends

of his family are deposited, appears the following inscription to their memory, from the same hand:—

Tears for the dead unprofitably flow,
  Their virtues yet a richer tribute claim;
Let emulation sanctify our woe,
  And raise a living trophy to their name.

Hither may still each kindred mind retire,
  If boisterous passions urge, or vice allure,
Muse o'er these sacred relics, and aspire
  To keep their bosoms pure, as these were pure.

That when to these their mouldering dust may come,
  O'er their cold urn may fall the filial tear,
Still fond remembrance dwell upon their tomb,
  And virtue find a fair example here.

* Sir James's epitaph on the death of Francesco Borone appeared in the volume of Tracts.

have sent him out some immediate supplies of necessaries. Pray write to me, and believe me ever

Yours,

J. E. Smith.

Mr. Davall to J. E. Smith.

Orbe, April 13, 1795.

The sight of your hand-writing yesterday evening gave me such pleasure as I could never express, and at the same moment the black seal occasioned great anxiety.

Believe me, my dearest and most excellent friend, I grieve with you sincerely on the cruel separation you have experienced, and the more so by reason of the interesting qualities of the good father you have lost. From the course of nature we cannot avoid these trials, and in my opinion one of our best comforts is the happy reflection he was a good and upright man. I love you more than ever, as you have done justice to my cordial affection to you by the detail of your sad communication.

Mrs. Davall and I hardly pass a day without talking of you. She was scarcely less affected than myself at your account of the loss of poor François.

It had been a favourite hope of mine to know him at last decidedly attached to and fixed with you, well conceiving how important his aid would be. I should not thus renew your regret, but François, —his gentleness, his intelligent eye, his pleasing air,—you—all that is yours—every moment I have spent with you,—is as present to me, and will be while I live, as if there were no intermediate occurrence whatever. I am delighted with your dedicating a genus to that excellent and unfortunate young man.

I am extremely impatient to have some fresh news of you: your health, all that regards you, is to me more interesting than anything else. Not one day passes without thoughts of you,—remembrance, regret!—such as diminishes my share of happiness. This very morning, taking a turn in my little garden, quite in ecstasy at the sight of four or five delightful stems of *Saxifraga mutata* in full lustre, I exclaimed quite loud, "Why is he not here?—at least the greater part of the seed shall go to him; he will admire plants, offspring of these; though not with me, he will not see them without thoughts of his most cordially devoted friend."

I really languish to see the continuation of English Botany.

May Heaven's best care attend you!

E. DAVALL.

In a letter dated 1795, noticed at the beginning of this correspondence, Mr. Davall says, "I too often found the want of a present friend. It was decreed," he continues, "that in this very place there was a person of the best merit, whom I had long known, and whose sentiments in most respects agreed with my own. We were married. Providence placed her by my side, and to her I owe all the domestic happiness I have enjoyed; and although I did not escape censure, I had learnt long since that the criticism of certain persons is praise. I am sure you are my friend. I know you have a heart open to sympathy. It is to you alone I can mention these circumstances, and alleviate no inconsiderable degree of pain to my mind. I have a fine little boy, who on the 25th of March will be two years old: he has a sister, now near the age of six months, who promises health and strength. You cannot conceive what anxiety I feel when I reflect that these interesting objects of my attachment are strangers on the earth, strangers in the country of their birth, and in the native land of their father;—how, with all the pleasure I experience from the gradual progress of my boy, I regret that his first ideas are not expressed in English. Hitherto it is in vain that I speak to him almost every day of his father's best friend. He will answer me with uncommon feeling, *Je l'aime bien ton bon ami, papa, où est il? Je voudrois bien le voir Mr. Smif.* It is always *Smif;* and perhaps I shall never hear him say Smith.

"When I look forward I am lost in sadness; the desolation of the times in which we live is fit also to inspire melancholy. It should seem (I wish I may be in error) that England is running headlong to her ruin. Almost all I have is in the English funds, and I am not quite easy on that head. This country, as every other, is so affected by these disastrous wars, that the price of every necessary of life is doubled, and moreover we have almost to fear a famine.

"Our bread is much inferior to what it is in less

miserable times, and it has become necessary to husband the corn, to forbid all bakers to admit to their ovens every kind of delicacy made of flour.

"The armies on the Rhine? and the cold procure us frequent visits from wolves,—charming company! we have more than usual. Last night (January 24) one was seen in the garden of a countryman at Montcheraud, a village very near us; and plenty of their tracks about the Grotto of Montcheraud, of which you have a view. There are societies of five or six together."

The next letter has the date of 1798, expressing too plainly the progress of disease in his enfeebled frame.

Mr. Davall to J. E. Smith.

Feb. 13, 1798.

For a long time,—a very long time past, I have not written to you, because there was no proportion between the fullness of my heart, all that I wished to say, and the strength I have had for the execution. I have been obliged to comfort myself with thinking almost without intermission of you; but, also, for some time past, my dearest and only hope has been that of meeting my friend in another and a better world.

This letter, which did not arrive in England till after Mr. Davall's death, was left unfinished. He died on the 26th of September, 1798, at the early age of thirty-five.

"To him the tenderest sympathies were given!"

He resembled the chosen friend of his heart in the warmth and devotion of his affections, but had less power to resist the ills of life. Many passages in the foregoing letters have disclosed the truth of what is here remarked. It is to show his worth, his tenderness, his rare attachment, his devotion to science, his love of nature allied so closely with the love of God, that so little reserve has been made of feelings which, in the writer's opinion, confer upon him the highest honour. Should they by any be considered too acute or reprehensible let it be remembered, "those tender longings which absorbed his soul, preyed on his spirits, undermined his health," were the sorrows of a banished man. Let those who can command both pleasure and society at will, contemplate with deep compassion the generous, disinterested Davall.

CHAPTER VIII.

Miscellaneous Letters to Sir James Edward Smith, from 1797 *to* 1807:—*Mr. Voght.—Mrs. Charlotte Smith.—Mr. Voght.—Professor Williams.—Mr. Johnes.—Sir Joseph Banks.—Bishop of Winchester.—Mr. Brodie.—Rev. Mr. Abbot.—Dr. Buchanan.—Rev. Mr. Talbot.—From the same.—Sir Thomas Frankland. — From the same. — Dr. Hedwig. — Mrs. John Taylor.—J. E. Smith to M. Ventenat.—M. Ventenat.—From the same.—J. E. Smith to Dr. Hedwig.— Sir William East.—M. Mirbel.—Duke of Bedford.—Mr. Johnes.—Professor Martyn. —Mr. D. Turner.—Mr. J. Lee.—Colonel Hardwicke.—Dr. Maton.*

From Mr. Voght.*

My dear Sir, Flotbeck, May 23, 1797.

OFTEN indeed I took up my pen to recall myself to the mind of my most respected master in botany, and my philanthropic friend, whom I most sincerely love and revere. Somehow or other I as often have been prevented by employments less pleasing, but of a more pressing nature. But today I will not be prevented; I locked the door of a little thatched hut on the bank of a rivulet that meanders through my meadows; and amidst the full chorus of

* Casper Voght, Esq., author of a pamphlet entitled, "Account of the Management of the Poor in Hamburgh, since the year 1788; in a Letter to some Friends of the Poor in Great Britain:" printed at Edinburgh in 1795.

our nightingales, accompanied by the murmur of a little rill, I intend to consecrate a few moments to one of the best men I ever knew.—Thanks then, my dear friend, for your most magnificent present: they are indeed fine and rare plants, unknown on the continent, and which will do great credit to my herbarium. Botany is still a study I am very fond of, but, like a man who loves an absent mistress, my several, perhaps too numerous, avocations prevent me from paying so much attention to her as I wish to do, and shall do hereafter. In saying so, I only speak of the scientific part of the systematical study of botany; for I am very busy as to the œconomical part. I try to find out plants for the use of men and of cattle, and so improve by culture those that are already in use. You will do me a great favour, if among your pursuits you happen to think of some plants, the use of which ought to be tried œconomically. I promise to attend to it, and, if you can give me some seed, to send you an exact account of the result. The physiology of plants continues likewise to attract our notice; and I am pointing out with M. Schmeisser a plan of experiments about some causes of the fertility of peculiar ground for peculiar plants, along with the influence of light, and of water, whose decomposition by an unknown cause is certainly an immediate food of plants; and of the different kinds of gases upon vegetation.

I arranged for my particular use a *herbarium pictum* in the way of Lord Bute's, which I find very convenient.

I am, thank God, well enough, which I ascribe particularly to my country life, exercise, and rising early in the morning. I renounced supper long ago. Thrice in the week I go to town to attend to the direction of the Institute for the poor, whose number since my publication still continues to diminish. We have not above 2500 poor now in Hamburgh, and of them, all those that choose to be industrious are comfortably situated. Our method has made proselytes amongst our neighbours: at Bern and at Copenhagen they are busy in founding an institution upon the same plan. I am called upon in the latter place to take an active part, which, to a certain degree, my connections with the Danish Court does not permit me to decline.

I give some time to chemistry, but the most of my leisure is engrossed by farming. I have now 500 acres of excellent land, three miles from town on the river side, in my hands, which by-and-by will be in the state of the richest culture. I teach my neighbours the use of marling, of oxen, and of better implements of agriculture. If they learn from me to do as much work, and infinitely better done, with one half of the powers employed till now; if they learn to keep half the teams they have now, and save one half of the seed they annually throw into the ground,—then the same ground will nourish thrice the number now living upon it, and *they* enjoy comforts of which they have not yet an idea.

So, my dear friend, in drawing into the circle of my family all that surrounds me, I have not yet perceived any regret of having none of my own; and it seems even to me that I would lose a great number of my old relations, were I to contract a more intimate one, that would more narrowly circumscribe my attention and my cares. Yet I know very happy marriages. I am intimate in some such fortunate families: I partake of their happiness, and believe it truly the most comfortable situation in life. I think you will find the life in Norwich more suitable to your taste than the continual hurry and bustle in a town like London, where for communicating and increasing knowledge you ought to spend a part of the year, and live the remainder to yourself and your friends.

Dr. Giesck is gone over to another world, where I hope he will not be punished for his plagiarisms and hasty productions in this.

The Germans are all mad now about Cryptogamy. It is vulgar now to pursue any other line in botany, and I really expect they will be great in these trifles.

M. Schmeisser and M. Wattenbach wish to be kindly remembered. As for me, I'll never cease to love you as long as I live.

Vale et fave,

Ever and sincerely,

VOGHT.

From Mrs. Charlotte Smith.

Sir, March 15, 1798.

The friendly politeness with which you honoured me when I took the liberty of addressing you almost twelve months since, has made on my mind too flattering an impression to allow me to forget, that in the letter you then favoured me with, you mentioned that in the months of April and May you were usually in London. As I am now settled there, permit me to hope that I may be so fortunate as to be favoured with your personal acquaintance when you this year make your annual visit. Though after a long and successless struggle I am compelled to leave the country, my passion for plants rather increases as the power of gratification diminishes; and though I must henceforth, or at least till peace, or something equally conclusive, dismisses me to the continent, (whither I will go if I have strength whenever it is practicable,) botanize on annuals in garden pots out at a window, it will be a considerable consolation to have an opportunity of being known to the principal of that delightful and soothing study; and who is, as well in science as in benevolence and cultivation of mind, an acquaintance so greatly to be desired.

I have not forgotten (being still compelled to write, that my family may live) your hint of introducing botany into a novel. The present rage for gigantic and impossible horrors, which I cannot but consider as a symptom of morbid and vitiated taste, makes me almost doubt whether the simple pleasures afforded by natural objects will not appear vapid to the admirers of spectre novels and cavern adventures. However, I have ventured a little of it, and have at least a hope that it will not displease those whose approbation I most covet.

A domestic occurrence, very much unexpected and very unpleasant, has made my heart ache and my hand tremble; but they are used to it, and both should by this time know better. It would be a prettiness, though very true, to say that while either are in existence it is a pleasure to say how much

I am, dear Sir,

Your obliged Servant,

CHARLOTTE SMITH.

From Mr. Voght.

My dear Sir, Flotbeck, Aug. 11, 1798.

It is a charming description you make of Hafod and its owner. I know of nothing that strikes me more, and awakens more pleasant feelings in my mind, than thinking of natural beauties enjoyed by active benevolence. Such a good man as your friend ought not to want good wine; and he shall not if I can help it. As to the yellow turnip, I know what you mean; it is the teller turnip, cultivated in the driest sands of the Mark Brandenburgh. It will not succeed unless you have such sands. As for the turnip itself, you may along with the hock get a barrel of them.

I have a great respect for a man who has planted 697,000 trees in one year. On the continent planting on a large scale is perfectly unknown, nor are there any nurseries for that purpose.

Among many future improvements of agriculture in general to which I was so fortunate as to lay the foundation there is a nursery of 1,500,000 plants begun this spring, which will I hope in a few years give the means to plant four fifths of Holstein, which after my example will perhaps some years hence have diffused the spirit of planting.

Yes, my dear Sir, I have received your valuable volume of Tracts relating to natural history, and also the pamphlet you sent me by Mr. Clark; but I will be obliged to you for a syllabus of your Botanical Lectures. However distant my present pursuits are from the track of this lovely science, I hope sooner or later to wander them over again, and would be sorry not to be guided by you. The exercise of practical agriculture and the promotion of this useful knowledge, together with our Poor Institute, take up all my leisure. I shall certainly revisit England as soon as Europe shall be restored to peace. You know, my dear Sir, that you stand foremost amongst the friends whom I shall feel myself happy to meet again.

Ever yours,

VOGHT.

From Professor Williams.

Dear Sir, Oxford, Aug. 1, 1798.

I passed a short time this morning in the Botanic Garden with Mrs. Kett*. At this season of the year a botanist is of rare occurrence in Oxford. It is singular on any occasion to meet a lady who is really conversant with plants. I am obliged to you

* Of Seething Hall, Norfolk.

for affording me an instance, and introducing me to an acquaintance so valuable. A very hasty visit, and under a vertical sun, could give little satisfaction to your friends; yet Mrs. Kett has kindly promised to supply me with any of my *desiderata* in her pos session. I beg you would assure her I shall have much satisfaction in sending to her, as I shall indeed to any of your friends, whatever specimens we can afford from the collection here. Your name could not but occur to us, and very frequently in our walk; for many reasons, both of private satisfaction and of botanical importance, I wish you could have been of the party. I am surrounded by many nameless friends,—by many at least of very doubtful characters. You are destroying, I am assured by the information of Mrs. Kett, a new and favourite *Verbascum:* I shall be glad when you can examine this genus, with which I have been lately not a little perplexed;—yet in fixing the Gramina and Mints you are better and more laboriously employed.

I am yours faithfully,

G. WILLIAMS.

From Mr. Johnes.

My dear Sir, Hafod, Sept. 2, 1798.

I was very sorry Mr. Gurney * made so short a stay, for I was never better pleased with any one on so short an acquaintance. I liked one of his daughters very much, but it was all in such a hurry I could scarcely speak to all of them.

* The late John Gurney, Esq. of Earlham, Norfolk.

They travelled in proper style, determined to be pleased with everything; the only mode that could have made their number any way comfortable.

Have you read Lord Orford's Letters? If not, you will I am sure have a great treat. I have but begun on them, and am more pleased than with any correspondence I ever read; it shows an excellent head and heart. They are very witty, and have excellent good *nonsense*, which is one of the most difficult things to write.

Mrs. Johnes is not well, but says she shall soon recover when her daughter does.

It is a great comfort to me, that as those from whom I had a right to expect kindness fall off, I find others who make ample amends; and the more flattering to self-love, because one hopes it may be owing to personal qualities. Among all, my good friend, I can never forget your great kindness and affection, "*dum memor ipse mei dum spiritus hos regit artus.*"

All our kind compliments attend you.

Truly yours,

THOMAS JOHNES.

From Sir Joseph Banks.

My dear Doctor, Soho Square, Jan. 6, 1799.

I shall feel myself much flattered if you finally determine to dedicate your *Flora Anglica* (*Britannica*) to me. It is a book much wanted, and one that cannot but receive advantage from the hand of the possessor of the Linnæan collections; from your hand it will also receive many other advantages of various kinds,—enough to make it far surpass all its antecessors.

Pray how low was the thermometer with you? and on what day was it the lowest? I began to think that the great severity of the frost has been in your parts and in Lincolnshire. On the 27th, my servant Roberts, whom you know, was valuing wood in a copse near Tattershall, where he heard noises several times repeated, like the loud smack of a whip, which proved to be the riving of the branches of trees by the intensity of the frost. I do not remember that this phænomenon, though common in the North of Europe, has been observed since the hard frost, as it is called, of 1739-40, when Miller mentions it. I believe that a combination of circumstances are necessary to this phænomenon, but intense cold below the 0° of Fahrenheit is certainly one of them.

My time has been of late so taken up by the Committee of Coinage that I have done little in botany. I regret it the less, as Dryander is employed in printing the Catalogue.

Dickson tells me that last market at Covent Garden was well supplied, and at reasonable prices, —a sign that the frost has done little mischief to our market-men. White brocoli is said to be the only plant that has suffered materially.

Pray offer to your wife and mother the best good wishes of the season, and for many happy returns; and believe me very faithfully yours,

JOSEPH BANKS.

From the Bishop of Winchester.

Sir, Farnham, Oct. 29, 1799.

Mr. Poulter has sent me your letter to him, and I inclose one to you from him. I am so persuaded that the fructification of the *Cycas* (*revoluta* as we suppose) is new in England, and probably in Europe, that I desired Mr. Poulter to write to you upon the subject, and also requested Mr. Brander to do the same.

I dare not aim at a description in proper terms, being rather an amateur than professor in botany, but I will relate what I know and have observed.

When I came to Farnham about the beginning of September, my gardener told me the *Cycas* had borne a singular appearance during the summer. As I did not see it earlier, I cannot speak of the flower; but on examining it I found a fine large *strobilus* formed upon the head of the plant, within the circle of the external leaves. The leaves of the *strobilus* itself were of a greenish gray colour, thick, pulpy, and beautifully laciniated: the *strobilus* hollow like a bird's nest, and filled with a quantity of *drupæ* (then green), about the size of a half-grown apricot, or large green almond; leaves growing mixed among the fruit. The only changes which the plant has since undergone, are, that at each point of the laciniated leaves there is a strong spine come forth, probably to guard the fruit; and the fruit itself, now probably ripe or nearly so, as the kernel is completely formed, is become of a very fine reddish orange colour, and covered with a fine gray down, and is very picturesque and handsome. I have cut open one of these drupes, the half of which Mr. Poulter has sent to you, but it is now so much faded as to give a very imperfect idea of the fruit upon the plant.

I shall have great pleasure in sending you two or three specimens of the fruit; but I should be much happier if I could have the pleasure of seeing you here, where I have beds for you and any one you may desire to bring to describe or paint the plant.

The painting alluded to by Mr. Poulter is now taking by one of my daughters; hereafter I shall hope to show it you. She is painting the fructification, that is, fruit and leaves of the *strobilus*, of the natural size, which is all that appears to be now necessary, for the external leaves are permanently the same; and this indeed is all that I conceive to be possible, for the circumference of the whole plant is much too large for any paper.

If you should not be able to give me your company, be so good as to let me know what you think can be best done for the information of the Linnæan Society or yourself; for it seems worthy their notice, as the whole which Linnæus says on the genus *Cycas* is, "De hujus charactere nihil etiamnum ab auctoribus traditum reperi." If you wish to send down any painter who will paint it at a reasonable price, (I say from five to ten or twelve guineas, or thereabouts, as I know not what painters of that branch usually charge,) I shall willingly be at the expense of it, and give a copy to the Society: but I am in hopes the painting of my daughter promises to be very accurate.

I am, with great respect, Sir,

Your faithful Servant,

B. WINCHESTER.

From Mr. Brodie.

Dear Sir, March 31, 1801.

Believe me, I shall always feel much pleasure in exerting myself to gratify your wishes, or those of any friend in whom you take an interest. My poor girl had the same desire,—that of rearing all sorts of birds from the egg; and succeeded with a great many, but was always baffled in the Ptarmigan, from the distance; for notwithstanding every precaution in packing, and their being carried by a man hired on purpose, swung in his hand, they were always addled. I would recommend their being sent by sea, for many reasons,—a principal one, that there is no mail-coach at present beyond Aberdeen.

As to the Black Game, they are become so scarce in Scotland that I am at a loss even how to advise you in procuring them. I remember in the year 1770, when I resided about ten miles immediately to the westward of Morpeth in Northumberland, they were in immense flocks on an estate in the neighbourhood, belonging to (I think) Greenwich Hospital, having been part of the Derwent Water estate. Perhaps this hint may be of service. The cock is a most beautiful bird, and becomes easily familiar. With the highest respect and esteem,

I remain, dear Sir, most faithfully yours,

JAMES BRODIE.

From the Rev. Charles Abbot.

Dear Sir, Bedford, Feb. 15, 1802.

The receipt of the last number of English Botany gave me a melancholy pleasure in retracing the happiness I enjoyed in a botanical acquaintance with the late Lady John Russell *; and whilst I thank you for your punctual observance of my request, in giving her ladyship's name to the *Rosa tomentosa*, permit me to say that no one, in my own personal knowledge, ever so well merited a posthumous attention of this sort. Her virtues do not sleep with her in the grave, for they are registered in the hearts of all who knew her, and her unexampled worth: Her noblest mausoleum is erected in the hearts of a grateful poor, who have lost their best friend. Lord John is not now in Bedfordshire, or this letter would not go unfranked; but I have long wished for your opinion on the subject of *Pyrus Aria*, which I expected to see figured in English Botany before now. The last fructification I sent you was most undoubtedly right, or the plant in my *Flora* is wrong: be it what it will, it certainly grows wild here.

I remain, dear Sir, yours very faithfully,

CHARLES ABBOT.

* First wife of the present Duke of Bedford, and mother of the Marquis of Tavistock and of Lord John Russell.

From Dr. Francis Buchanan.

My dear Smith, Bassaria, March 3, 1802.

I was favoured a few days ago with yours of the 17th of May, 1801. During a great part of the years 1800 and 1801, I was employed by the Governor-general in investigating the state of agriculture, commerce, and arts, in the dominions of the late Tippoo sultaun. I had just returned to Calcutta with a considerable collection, and had commenced to arrange my papers, when I received an order to proceed to Patna, and join the resident going to the court of Nepal. I am now busy in digesting the journal of my proceedings during the trip to Mysoor.

Nepal is a country extending south from the snowy mountains called Himaleh by Major Rennell, but Emodu Ghiry in the Sanscrit, and no doubt the *Emodus* of the ancients. Its southern boundary is the territory belonging to the English, so that its breadth from north to south is not great; but on the west it includes what Rennell calls Almorah, and on the east what he calls Morung. About sixty years ago this great extent of country was subject to a number of petty chiefs, who all soon after were subdued by Pritty Narain, rajah of Gorka. Dissensions have lately broken out in this ill-united kingdom and turbulent unprincipled people; and the principal persons of the reigning family have been obliged to take shelter in the Company's dominions. Although cruelly afraid of European interference, all parties among them, distrustful of each other and terrified at the thoughts of the return of the exiles, have wished for an English resident to preserve the peace. Captain Knox has been appointed for this office, and I go as his surgeon. I have a painter with me, and all requisites for making a collection. We have just entered the dominions of Nepal, but are still in the country below the mountains. We however expect in a few days to ascend the lower range that elevates Catmandu the capital, high above Hindustan. From its appearance, this ridge will be about four or five thousand feet high, and I expect to meet with little else but plants resembling those of Europe. Even here I find many common European plants, such as the *Anagallis arvensis*, *Ervum hirsutum*, *Sonchus palustris*, *Fumaria officinalis*, *Saponaria*; and the common produce of the country is wheat, barley, mustard, and flax. The *Cannabis sativus* grows here wild.

We have had a fine view of the snowy Emodus, and got several good altitudes of it, from which, if its latitude is not greatly erroneous, it must be considerably above five miles in perpendicular height above the level of Bahar.

Two very badly stuffed skins have been presented to Captain Knox from these mountains: from the horns and teeth there is no doubt of their being two species of wild sheep. The one is about the size of the *Argali*, and may be that animal, but it has hair exactly like the musk deer: this hair, although soft, is analogous to the prickles of the hedgehog, as it is hollow and inserted by an acute root. I should be glad to know if this be the case with the *Ovis* or *Capra Ammon.* The other may perhaps be the *Capra caucasia.* The length of my stay in Nepal is very uncertain. Whenever I am able, I intend returning to England, and shall have amusement enough for my old age in putting my papers into some order. At present I write my descriptions at full length, but I intend afterwards to make out a *Flora Indica* with short characters of the species, as much as possible resembling those Jussieu gives of genera as I am able.

In Mysoor I found another species of the *Vatica,* and plenty of the *Vateria,* which is one of the most ornamental trees I have ever seen. The *Dua Bunga* may be a *Sonneratia,* but this must be determined by the fruit. The unripe fruit did not seem inclined to become pulpy, but rather to be capsular: at any rate it is quite different from the *Sonneratia acida,* with which I am as well acquainted as with oats.

The plants I proposed calling *Hopea* are found in the woods of Chittagong and Tippera. Two of them produce an oil of turpentine, or at least one analogous to it, and which is called wood-oil by the English of Bengal. The two last species are valuable timber trees.

Roxburgh has since called another genus after Hope: its fruit is the same almost exactly with that I have now mentioned, but the stamina are definite. He found one species in a garden at Calcutta with ten filaments, alternately simple and bifid; the anthers are therefore fifteen. I found another in Malabar, where it is called iron-wood: it has ten simple filaments and ten anthers.

The *Eugenias* are worse than the Willows or Roses of Europe, especially those that resemble the *Myrtus Cumini*, which Roxburgh makes an *Eugenia*. I am sick at the sight of one of them. I am still sicker of the *Conyzas*, of which in this very camp I have been obliged to describe eight, all as difficult to define as any *Eugenia* in the world.

It is fortunate that Sibthorp's materials fell into your hands. I promise myself great pleasure in looking over my old friends the Lichens and Mosses, with the assistance of your *Flora Britannica*. They are the only resource for a botanist in the bleak hills among which I intend to pass my old days.

Yours very truly,

FRANCIS BUCHANAN.

I have left a number of drawings and descriptions of fishes with my good friend Mr. Fleming, President of our Medical Board, a very great encourager and promoter of natural history. You will find him a frank, pleasant, and learned man, with a greater knowledge of man and manners than usual.

From the Rev. Thomas Talbot.

My dear Sir, Paris, 23 Pluviose, l'an 10 (1802).

Soon after my arrival, I delivered your letters to Mr. Thouin and Mr. Millin; the former is extremely well, and was very happy at hearing from you. My time has been so much occupied, and the Jardin des Plantes at such a distance, that I have not yet had time to make him a second visit.

Mr. Millin has been so polite as to invite me to the evening meetings at his house every septidie of the décade, which I find extremely interesting and pleasant. The evening I spent there we had perhaps thirty people,—several foreigners, but no Englishman but myself. We have the good fortune to be extremely well situated in the Place du Carousel, or what was formerly St. Wicaise, one side of which has been pulled down, with all the old houses that encumbered the Carousel; and the square is now completed, and a most magnificent thing it is. We are directly opposite the Thuilleries, and St. Mark's Horses are placed just before our windows. We have seen, or rather looked at, most of the curiosities of Paris. It would be impossible to attempt giving you any thing like a description of what is to be seen among the statues and pictures of the Louvre. To you it is only necessary to mention, that among two hundred of the choicest models of statuary, are to be reckoned the Apollo Belvedere, the Torso, the Antinoüs of the Capitol, the Laocoon, the Venus, the Juno, the Bacchus of the Capitol, the Gladiator moriens, the delicious Venus sortant du Bain, the Cupid and Psyche, and many little if at all inferior.

Among the pictures which the Louvre Gallery (1500 feet long) will not contain, are the Transfiguration, and other of the greatest works of Raphael; Communion of St. Jerome, Martyrdom of St. Agnes, &c. by Domenichino; Vierge du Rosaire by the same; St. Jerome by Correggio,—the first and last esteemed the masterpiece of those two painters: la Mère de Pitié, the *chef d'œuvre* of Guido, that of Titian, of Tintoret, of Rubens, the Descent from the Cross (for which the inhabitants of Antwerp offered 40,000*l.*), Vandyk's Dead Christ, la Femme Hydropique of Gerard Douw, Poussin's Deluge; &c. The collection is particularly rich in Guidos. There are perhaps forty Guidos, and certainly not so few as a hundred pictures of Rubens; every church in the Low Countries being pillaged. Any attempt to express what one feels on entering this extraordinary place would be absurd; I felt all the time quite bewildered with wonder and delight. We have received great civility from Mr. Jackson, and have lately been introduced to some Russian ladies,—the Princess Galitzin and her sisters, who are most delightfully pleasant people.

You shall hear all our motions; and with our united regards to yourself and Mrs. Smith,

Believe me, ever yours,

THOMAS TALBOT.

From the same.

My dear Sir, Paris, 19 Ventose, 1802.

It is really almost setting about an impossible task to give you an account of what is now to be seen in Paris which did not exist when you were acquainted with it; though at the same time it must be confessed that in many points it has experienced a change much for the worse. The churches, stript of their pictures, their fine altars, and rich marbles,—which have, since the last Revolution has restored tranquillity to France, been replaced in a manner calculated only to make the loss of the originals more conspicuous. The church of the Invalides has suffered nothing, and is now converted into a temple of Mars, and filled with the trophies of the present war. The body and tomb of Turenne have been brought there. St. Geneviève, now the Pantheon, intended as a burial-place for illustrious characters, is getting very forward under the original design, and will, when completed, be a magnificent national monument. Voltaire and Rousseau are already removed there.

The Museum of Ancient Monuments is wonderfully curious and interesting: they are placed in the late convent of the Petits Augustins, which is somewhat too small for their reception; it is at present unfinished. All the monuments from St. Denis, and indeed throughout France, have been brought to enrich the collection: those of Richelieu, of Mazarine, of Abelard and Heloise from the Paraclete, Anne de Montmorency, Diane de Poitiers, &c., most of which are arranged in large halls appropriated to each century, beginning with the eleventh. The catalogue is rather too large to send you.

The library, which the Government proposes to transport to the Louvre (at present inhabited by artists, who are to be removed,—the professors to the Eglise de Quatre Nations, and the students to the Sorbonne), is beyond measure rich in manuscripts, (the gleanings of Italy,) in the complete collection of engravings, and the printed books composed of an assemblage of all the best libraries belonging to the Sorbonne and other public institutions, independent of the original collection. The medallions, as well as other objects of antiquity, and above all, the gems of the Cabinet de l'Antiquité, are matchless; the latter, for their immense number and extraordinary beauty and perfection, are far superior to any conception I could have formed. We have lately seen the Gobelins, which is exceedingly improved of late years: when you see the tapestry placed by the side of the pictures, it is almost impossible to tell the one from the other.

We very happily had taken a pretty wide range in seeing the curiosities of Paris, before we found ourselves immersed in engagements as we are at present; so that we have had an opportunity of enjoying these different modes of amusement both in perfection; and great as has been the pleasure we have received from the one, yet I hardly know whether it is not equally as much enjoyment that we derive from the other. After a time, however we may be surrounded by interesting objects, yet it must be from agreeable and rational society that we are to look for the pleasure of so long a visit as this of ours; and we have been happy enough to find it. The state of society here is of course materially changed since the Revolution. The people who keep the best houses, and receive in the most showy style, are for the most part the bankers, and those whose fortunes have risen since that period. Those of the old régime who remain here, are for the most part so reduced in fortune as to be unable to receive anybody in their houses; but we see the greater part of them among those with whom we visit, and have begun to get a footing among some of those who still see company. But the persons whom we chiefly spend our time amongst, and who actually seem to be the only set in Paris that understand real hospitality and pleasant living, are the Russians, with whom I believe I mentioned to you that we had formed an acquaintance, and are now perfectly naturalized. Every one of their houses is open every night; and at some or other of them we are certain, if they are going to no ball or regular party, to meet with those who have something besides stars, ribbons, and titles to recommend them; namely, good education, good breeding, and good sense. We contrive to fill up every evening with engagements to parties, after going to some of the spectacles, which we do most evenings, with a most charming woman, the Princess Galitzin, whose beauty, whose manners, and whose character I feel myself unequal to describe. We have been at a great many very fine and elegant balls, particularly one at the Comtesse de Gerebtzoff's, who with the Princess were the first of our acquaintance, and have been infinitely kind in pushing us forward. Three nights ago we were at an exceedingly elegant entertainment at the Princess Hohenzollern's, and last night at her sister's the Princess of Courland, who is a most beautiful and pleasing woman. Her parties, which are the finest and the most select, as well as the most easy and pleasant things going forward here, occur every octodie of the décade: we have a general invitation to them, and to her house at all times; and every morning when the weather is good we assemble there to breakfast, and ride on horseback to the Bois de Boulogne.

Prince Hohenzollern and Mr. Dernidoff furnish horses; and Fanny* is in such request as a horsewoman that she generally has the choice of the stables. Tonight Fanny means to keep back for a ball; and tomorrrow at Mrs. Lemaistre's, an English lady to whom Lord Henry Petty has introduced us. I intend going to the bal masqué at the opera, which commences about one in the morning, and finishes about six or seven; indeed the Parisian hours are very terrible. We never break up at Madame de Courland's before five, and at most other parties three or four.

The carnival has afforded us great amusement: it was the first regularly observed here since the Revolution, and, as you may suppose, not wanting in gaiety.

I was presented to the First Consul by our Minister at the audience on Saturday. His manners are extremely mild and interesting: he talked a great deal to us, and paid many compliments to the English nation.

* Mr. Talbot's sister, now Countess of Morley.

I have been this morning with a Pole, who knows a great deal of pictures, to see an immense salon, opened yesterday at the Louvre, filled with the choicest morceaux of the whole collection, of which the small appendix to the catalogue gives you an account.

We see a good deal of an extremely pleasant Italian family, whose name is Gallo. The Marquis de Gallo is ambassador here from Naples, and his wife a very accomplished woman. We have been introduced to Lady Mount Cashel, who called on us yesterday and invited us to a ball, the 1st Germinal. We dined yesterday with the Rohans at the Bagatelle, and passed the soirée with the Hohenzollerns, and have abundance of balls in view for next week; one of which is to be given by Dernidoff, the richest man in the Russian empire, and who prides himself a little on his magnificence, such as wearing a diamond of 19 carats for a shirt-pin, &c., and is intended merely to astonish the people here. His house and his wife's which adjoins it, and both of which are very immense and fine, are to be laid open for the night. Adieu! You will think our heads run only on dissipation.

T. Talbot.

From Sir Thomas Frankland.

My dear Sir, Chichester, April 25, 1802.

As you probably know the sad event of my endeavours last autumn, I shall only say that I am in good health, supported by hope of re-union with those whom I most loved, in a happier state!

Our plans are to pass the summer at Brighton: no place is so agreeable to me as my own home, where I have been seventeen years collecting sources of pleasure, though it does not seem to be the will of God that I should enjoy it. Soon after my arrival here I was informed that two of the Duke of Richmond's horses had been poisoned by some plant in their pasture. I remembered the *Œnanthe crocata** by the side of a rivulet in 1782, and pronounced it to be the evil in question, and was then told that the pasture is by the side of this very stream.

I have received the last seven numbers of English Botany. I observe a good figure of *Conferva rosea* (a good name): I found it at Worthing in 1782, and showed it to Hudson, who thought it his *purpurascens*.

THOMAS FRANKLAND.

From the same.

Dear Sir, Chichester, June 20, 1802.

As Parliament will be dissolved in a few days, I take the opportunity of writing a few lines by my brother's *fiat*, whom however I purpose to bring in again, together with his present colleague.

I received your kind letter, and here conclude my

* Hemlock,—Water Dropwort. See English Botany, fig. 2313; not the Common Hemlock, *Conium maculatum*.

account of the *Œnanthe crocata*, which I have now convicted of poisoning two mares in foal belonging to the Duke of Richmond last November. I have viewed the field in company with the gardener and groom, who found one mare dead and the other dying. The stream (Lavant), which bounds one side of the field for a considerable length, was quite dry, and I was shown one plant of which the mares had evidently dug up the roots with their hoofs. Their stomachs were full of a white substance like saw-dust, agreeing exactly with what we observed on examining the roots this year, which were intermixed with those of last season in decay. The bank of the stream was abundantly full of the plant, and more decisive evidence of murder could not be pronounced against it. A few days ago I observed a horse in a ditch cropping the tops of the plant, which I suppose are comparatively innocent. The brood mares possibly might have depraved appetites, which induced them to dig up the *roots*.

Yours,

THOMAS FRANKLAND.

Viro gravissimo et experientissimo Naturæ scrutatori oculatissimo, D. Smith, Præsidi Societatis Linnæanæ Londinensis celeberrimo, fautori ac patrono summo, pietatem, observantiam, obsequium spondet Hedwig.

Lipsiæ, Aug. 2, 1802.

Quum amicissimus Dawson Turner me per literas certiorem reddiderit, te, vir gravissime, suscepturum esse non tantum pietatem meam, sed et animi vota indulgente tua et propria humanitate, suscepi hanc tibi conscribere epistolam, et quæ fovet pectus sensa tibi omni probitate proponendi. Fervet enim adhuc atque exsultat ante omnia animus meus in tantam tuam benevolentiam, qua olim patrem meum desideratissimum Societati tuæ Linnæanæ gravissimæ socium esse jussisti; honorificentissimæ sane et præclaræ Societatis, quam amant et suspiciunt quicunque in botanices studiis res suas honestissimas vere agunt. Quid igitur? dum honorem quem patri optimo parasti, respicio, non possum non, quam gratias non tam tibi agere, quam referre.

Neque enim Johannis Hedwigii faciem unquam vidisti, scripta tantum nosti! Jam ut et vultum ipsius intueri valeas, annulum tibi cum effigie patris (quem ipse per paucos dies in digitis meis habui annulum) mitto, humanissimis additis precibus, ut hunc annulum digito tuo applicare velis, et ut patris immortalis faciem agnoscas probi illius et integerrimi viri, et animi mei grata sensa clariora suscipias.

Quæcunque enim patri meo optimo acta sunt, quam ipse humanitatem ab aliis suscepit, in me collocata quasi sentio, et ad finem vitæ usque grata mente et sensu habebo. Sed hæc hactenus. Sufficit si a te, quæ hoc annulo exprimere volui, animo hilari et benevolo suscipiantur.

Quod jam ad ea pertinet, quæ Dawson Turner, vir æstumatissimus, tecum mei gratia loquutus est, pertinent ad Filices. Tenes, enim, me in tractandis, depingendis, et microscopice et analytice describendis his stirpibus versari. Dispositionem tuam sequuturo multæ adhuc desunt filices, quæ non nisi benevolentia aliorum, imprimis tua, suscepi possunt. Si ergo quædam ad labores meos quæ vel communicari et tum remitti possint, habes, vel a te humanitate singulari in me conferri, grato animo et domi et palam pronunciante suscepturus sum. Nam quis diffidere posset quod penes te sint multæ et novæ filices ex variis terris et regionibus allatæ, qui possides plurimorum regionum amicos, qui tenes Linnæi immortalis herbarium et quæ sunt reliqua. Si ergo humanitate tua singulari meis studiis succurrere velis, persuasum te habeas, non ingrato dedisse.

Mrs. John Taylor to J. E. Smith.

My dear Friend, Norwich, May 6, 1803.

I should not do justice to my own feelings if I omitted to send you a few words,—not to express my sense of your kindness, for that is impossible, nor yet to tell you I am better, for this you will learn without my wasting any paper upon the subject,—but to keep up in some way the communication which is suspended by your absence. I have often told you that the consciousness of yourself and dear Mrs. Smith being within my reach, the recollection that I can at any time see you and converse with you, forms one of the charms of my existence; and time and habit (which are said to lessen our sense of the blessings we enjoy) are continually increasing this feeling in my mind. Indeed, what can time and habit do, but add to the stores of recollection repeated proofs of friendship, or make these proofs more necessary and dear to us? When you are away, I think it right, because those who deserve your society in other places should have a portion of it, and because you are certainly varying your enjoyments, and enlarging the sphere both of your usefulness and of your gratification; besides, when you are able to do this, I am easy about your health. In order to answer the calls made upon you for exertion (in London particularly) your best health is necessary. How gladly would I encounter much more bodily inconvenience than I have lately endured, to have my mind free from anxiety concerning my husband and you! Indeed I have scarcely thought what I felt was an evil, so many interesting and pleasing sensations accompanied it,—even those speculations upon the uncertain tenure of our mortal existence, which a weak state of health naturally leads to, were wholly unattended on my own account with agitation or alarm; but for those I love, I am full of terror and dismay. Well! let me indulge the pleasing idea that you will meet happily in London, congratulate each other upon looking well and feeling well; and that, after the time of our separation is past, we shall all join and rejoice together, and renew our delightful intercourse. I had another kind visit from Mr. Trafford lately, when we talked of you particularly, of our little circle in general, of our Conversations, and the hope of their renewal.

How many fine things could I say to Mrs. Howorth and to Mr. and Mrs. Lane, if I were to indulge my own feelings! and all right, true, and sincere. I shall always think of them with the respect and affection their heads and hearts deserve, and as some of my most intimate friends, though we only became acquainted last summer.

Notwithstanding my having just filled two sheets to the Aikin family, and that I have reached the top of my third page to you, I know not how to cease talking to you; remember the restriction which has been laid upon my tongue, and consider that even the pen is a welcome substitute. Lucy Aikin is suffering under the same prohibition, and we have some joking upon the subject,—yet I am very serious about her. Her letters contain the sound sense and just criticism of a healthy mind, whatever may be the state of her body. Our attention has been much drawn to Cowper lately,—hers by the perusal of the new edition of his Iliad, and mine by his Life and Letters. Whether his interesting and unhappy malady was purely physical, or whether gloomy reading and conversation contributed to it,—if I had been his biographer, I should have freely expressed my feelings respecting Calvinism: you and I cannot bear such representations of the Father of the *whole* human race*. That I may engross no more of your

* This alludes to the doctrine of election and reprobation: yet the kindness and christian charity of Sir James's heart cannot be more clearly discerned than in his mode of reflecting on certain doctrines which he considered injurious to the impartial justice of God, and as tending directly to immorality.—In speaking of these he often observed, that the goodness of human nature, which they who hold these opinions so much depreciate, overcomes a bad system of religion; and that the alembic of an honest heart and sensitive conscience transmutes a poisonous doctrine into wholesome food.

valuable time, I will beg His blessing upon you wherever you go. Our children, who have so much reason to love you, desire their grateful remembrances :—you will frequently be the subject of our conversation, still more frequently of the meditations of your faithful friend,

SUSANNAH TAYLOR.

J. E. Smith à Mons. Ventenat, de l'Institut National, Bibliothecaire du Pantheon à Paris.

Monsieur, Londres, Juin 20, 1803.

Je viens de recevoir votre lettre obligeante du 30 Germinal, accompagnée des deux superbes livres de botanique, dont Madame Bonaparte m'a fait l'honneur d'enrichir ma bibliothèque. Ayez la bonté, Monsieur, d'offrir à cette illustre dame mes plus humbles remerciments, et de l'assurer que ces livres et leurs suites feront l'ornement de ma collection, par leur propre merite, et comme un temoignage de son estime. Je ne manquerai pas, aussitôt que les circonstances le permetteront, de faire passer à Madame Bonaparte quelques temoignages de ma reconnaissance, surtout des plantes sèches, qui par leur rareté ou leur beauté pourront l'interresser, ou même des graines pour son jardin. J'attendrai avec impatience les Numeros suivants du magnifique *Jardin de Malmaison* et de *Plantes Liliacées,* aussi bien que votre ouvrage sur le jardin de mon bon ami Cels, que je n'ai pas pu encore me procurer, mais que j'ai longtemps desiré d'avoir. J'embrasse avec joie cette occasion de profiter de votre correspondence, et je serai très flatté de vous être utile à quelque chose. Peut-être vous avez vous-même un herbier, et que j'y pourrois ajouter quelques plantes nouvelles. Permettez moi d'observer que votre *Melaleuca gnidiæfolia* me paroit ma *M. thymifolia*. Pour ce qui regarde l'opposition des feuilles de quelques *Metrosideros*, je puis vous assurer que c'est notre ami Cavanilles qui a tort.

J'ai donné ordre de faire mention honorable de vos publications dans nos journaux, quoique je me fais scrupule de n'y rien écrire moi-même, afin d'eviter les disputes literaires. Croyez moi, Monsieur, avec les sentiments de la plus parfaite consideration,

Votre très humble et très obéissant Serviteur,

J. E. SMITH.

From Mons. Ventenat.

Monsieur, Paris, Octobre 10, 1803.

J'ai fait connoître à Madame Bonaparte combien vous aviez été sensible à la marque d'estime qu'elle vous a donnée, et qui vous etoit due à plusieurs titres que votre modestie ne me permet pas d'exposer. Elle se fera un plaisir de vous envoyer la suite des deux collections ; et elle a recommandé devant moi à son intendant, s'il ne trouvait pas moyen de vous les faire parvenir en le moment, d'en garder le dépôt jusqu'à la première occasion favorable. Elle recevra, Monsieur, avec la plus grande reconnoissance les graines que vous voudrez lui envoyer. Pour ce qui concerne les plantes sèches, je vous avouerai franchement qu'elles m'étoient destinées; et puisque vous avez la bonté de m'en offrir, il est inutile que vous en fassiez deux pacquets: je recevrois deux fois les mêmes exemplaires, et je priverois des effets de votre générosité quelque botaniste qui en sentiroit le prix aussi bien que moi. Je vous remercie, Monsieur, de ce cadeau que vous voulez bien me faire; vous enrichirez ma collection, et vous me mettrez à même d'éviter des erreurs, ayant sous les yeux des objets de comparaison. Si l'envoi précieux que vous aviez fait à M. de Jussieu fût arrivé avant la publication de ma première livraison, je n'aurois pas donné le nom de *M. gnidiæfolia* à la plante que vous aviez déjà nommée *thymifolia:* c'est un de vos compatriotes, Mr. Kennedy, qui m'induisit en erreur, en m'assurant que cette espèce étoit parfaitement nouvelle: voilà une preuve de plus pour moi, que l'autorité des cultivateurs n'est pas toujours d'un grand poids en botanique.

Je serai très flatté, Monsieur, de vous faire hommage du *Jardin de Cels:* il m'en reste encore quelques exemplaires de choix et sur papier vélin. Vous me ferez plaisir de me procurer une occasion favorable pour vous en faire parvenir un; j'y joindrai les livraisons qui auront paru d'un nouvel ouvrage qui fait suite au Jardin de Cels, et qui est du même format: quoique les deux ouvrages ne soient pas coloriés, je crois qu'ils pourront être plus utiles à la science que celui de la *Malmaison,* où je me trouve forcé de faire entrer des plantes déjà publiées. Oserois-je, Monsieur, vous exprimer le desir que j'aurois de posséder quelques uns de vos ouvrages, que je n'ai jamais pu me procurer chèz nos libraires de Paris; savoir, vos *Icones pictæ*, votre *Spicilegium*, et *Tracts relating to Natural History*.

Soyez assuré, Monsieur, que j'apprécie tous les avantages que je puis retirer de l'honneur de votre correspondence. Si j'étois assez heureux pour pouvoir vous être ici de quelque utilité, j'en saisirois les occasions avec bien de l'empressement.

Agréez, je vous prie, Monsieur, les sentiments de la plus haute estime et de la plus parfaite considération.

VENTENAT.

From the same.

Paris, 28 Nivose, An xii,
19 Janvier, 1804.

Monsieur,

J'ai reçu avec d'autant plus de plaisir la lettre que vous m'avez fait l'honneur de m'écrire, que je craignois dans les circonstances où nous nous trouvons de ne point en recevoir:—je vous remercie de m'avoir indiqué un moyen de correspondence aussi simple que commode: j'en profite, Monsieur, pour vous faire passer le *Jardin de Cels*, et les 4 livraisons qui ont paru du *Choix de Plantes*.

J'ai communiqué votre lettre à Madame Bonaparte: elle m'a prié de vous témoigner combien elle étoit flattée de l'intérêt que vous preniez à son jardin. J'aurois désiré qu'elle m'eût fait remettre les fascicules de la *Malmaison* qui ont été publiés, et que vous n'avez pas encore reçus, pour les réunir à mon envoi; mais il paroît que le nouvel intendant de la Malmaison revendique comme un droit de sa place, l'avantage de vous les addresser ainsi qu'aux autres botanistes que j'avois désignes à Madame Bonaparte; je me suis contenté de donner votre addresse, et d'indiquer le moyen de vous faire parvenir la suite.

Je récevrai, Monsieur, avec de la reconnoîsance, les objects que vous avez la bonté de me destiner. Les graines surtout me seront d'une grande utilité. Les circonstances ont empêché Madame Bonaparte de se pourvoir chez Messrs. Lee et Kennedy; et les envois de graines et de plantes qu'on lui faisoit, ont été capturés par vos compatriotes.

Je prends la liberté de vous consulter sur un *Echium* que j'ai fait dessiner, et dont je vous addresse une feuille avec un epillet. Cet *Echium* est originaire des Canaries; je crois bien que c'est la *giganteum* * d'Aiton, mais les *strigæ* ne sont pas très sensibles sur les bractées, et sur les calices; de plus, les nervures des feuilles sont rougeâtres.

Monsieur de Jussieu prétend que c'est une espèce nouvelle. Veuillez bien, Monsieur, me faire connoître votre avis, qui sera pour moi une décision arrêtée. Je désirerois aussi savoir si *l'Echium candicans* de Jacquin est le même que celui d' Aiton †. Messieurs Thouin, Jussieu, Cels, et Desfontaines, m'ont prié de vous témoigner combien ils étoient sensibles à votre souvenir.

Si j'étois assez heureux pour pouvoir vous être ici de quelque utilité, employez, Monsieur, avec

* In a marginal note in the original letter—"right":—J. E. S.

† "Different":—J. E. S.

assurance celui qui seroit très flatté de pouvoir vous donner des preuves de sa profonde estime.

VENTENAT.

P. S. L'Institut nomme dans ce moment ses Correspondants; et sous peu il aura l'occasion de rendre justice au mérite de Monsieur Smith.

J. E. Smith to Dr. Hedwig.

Viro clarissimo, doctissimo, illustrissimi patris filio dignissimo, D. Romano Adolpho Hedwigio s. p. d. Jacobus Edwardus Smith.

Literæ tuæ gratissimæ, vir summopere æstimande, quamvis Augusto mense anni 1802 datæ, haud ante Maium anni præteriti ad manus meas pervenerunt. Responsionem detuli ac procrastinavi, occasionem exoptans qua fasciculus plantarum, ex desideratis tuis, ad te mitti potuerit; sed adhuc frustra. Commercium solummodo epistolarum tempora bellicosa ac portentosa nobis non vetant. Inter arma silent artes, at diutius non sileat amicitia.

Annulum aureum, et imaginem auro longe cariorem, digito jam iteratis vicibus imposui, numquam meritorum parentis tui immortalis, neque benevolentiæ tuæ, immemor.

Præclari viri laboribus et doctrina plus plusque indies fructus ac dilectus sum, cum in *Fl. Britannicæ* vol. tertio mox prodituro muscos nostrates ad normam ejus disponebam. Societati Linnæanæ magno sane ornamento fuit; sed oculos in filium ejus jam vertimus, nomen tuum, vir clarissime, inter Socios nostros exterraneos, quam proxime ut eorum numerus legitime augeri possit, facile inscripturi. Interea tu valeas, vir optime, mihique semper faveas. Amicissimus D. Dawsonus Turnerus te ex corde salutat. Libellum doctum et elegantem de Muscis Hiberniæ jam typis mandavit.

Dabam Nordovici, die 22 Februarii, 1804.—Literis tuis inscribas, "J. E. Smith, M.D., President of the Linnæan Society, Norwich."

Sir William East, Bart. to J. E. Smith.

My dear Sir, Hall Place, July 25, 1804.

Accept my most sincere thanks for your very kind remembrance of me, not only by the favour of your very flattering letter, but for the present which accompanied it, and which I shall always set a high value on, as a gift from a person whom I so much esteem; and permit me to say, that as there are few days, perhaps hours, of a man's life, that are passed with gratification to our feelings, so shall I look upon the short time, which you and Mrs. Smith favoured this place with your company, amongst the number of our pleasantest moments. But, my good Sir, you must not, by the rules of plain dealing, allow us barely to taste of the agreeable,—you are bound in conscience, after having once indulged us, to repeat the favour. May we not then be allowed to expect, when you find yourself inclined to revisit this part of the kingdom, to recollect those whom you will make so truly happy, by passing as much time with them as you can afford from your pursuits in a science the most noble, and the most worthy of an enlightened mind. I should have been truly fortunate, if in the younger part of my life I had fallen into your line of study; for I perceive those who have done so, by their unwearied attachment, meet with charms which I am absolutely forbidden to enjoy;—and now, alas! it is too late: the imagination is no longer vivid enough at sixty-seven to become enthusiastic; and without some degree of enthusiasm everything in nature becomes dull and torpid. Is not this true, my dear Sir, whatever the philosopher may urge to the contrary? I now expect, my good friend, you will exclaim—Enthusiasm! —what but this passion could reconcile you to a red coat at your time of life? And this charge will at first sight appear but too well grounded, more especially so when I declare that the military character is the most foreign of all my inclinations;—but dread of falling an unarmed victim into the hands, probably subjected to the persecution of an inveterate enemy,—with this further incentive, a desire natural to an Englishman, to offer, at such a period as the present, all that was in his power to offer, to his king and his native country,—this led me to the adoption of my scheme for the immediate defence of our own neighbourhood. Had I been a younger man, I would have offered at this crisis my services in a more extensive way. Should our threatening enemy succeed in landing a powerful force, dreadful will be the consequence, let the conflict end which way it will: but those unemployed, unengaged in the contest, will be treated with unexampled severity; and should the enemy prevail, I could not bear the thought of seeing those I loved treated with inhumanity. My witnessing such shocking scenes could avail them nothing. How much better then is it, if necessary, in trying to defend them and yourself to the last, to end your life in the most glorious way in defending all that is dear to you?—What will Mrs. Smith say to my long declamation? I think, however, and I beg you will tell her so, that an old captain, a soldier at sixty-seven, deserves at least, by way of encouragement, the approbation of the ladies; and you may assure her there is *but one* amongst them whose good opinion I should set a higher value upon than hers.

Having now intruded upon your patience, it is time that I should release you, by assuring you how sincerely Lady East and her sister join in every kind wish with your faithful friend,

W. East.

From Mons. Mirbel.

Monsieur,

Sa Majesté l'Impératrice m'ordonne de vous addresser la suite de *la Flore de Malmaison* et des *Liliacées*. Je saisis cette occasion pour vous offrir mon traité d'*Anatomie et de Physiologie Végétales*. Je vous prie, Monsieur, d'agréer cet ouvrage comme une marque de ma profonde estime pour vous, et du prix que j'attache à votre jugement. Je serois pénétré de reconnoissance si vous daignez m'honorer de vos conseils.

Je suis, avec une haute consideration,

Monsieur,

Votre très humble et très obeissant Serviteur,

B. MIRBEL.

De Malmaison, à 13 Prairial, an 12, 1804.

From the Duke of Bedford.

Sir, Woburn Abbey, Nov. 11, 1804.

I have to acknowledge the receipt of your letter of the 7th instant, inclosing one for Dr. Abbot on the subject of the *Salix Russelliana*. I have derived much gratification from the distinct and satisfactory manner in which you draw the line between his Willow and the *S. fragilis* of Linnæus; not merely from the pleasure I feel from everything that tends to elucidate the science of botany, but because I have a peculiar interest in this Willow, from the circumstance of its having been cultivated here by one * whose most anxious and warmest pursuit was a knowledge and improvement in every branch of agriculture, and the consequent benefit to the community at large. It remains for me only to return you my best thanks for this obliging communication of your opinion, and to assure you that I am, Sir,

Your very faithful

BEDFORD.

I have lately discovered accidentally, that my

* His Grace's brother, Francis Duke of Bedford.

grandfather introduced the same Willow into his county forty or fifty years ago, and distributed plants of it among his tenants; which perhaps may give it some additional claim to the specific name you have been pleased to bestow upon it.

From Mr. Johnes.

My dear Friend, Hafod, Dec. 6, 1804.

I have read Mrs. Barbauld's Introduction to Richardson's Letters. It is most excellent; but there is a Mrs. Klopstock who goes to your marrow, for she pierces through the heart. It is a charming collection.

I am very proud you continue to like Froissart, and thank you very kindly for your expressions of him and me; for I shall tag myself to the old gentleman's skirts, in hopes he will make me mount to fame with him. I shall go on with Monstrelet, who has never had an English coat on; and probably Joinville, by way of *intermezzo*, and to give time to collations of Monstrelet. Let me know how you like this plan, for I am not quite determined as to Joinville's Memoirs*. You are among the few who

* Mr. Gray, in a letter to the Rev. N. Nicholls, speaks thus of these old historians:—"I rejoice you have met with Froissart: he is the Herodotus of a barbarous age. Had he but had the luck of writing in as good a language, he might have been immortal! His locomotive disposition, (for then there was no other way of learning things,) his simple curiosity, his religious credulity, were much like those of the old Grecian. When you have *tant chevauché* as to get to the end of him, there is Monstrelet waits to

really know me. I will tell you all about Froissart when we meet. Talk to me of a cart-horse if you dare! Grub-street is a farce to me. I am hard at work; and you will say hard at work, when I keep two compositors and one pressman fully employed.

I forestalled the puns, or I should have been borne down by them. I have done with you, when you condescend to explain them, for a commentary on a pun is the devil.

I have a bridge—beautiful,—through the kindness of Mr. Shepherd, whom you know. Poggio—an excellent fellow, with more abilities than one treble his size. I want to show you the new walk. It is *superbe*, and as yet a nondescript.

THOMAS JOHNES.

From Professor Martyn.

Pertenhall, near Kimbolton, Jan. 11, 1806.

Dear Sir,

I thank you for your letter, though I should have been much better pleased, had you scourged rather than complimented me.

I was grieved to hear that you had been in a state of suffering, and that so much of your precious time had been lost to the public.

I shall take the liberty of continuing the dedication* to you; for though you have migrated east,

take you up, and will set you down at Philip de Comines; but previous to all these you should have read Villehardouin and Joinville."

* The Language of Botany; being a dictionary of the terms.

and I north-west, yet I still retain the favourable impression that your agreeable manners in general, and particular civility to me, made upon me long since, and beg still to be considered as your friend. I am persuaded that we shall never quarrel, however we may differ in opinion. I consider you as the Establishment, and only desire toleration so long as I am a quiet subject.

Respecting *plantæ acifoliæ* and *ordo acifoliarum*, I cannot help you. *Drifted* might well express *secundus*, but I dare not hazard it, any more than you. I thank you for your hint on *propago*, which I had not attended to. I have long thought *general* umbel better than *universal*. I am happy you approve my distinction between *labiate* and *personate*. Your sanction was wanting to satisfy me entirely. *Downy* best expresses *tomentosus*: I only hesitated on account of *pappus* being *down*; but I adopt *seed-down* for that, which I have taken from you. I have no objection to *anthera*, if you will use *antheras*, and not *antheræ*, in the plural. I use *corolla* and *corollas*. So *nectarium* sounds better to me than *nectary*; but then it should be *nectariums*, and not *nectaria*, in the plural. My only difficulty about the final *a* is, that ignoramuses will use *stamina* in the singular, and *stipula* in the plural. But why do I torment you with these trifles? I hasten to the best part of my letter,—my sincere good wishes for your health and comfort; for I am, dear Sir, with much regard,

Yours most faithfully,

THOMAS MARTYN.

*From Dawson Turner, Esq.**

Yarmouth, Feb. 5, 1806.

Knowing, my dear Sir, how frequently you heard of us by Dr. Rigby, I was willing to put off the painful task of writing to you, as long as I could. A painful one indeed it is, and yet a most sweet one; for without experiencing equal affliction, you cannot conceive the sensations of pleasure which the affectionate kindness of my friends has occasioned; and to you I may say, what I should be sorry to say to any other, lest I should be regarded as affected or foolish, that the reading of your letter drew from my eyes the most delightful tears they ever shed.

You knew our poor babe, and you always treated him with a kindness which his mother and I shall never forget so long as we live; for those who were kind to him have now the strongest claim upon our gratitude. You know, therefore, what a loss we have sustained in being deprived of him, even without considering those dreadful circumstances attending his death, upon which I cannot at this time reflect, without shuddering with horror. And yet in the midst of this calamity, the mercy of Divine Providence was so striking in the preservation of my wife and of my remaining children, that, if it do but please the Almighty that their lives should be spared, and her health and spirits restored, I feel

* This beautiful letter was written upon the loss of his eldest son, a most engaging child, who was, unhappily, burnt to death.

that I can in time regard even this visitation in the true light which I believe that duty to my Maker requires.

I endeavour again to turn my thoughts to botany; and though my mind has not yet regained its spring, I find in it great amusement and comfort when I walk out alone. I cannot attempt to say, my dear friend, how much your expressions of affection and friendship have moved me: it gives me the liveliest pleasure to feel that I have a heart capable of valuing them, and I trust in God that you will always find me most anxious to strive to merit them. If you see anything amiss in my conduct, my disposition, or my temper, I entreat you, by your friendship for me, never to let it pass without notice, and I assure you that you shall find me grateful. Conscience indeed speaks, but we learn in time to disregard her like Amurath's ring. The words of a friend are much more sure of being attended to; and when that friend has such a heart, and leads such a life as you do, it is hardly possible that they should fail of effect; at least I would fain hope so, for the credit of my nature. That I may, as long as we both live, enjoy and deserve your friendship, is, I assure you, one of the most earnest wishes of

Your truly affectionate and obliged friend,

Dawson Turner.

From Mr. James Lee.

Nursery, Hammersmith, March 11, 1806.

Dear Sir,

We cannot procure any *Fuchsia* seed, and *Dodecatheon* does not ripen in England, and it is now impossible to send anything to Naples. We should have wrote to you about this before, but at this season are so overpowered with business that you must excuse me. You may depend on it that no one will be able to make an impression on our minds unfavourable to you: we have such high esteem for your abilities, such gratitude in store for what you have done for science, for the public, and for us, that we shall always be willing to defend you on all occasions, and at all points.

We are sorry to have to communicate to you the death of our dear friend Masson, who died at Montreal in January last. We lament his fate most sincerely. He was hardly dealt by, in being exposed to the bitter cold of Canada in the decline of life, after twenty-five years services in a hot climate,—and all for a pittance. He has done much for botany and science, and deserves to have some lasting memorial given of his extreme modesty, good temper, generosity, and usefulness. We hope, when opportunity serves, you will be his champion*.

I am, dear Sir, &c.

James Lee.

* "Francis Masson," Sir James informs us in his biographical memoir of this excellent naturalist,—"a name which ranks very

Lieutenant-Colonel Hardwicke to J. E. Smith.

My dear Sir, Calcutta, Nov. 4, 1807.

I was a few days since highly gratified by the receipt of your friendly letter of the 12th of last April. It is a pleasure I had hardly promised myself; and I will assure you if my writing now and then will induce you to spare me sometimes a leisure hour, to keep up the correspondence thus begun, I shall on my part often give you a letter,

high among those who, by encountering personal difficulties and hardships with the most indefatigable and disinterested zeal, have promoted botanical knowledge,—was born at Aberdeen in August 1741.

"The writer of this," he continues, "well recollects the pleasure which the novel sight of an African Geranium in Yorkshire and Norfolk gave him about forty years ago. Now, every garret and cottage window is filled with numerous species of that beautiful tribe, and every green-house glows with the innumerable bulbous plants and splendid Heaths of the Cape. For all these we are principally indebted to Mr. Masson; besides a multitude of rarities, more difficult of preservation or propagation, confined to the more curious collections.

"During his stay at the Cape, he entered into correspondence with the great Linnæus. Having discovered a bulbous plant of a new genus, he was not only laudably ambitious of botanical commemoration in its name, but he was particularly anxious to receive this honour from no less a hand than that of his illustrious and venerable correspondent. This indeed was the '*unicum præmium*,' the only reward to which he aspired for all his labours. He obtained the honour to which he aspired. The specimen of *Massonia* in the herbarium of Linnæus, named by his own trembling hand, near the close of his life, proves that the name had his sanction."—*Rees's Cyclopædia.*

however little important matter I may have to communicate. I am quite out of the way just now of adding to my collection of drawings, unless I were to take my subjects from the botanical garden; but as that would be a species of robbery or encroachment upon Dr. Roxburgh's province, I must wait with patience till my professional duties call me to the upper provinces of Hindostan, when I shall hope once more to range in the same field I have hitherto found so interesting and productive.

Since my arrival at Calcutta, my attention has been more directed to the collecting of insects than anything else, and in which I have been very successful; for I am just now in possession of a very splendid assortment, from which I hope some day or other to add useful varieties to your cabinet.

I am sorry, my dear Sir, to find you have been less successful in the circulation of that beautiful work Exotic Botany, than its merits deserve; and I feel it an addition to the disappointment, that every number I received from your bounty was swallowed up by the merciless ocean. It makes me melancholy almost, when I reflect on the valuable books and papers I lost on that unlucky event, the loss of the Lady Burgess; and the drawings of insects which you and Mr. Marsham admired so much, unfortunately were with me.

I am very glad Dr. Buchanan preferred leaving his collection of dried plants and drawings with you. He set out on his statistical mission about a month gone, and the early part of his route is through a country little explored, and rich in vegetable productions. His views have met the aid and encouragement of Government, and he travels with more useful advantages than an individual would be likely to obtain without such powerful assistance.

Mr. Fleming and myself have often indulged in conversation about you. He has a great desire to be better known to you; but how soon he may have that opportunity I cannot say. He is here in the full enjoyment of health and all the ease and luxury a man can wish for; consequently he is very indifferent about leaving the country.

Dr. Roxburgh arrived here a few weeks since, and is as busy as ever. I have this moment a note from him, wherein he says, "I am surrounded and in confusion with the various parts of a new palm from the Moluccas, which is now in flower." In my next I may be able to tell you something about this said palm, as I mean to go down to the garden in a day or two.

I find the plant which produces that valuable drug in the Materia Medica, *Columba Root*, is at length ascertained. It proves, I believe, to be a species of *Menispermum*, just what I some years ago predicted, when I sent Dr. Roxburgh specimens of a root bearing close resemblance with *Columba Root*, and used by the natives of Hindostan in the same cases you use that plant. This was also a *Menispermum*.

Pray, my dear Sir, make my best respects to Mrs. Smith. How often do I reflect with feelings of pleasure on the few happy days I passed in the society of your family and friends! And may I add, without being suspected of studied compliment, I shall never lose the remembrance of those friendly attentions you favoured me with, and the engaging manners which made your fireside so desirable. Be assured you will soon have another visit from me after I return to Old England. Let me know what I can procure you in India, in the vegetable kingdom. In September last I dispatched a parcel of seeds for the Marquis of Blandford; and the half of this packet I intended for your friend Colonel Johnes.

Let me repeat, I shall always be highly gratified with your correspondence, and that I am, dear Sir,

Your obliged and faithful

T. HARDWICKE.

From W. G. Maton, M.D.

My dear Sir, Spring Gardens, Dec. 19, 1807.

The specimen sheet of our new *Pharmacopœia* is just gone to the press, and I hope to be able to send you a copy for your perusal very soon. I have named the new *Melaleuca, Cajeputi;* and had I all the requisite specimens of that tree before me, I believe I should have been tempted to draw up a description of it, under the above name, to be laid before the Linnæan Society; but as it would be done so much better by yourself, I trust that you will some day give it, by way of supplement to your observations on the genus, in our third volume. I had not the pleasure of hearing your paper on the Vitellus of Seeds, which I regretted much. It has been talked of a good deal among our botanical brethren; but few of them are competent to judge of its merits, and those who are disposed to be critics-in-chief on all occasions, have not ventured to arraign any part of your positions.

Your late publication is lying on my table, and I greedily devour its excellent contents, though, from the professional interruptions I experience, it is only by fits. I do not misinterpret, I think, an allusion in the preface on the subject of Paradise*. When may we hope to see the Lapland Tour in its English dress? I am quite impatient to peruse it, as indeed everything that is sent into the world by you.

Yours most truly,

W. G. Maton.

* This alludes to a passage in the Introduction to Botany, in which the author observes, that "None but the most foolish or depraved could derive anything from the pursuit of this science but what is beautiful, or pollute its lovely scenery with unamiable or unhallowed images. Those who do so, either from corrupt taste or malicious design, can be compared only to the fiend entering into the garden of Eden."

CHAPTER IX.

Correspondence of Andrew Caldwell, Esq.;—and a few other Letters relative to Ireland.

THE correspondent whose letters here appear was an Irish gentleman who represented Downpatrick in parliament in 1788; and when the compiler of these memoirs had first the pleasure of being made acquainted with him, in the year 1799, was verging towards sixty. Of the mildest aspect and highly polished manners, a tinge of melancholy cast a shade over his dignified deportment, which at once engaged the affections on his side, and broke every barrier of formality and reserve. He lived much among persons of high birth, and his conversation was replenished with anecdote and entertainment, enlivened with the peculiarly pleasing manner and dialect of the superior ranks in the sister isle.

He had a noble generosity of spirit, and his compassionate heart revolted at oppression in every form; but especially did he deplore and execrate the merciless traffic in negro slaves. He read with remarkable grace, and recited poetry with more than common tenderness; he was altogether a being of a higher order. "How is it," asks Burns, "that in the short stormy winter day of our existence, when you now and then in the chapter of accidents meet an individual whose acquaintance is a real acquisition, there are all the probabilities against you that you shall never meet that valued character more?"

This amiable man regrets his separation from the friends he so greatly honoured with his regard, in terms not unlike those in which Mr. Davall expresses himself. "Often," says Mr. Caldwell, "do I lament being deprived of more frequent enjoyment of your presence and conversation; but such is the condition of life, that variety of advantages cannot be had at once; we must learn to be content with them separate and occasionally."

The first letter in this correspondence, from A. B. Lambert, Esq., is inserted among Mr. Caldwell's, in consequence of its relating, like a few others in this selection, to Ireland only.

Mr. Lambert to Sir J. E. Smith.

Dear Sir, Castle Bourke, near Tuam, May 1790.

I hope you received the letter I wrote just before my leaving England. Ever since my arrival here I have been taken up with business, so that I have not been able to pay that attention to the natural history of this country I could wish. I have been this last month in the county of Mayo, at Westport, the seat of Lord Altamont, surrounded by the Hibernian Alps,—the most mountainous country I ever saw; near it is the famous mountain of Croagh Patrick, reckoned one of the highest in Ireland, which seems to promise a fine field for a naturalist. It took me up two hours walking from the foot to the top of it. I found the *Andromeda Dabœcia* * growing in great abundance on the sides of the mountains; the *Empetrum nigrum*, and some other plants which were new to me, and not being in flower I could not determine them; and within a few yards of the top, the *Saxifraga umbrosa*, the London Pride, not in Hudson. If any one should wish for any plants of the *Andromeda Dabœcia*, I could send him plenty. I hope this summer to

* Figured in the first volume of English Botany, p. 35, under the name of *Erica Dabeoci*, Irish Heath. This genus is now removed to *Menziesia* by Jussieu and Swartz. See English Botany, vol. xxxv. p. 2469.

The following letter to Sir J. E. Smith, from R. Duppa, Esq. of Lincoln's Inn, dated July 1827, may here furnish a note respecting this genus.

Dear Sir,—Somewhere or other I think I have met with an assertion that there was but one Heath indigenous to Ireland, and that is now removed to the genus *Dabœcia*[1]. My question is, whether, of the four species of English Heath, any one is found wild in Ireland? and also, if our English Heaths are not found in Ireland, whether the bogs of Ireland are composed of the roots of the *Dabœcia*, the *Salix herbacea*, or what?

The *Erica vagans* is peculiar to Cornwall.—Query, are the *Erica vulgaris*, *cernua* and *Tetralix* found in Cornwall as well as the *vagans?* and is the *Dabœcia*, which is peculiar to Ireland, never found in any other country that we are acquainted with?

R. Duppa.

---

[1] Mr. Duppa probably meant *Menziesia;* there is no genus *Dabœcia*. The plant was called *Andromeda Dabœcia* in *Linn. Syst. Veg.* It is named *Dabeoci* after St. Dabeoc, whence the Linnæan trivial name has been corruptly taken.

be able to give you a good account of the plants of Mayo ; and nothing would give me greater pleasure than to see a brother member of the Linnæan Society at Castle Bourke where I reside, which is situated on Lake Carra, joining Lake Mask and Lake Corrib ; being surrounded with immense high mountains and bogs, which seem to promise much gratification to a botanist, and have been, I believe, but little examined. Dr. Brown, the author of the Natural History of Jamaica, I heard chiefly resided in lodgings at Ballinrobe. I paid him a visit one morning, and found him in bed quite a cripple with old age and the gout. He showed me a copy of a *Flora Hibernica,* which seemed not much more than a catalogue, and very imperfect. Some old plants he has mentioned as new species, and showed me a specimen of the *Juncus sylvaticus* for one. The copy that was in London is coming over here for correction, which I rather think he will have some difficulty in doing. He talked to me a great deal about the Jamaica plants, and the number he had formerly sent to Linnæus, who he told me corresponded with him above twenty years ago.

I saw at Lord Altamont's the true Irish wolf-dog ; he has seven of them, and the only ones now in Ireland. His brother has given me a very good painting of one of the largest, done the natural size.

In a letter from Dr. Pulteney, he writes me word you mention some sheets of a botanic work which straggled to London last year, nobody knew to what they belonged. I believe I can inform you ;

they came from hence, being part of an intended *Flora Dubliniensis,* published by Dr. Wade, a physician here, and a very good botanist. He has been the means of establishing a botanic garden here. Having very lately applied to parliament for that purpose, they have granted three hundred pounds for the first year to begin it; and I believe Dr. Wade will have the direction of it. He was with me yesterday, and wishes very much to become a correspondent with the Linnæan Society, and likewise to establish one in this country on the same plan as ours, and he will be much obliged to you to send him the rules, &c. I am very glad to hear from Dr. Pulteney that the Linnæan Society will soon publish a volume of *Acta.* Any botanical news will be very acceptable to me, as I am now in a kingdom that is not very famous for science, and I do not find it very easy to get information.

I am, dear Sir, yours, &c.

A. B. Lambert.

P.S. The following account is in a book lately published in Ireland, which perhaps you may not have seen: "On the see lands of the Bishop of Dromore were found in 1783 a pair of elk horns, which measured from tip to tip fourteen feet four inches, as also almost the entire skeleton, in the most perfect preservation, of the enormous animal that wore them. From the length of the bones of his fore-legs he is judged to have been almost twenty hands high.

I inclose you a drawing of an insect called the Borer, which has occasioned more destruction among the sugar-canes in the West Indies than the Hessian Fly has in America.

From Andrew Caldwell, Esq.

Sir, Dublin, Sept. 23, 1793.

Your polite attention, when I took the liberty of writing to you before, gives me courage to intrude on you again. I hope you will have pleasure in being informed that a garden for indigenous botany, under the patronage of the Dublin Society, is a measure determined upon. A committee, at the head of which is Mr. Forster, speaker of the House of Commons, and of which I have the honour to be a member, is vested with proper powers and a sum of money for purchasing ground. We have actually agreed for the house and garden formerly belonging to Dr. Delany. It is within a short mile of town, the soil excellent, the ground well varied with high and low, a small stream running through; it seems well adapted in every respect. I mentioned to the Speaker that I had hinted this already to you:—it gave him great pleasure; and whenever we get possession, and are enabled to begin the garden, you shall be principally consulted, and your advice and assistance chiefly relied on. The ground consists of eleven acres, which is more than seventeen English acres. It is at present too much covered with trees, but I shall be for cutting down sparingly and with caution. Much of the old-fashioned garden remains, and is pretty in its kind. It is a sort of classic ground, having been admired and frequented by Swift, and all the people of literature of that day. The Speaker proposes that this should also be a garden for agricultural experiments: I cannot say that I agree with him perfectly in this scheme. There was a similar project undertaken by the Dublin Society several years ago, and it turned out very unfortunately. They brought over a gentleman from England, a Mr Wynne Baker, rented a large farm for him, and allowed him a salary of 300*l.* per annum. But this did not satisfy him; he was for ever bringing in bills for contingencies, and carrying them by his influence. He was a man of talents, agreeable conversation, and convivial: an excellent farmer at his desk and over the table, but execrable in the field. When I first became a member of the Society, there was regularly a pitched battle twice a-year between him and the real public-spirited gentlemen; but Baker always brought down a crowd of jovial squires and members of parliament, and outvoted us. A fever at last carried him off, and relieved the Society; and had he lived on, he would long before this have swallowed up the Dublin Society and their funds altogether. I confess, after this experiment, I am dreadfully afraid of anything like a renewal.

I cannot help saying I am charmed with the accuracy of the figures, and the clearness and elegance of the descriptions in the English Botany: nothing can be more complete.

I have not seen Dr. Martyn's Botanic Dictionary. That, if well executed, would be a very convenient and desirable work; but dictionaries are generally loaded compilations, in which you meet with a crowd of things, except the very thing you want to find. It would be excellent if he would refer to the particular flower, leaf, plant, or root, that might illustrate his article: he should mention three or four of the sort, that if the reader could not procure one, he might another. A collection of dried specimens, ranged in the order of the dictionary, would explain the terms infinitely beyond any plates or verbal descriptions that could be given. The reader should be impressed with a particular caution as to terms that resemble and might be easily confounded; for instance, *pinnatifid* and *pinnate*,—the one seeming the substantive, and the other the adjective. I could wish Linnæus had dropped one of those terms, and invented a new word. The difference between *Ranunculus Flammula* and *R. Lingua* is not very satisfactory: the *Flammula* is not procumbent, and difference as to size is vague and uncertain: if it be decided that the bottom leaves of one are always ovate, though concealed in the grass, and the other never so, that would be something to depend on. I wonder it is not mentioned that the leaves rise each from a sheath that envelopes the stalk. There is often occasion to lament that, in describing a flower, the student is not sufficiently cautioned against confounding it with another to which it may bear a striking resemblance; for example, *Stachys* and

Galeopsis, and many flowers of that class. The most of the umbelliferous weeds are subject to the same perplexity. If there could be a precise positive character pointed out to refer to at once, it would save much time and trouble.

I am, Sir, your most humble servant,

A. Caldwell.

From the same.

My dear Sir, Dublin, Nov. 17, 1795.

I have a long time owed you some account of myself; I flatter myself at least that you expect it from me, and think it your due. Nothing entertaining occurred on the journey;—the weather pleasant, and North Wales I had often traversed before. We had a good passage from Holyhead, and the ladies under my charge landed safe in Dublin. My stay there was but for a few days, being obliged to go to the country on particular business. Though I travelled many miles, I cannot say the landscape anywhere was as beautiful as what you showed me in the neighbourhood of Norwich; if truth must be uttered, it was more the reverse than you could conceive. The posting too, I think, would not have put Dr. Smollet into good humour. The first chaise, of a venerable antiquity, broke all to pieces and overset, just as it came to the end of the pavement; *et pour comble de disgrace*, a fair stranger, a cousin of mine from Shrewsbury, was under my care: this was her first adventure in a foreign land; but we escaped with whole bones;—the journey altogether a tragi-comedy. Having set down my English girl at a friend's house, I proceeded further. You shall be spared an exact recital of all the varieties: I shall only mention, I was obliged to get out of the chaise at half-past seven at night, on the top of a mountain, and walk five miles to the house to which I was bound; the chaise, meanwhile, left amongst the wild natives to be dragged on till horses could be sent to meet them, the hacks positively refusing to stir a step,—I believe, indeed, perfectly unable. The natives little dreamed two thousand guineas were shut up in the seat of the chaise. Had they known the secret, such is their simplicity, I am not sure but the shiners might have been as safe as near the police of your great metropolis. I thought it, however, as well not to put the moral sense unnecessarily to the test. The night was fair, and a little moon out, but such a storm of wind, and in my face, that it retarded my march excessively. It was the night, I believe, that did so much damage in England. I often involuntarily recollected my late travelling companion,—not that I wished him to share the distress, but if by chance he had been there, his presence would have cheered me.

I presume you did not forget the engaging bird at Exeter 'Change, whether or not it be the *Turdus nitens*.

Dr. Wade last summer had the ardour to make a stolen visit to the Giant's Causeway, to search for the *Scilla verna*, having met in a manuscript note in an old book that it grew there. He found it in abundance, and brought away some roots.

Yours most sincerely,

A. Caldwell.

From the same.

My dear Sir, Dublin, April 5, 1796.

I am highly gratified by the interesting communication in your last letter. I hope one day or other, at no very distant period, the lady of your choice will honour me with a place amongst the number of her friends. I claim that from partiality to you, and generous resignation to the formidable rival of botany and me. It is true I can no longer flatter myself with the imagination of exploring together this Western Isle, or still less of visiting Italy with one whose taste and mine are so congenial; yet, notwithstanding, I willingly raise my voice.

Tu festas, Hymenæe, faces; tu, Gratia, *flores*
Elige; in geminas, Concordia, necte coronas.

My warmest wishes are for a long continuance of health to you and yours. You have too just a sense of worldly prosperity to look for happiness on any other principle than your own moderate desires, goodness of heart, and ingenious occupations.

I have little to say about myself; it is unavoidable not to share in some degree in the dissipation of a great town: that so ill suits my disposition now, that I am never so contented as when alone, surrounded with my books.

A gentleman of *your name* is come over; he has brought several valuable things he purchased at the Duchess of Portland's sale; he told me he laid out there upwards of two thousand pounds. Perhaps you may know who he is?

A friend at Liverpool is to lend me Lorenzo de' Medici: I wish much to read, since you recommend it.

Yours ever,
A. CALDWELL.

From the same.

My dear Friend, Dublin, Jan. 6, 1797.

You might accuse me of the spirit of retaliation, but I believe you think better of me than to suppose such a motive. The influence of procrastination is too powerful, and I fear I am as little able to resist as those who are much my superiors.

There was a momentary feeling of regret at finding you intended to remove,—an involuntary impulse towards self; but I soon coincided with your ideas, and allowed that you judged right.—How things come about! My station is so distant from Norwich, it is one of the last places I could have thought of. I now think myself connected there.

Let me now give you my own history. The months of June and July being incessant rain, almost confined me to the house; August clearing up and promising well, the knight sallied forth. The first excursion was to a place called Ross Trevor, situated on Carlingford Bay, a high hill with oakwood down to the strand; lofty mountains all round, except the mouth of the bay, which is broad and due east. It is the prettiest place I have seen in this country, and wants nothing but English taste and opulence. I thought it was hardly ever visited, but found it a noted resort for bathing and dissipation, and the second evening met a multitude of acquaintances; so next morning I packed off, and strayed on to Tullymore Park, Lord Clanbrassil's, a fine retired spot,—a mountain his lordship has planted almost to the summit, a very rocky glyn with a fine stream terminated by the sea. His lordship has naturalized *Antirrhinum Cymbalaria*. It seems to grow spontaneously on bridges and rocks.

I then moved forwards to Donoghadee, to visit my friend Mr. Arbuckle, who is collector there. He was just gone over to Port Patrick, but expected to return immediately. I waited three days, having acquaintances, and Mr. A.'s elegant botanic garden, where several plants of *Veronica decussata* formed great bushes in the open ground, covered with profusion of blossoms. It is strange it will hardly flower here, nor at Lord Clanbrassil's at Dundalk. Growing tired of waiting for my friend, I determined to join him at Port Patrick;—we sailed the same morning, and crossed in the middle of the channel; so on landing in Scotland I found myself alone. I resolved however to proceed, and went the first ten miles to Castle Kenedy, Lord Stair's. There is a rising ground very nearly surrounded by two fine lakes, with some wooded islands; the ancient castle, burnt in the year 1716, but the walls strong and good, stands elevated on the summit of the ground, a fine platform round, and the sloping sides thick-planted, with many avenues cut in the stairway down to the lakes. It looks like a magician's castle in romance, shut up by dark woods and lakes. It strikes me as a fine place, and wants only a long range of hill on the opposite side of a lake, and to have a thick mass of planting. I resolved to push on to Glasgow, to review early scenes of happiness in my college days; —went in my way to look at Cullean, Lord Cassilis's. It is a magnificent castle, lately built. It stands on the edge of a precipice 150 feet above the sea; fine views of the Arran mountains at a great distance; the plantations extensive, and all thriving in the most exposed situation that can be imagined; it is a prodigious fine place.

Glasgow has double the inhabitants since I first knew it. I spent ten days there with an old friend. The town is now elegantly built, but they assured me was nothing to Edinburgh. I returned to Port Patrick by another road, Paisley, Port Glasgow, and Greenock. The view when you come to the Frith of Clyde is beyond anything seen in Wales: the vast expanse of water; Dumbarton, with its double summits, on the opposite side; Ben Lomond's vast pyramid rising behind; the mountains of Bute and Arran; the numerous cultivated seats and plantations on the low ground near the water,—form the noblest assemblage that can be imagined.

The remainder of the journey was coastways to Port Patrick; very charming, but the weather was bad. Ayrshire and all about Glasgow is richly improved; the trees only want age, and twenty years hence the country will look as well as England.

I believe August is a bad month for botany; the fructification of most plants is over, and the *Crypts* not come on. I found a profusion of *Parnassia*, larger and more luxuriant than I had ever seen in England or Ireland. *Campanula rotundifolia* covers the fields, but no *Echium* or *Verbascum*. *Rhodiola rosea* is found on one side of Port Patrick, but I took the wrong side and missed it.

I spent a few days with Mr. Arbuckle. Lord and Lady Clanbrassil landed and staid with us. I then moved slowly on, having visits to a numerous cousinhood. I gave three days to each, and seldom more than five miles to go at a time. My last stop was at Dundalk, forty miles from Dublin; a few days there with Lord Clanbrassil. He goes first there when he comes from England. The house is an old patched one close on the street, but the grounds behind quite in the country, and beautiful. Here he has his garden and hot-houses. He is very worthy and a great *amateur*, but not scientific; everybody needs not be so. I have not room, I see, for all my travels. When I came back to Dublin I was obliged to set out for the south, chiefly on business. I am home, and settled about a month.

A. Caldwell.

The two following letters relate to the state of insurrection and rebellion in which Ireland was involved at the period at which they were written. In the first of these, dated May 16, 1797, Mr. Caldwell addresses his correspondent thus:

"It is difficult now to think of any subject but

the alarming state of this country. A secret committee of the two Houses of Parliament have just published their Reports. I wish I could send them to you; they are worth your looking into, and are curious historical documents. The Lords state that it was calculated by one of the leaders of the conspiracy, that 30,000 persons must be massacred. I understand, however, this was only from oral testimony, and therefore to be received *cum grano.*

The Report of the Commons is satisfactory, and seems well supported by evidence. Great ability and contrivance is shown by the conspirators, and yet many of the managers perfectly illiterate, and not able to spell common words. If the French could land a tolerable force here, the country is lost. I perhaps, if I escaped with my life, should be a poor emigrant; but I have not strength to work in a garden like Mr. M'Mahon;—I don't know what I am fit for. Alas! I have this moment received very melancholy news,—the death of a brother-in-law, a most worthy man, and of great consequence to his family; and a niece, an amiable young woman, very happily married in Yorkshire. Between the public and private distresses, every thing around me wears gloom and melancholy."

From the same.

My friend, Dublin, July 11, 1798.

What have I done, that you should forget me entirely, and in the hour of peril and distress? I am, it is true, alive and safe; and that probably I owe to having command of myself enough to submit to confinement in Dublin. All this fine summer, unusually so, has been lost in the smoke and dust of a populous city. We seem secure here, and that is the utmost can be said;—but we live in a state of siege. All the avenues into town are secured with guards and palisades; we must keep strict hours, and be at home at nine o'clock:—that perhaps will not appear to you a hardship, yet it is to sober people like me an inconvenience, and sometimes debars one from little indulgences. The pleasures of rural excursions and country life are totally prohibited. There is no safety, except military are quartered at hand. What a misfortune is civil war! You can form no idea of it: and though it may be supposed that, with the great force now collected here, a rabble, undisciplined, without leaders of eminence, or concerted plan, must be supprest; yet it will, their numbers being very great, require time; and we fear that when the days grow short, and winter comes on, there will be numbers of irregular banditti in various directions; so that it will be extremely dangerous to stir from home. I don't choose to mention particular facts; it would be a long history, and not agreeable enough to compensate for the trouble of reading it. I can say no more than you will find in the newspapers; and I am the less inclined, as reports are spread every hour, and the last always contradicting the preceding:—we are confounded, and live in a state of constant suspense.

The night before last, the 9th, a sudden engagement happened at a place within two miles of town.

A small body of rebels came down from the mountains; a few were killed on both sides; four prisoners were immediately shot, and five others brought into town, tried and condemned at a court martial, and they are executing at this very moment. Such things are so frequent, nobody wonders or seems concerned.

I fear you have lost your lively friend Dr. Gwynn. You may imagine we have had little temptation for botanical researches; yet it was a solace to turn one's thoughts to the subject: it served at least to divert the attention from painful images.

I should have taken a jaunt to England, but it was neither decent nor safe to leave home and all my little concerns, when the country was in a state of confusion and trouble.

I beg my most affectionate remembrance to your mother and all your ladies, and am,

Dear Sir,

Yours sincerely,

A. Caldwell.

From the same.

Christus natus est. 1798.

All the people of my neighbourhood are at their devotions in church; the streets are as silent as midnight, and I seize the quiet hour to write to my friend. It is a long interval since the date of your last favour; but procrastination is not altogether the fault;—I sometimes restrain myself, fearing to appear a troublesome, importunate correspondent.

Nothing better occurs than to give you a history of my life and labours;—very unimportant, but the partiality of friendship will receive the account with indulgence.

The very day after the news that the French had surrendered at discretion, I went into the packet for Holyhead, the 5th, I believe, of September. The scheme was to ramble for three or four weeks, to visit several places that we have long been in the habit of passing by, and never seeing.

The first place was Conway. It was too late for much botany, but the sea prospect and the country is beautiful. I then turned back and stopped at Aber. I went up there to the top of a mountain, but met with no plant worth notice; the dry, barren summit of a mountain is not the place to expect botanic curiosities. Hearing at the inn of Mr. Davies, I immediately wrote to him; he came directly, and in two hours you would have thought we had been acquainted all our lives,—such is the liberality and advantage of science. He is a most pleasing, well-informed companion: his kindness and attention were such that I staid six days. We were every day together *.

There is a romantic glyn two miles long adjoining the inn; it is closed at the end by a deep rock, down

* The amiable naturalist here mentioned is the Rev. Hugh Davies, a correspondent and friend of Sir James.

"Surely no chemical affinity is stronger than that of congenial minds!"—and these two admirers of nature, Mr. Caldwell and Mr. Davies, were remarkably similar in character, if we may judge from the similitude of their feelings, as expressed in their letters.

which a torrent rushes, and rolls through the glyn over a rugged channel to the sea, which bounds the other end, where the entrance of the glyn is. We found *Polypodium fragile, cristatum, Filix mas* and *fem., Asplenium adiantum nigrum;* and I gathered the *Papaver cambricum,* some of which was still in flower. There is plenty of *Lichen geographicus* and *concentricus, Sedum Telephium* and *reflexum.*

All over that part of Anglesea is the greatest abundance of *Lychnis flore rubro,* which I remark because scarcely to be found here, though the coasts are opposite, and the soil similar in all appearance. It seems difficult to account for the predilections of plants.

Bangor was the next: I prefer it to most places; you have prospects of sea, mountains, islands, woods, and can never want a varied entertaining walk. I went to visit Lord Penrhyn's slate quarry, rather unwillingly: it is well worth seeing, and I had like to have missed it. I was at Beaumaris and Caernarvon, towns worth seeing for their pretty situations and fine old castles, now picturesque ruins. I went to the famous copper-mine at Paris Mount; they have now got to a prodigious depth, but, contrary to most cases, the ore is not so good as near the surface. There is so much sulphur produced, you might think yourself at Solfatara. I came home after an excursion of five weeks: my return was unlucky;—incredible as it may seem, I was five days getting from Holyhead, when the passage ought to be in twelve hours.

Mr. Templeton has been in town for a few days.

I never saw him before ; he is an alert, active botanist ; knows every thing at sight. We went together to a friend's about ten miles off on the sea-side; it came on a storm, and so tremendous a surf it was impossible to walk on the beach. He picked up a *Chlora perfoliata* withered, but with ripe seed ; it don't grow in this country ; it seemed as much joy as a good prize in the lottery.

You hear, I suppose, of the project for a Union. It occasions much agitation here ; the majority are strongly against it, but the country appears totally indifferent as yet, and to take no part. A great deal can be said for and against it ; future consequences must be conjectural and uncertain ; that is one argument against a change. "Better, perhaps, to bear the ills we have, than fly to others that we know not of."

I must own, England appears in a desperate situation, notwithstanding the showy victories. The worst of all, public spirit is gone ; the country has not energy to make resistance; the Constitution is no more than a fiction, but everybody has entered into a tacit consent to suppose it a real existence. The power of the Crown is become irresistible. The new scheme of inquisition into every man's private circumstances is beyond any attempt I ever heard of under Louis XIV.

The disturbances here are so far at an end, that the people don't assemble in large bodies ; but the roads are dangerous, and houses in the country are repeatedly attacked. Discontent and disaffection prevail as much as ever.

Your volume of Tracts I have seen, but would not read till I get my own copy: all that came over are already sold. I have got an opportunity of sending your order to White;—bid him give me a choice copy; but I am sorry there cannot be "*ex dono*" in your *own hand*. You ought by this time to be thinking of the new edition of your Travels. I am anxious for one improvement,—that you should mention where good plates of the several articles of botany and natural history are to be met with. I mean to begin reading the Travels again immediately.

I have just had a letter from Mr. Davies. He laments the decease of Mr. Pennant; so must every friend to science and a respectable character.

Yours sincerely,

A. CALDWELL.

From the same.

My dear Sir, London, August 30, 1799.

The agreeable manner in which you contrived my time should pass,—your kindness,—Mrs. Smith's, and all your friends'—was more than sufficient to make me regret leaving Norwich: we must, however, submit to destiny, and often practise self-denial. Yarmouth was so full, the landlord of the inn where I stopped looked frightened at seeing me:—"Sir, if you were to give fifty guineas I could not let you have a bed." He calmed on my assuring him I wanted nothing more than the loan of a room for half an hour to dress myself. All I could get was the common powdering room.

The Greek botany I shall never live to see the end of; I had not even time to look through the thousand drawings. With what pleasure I recollect the quiet studious hours in your library! It put me in mind of college days; young men then frequently study together. It is not in every one's company now, that I can read with attention; but you never were a discomposure, nor I flatter myself was I to you: I was sometimes impelled by curiosity to ask a question, but I believe not too often.

I could not find the *Arabis stricta* at St. Vincent's Rock: the plant is too small to be easily seen, and I had no one to direct me to the particular situation. I observed the *Gentiana*, I believe the *Amarella*, on the other side of the water; plenty of *Geranium sanguineum*, looking beautiful; both the *Lychnis dioica*'s; and, in a woody glyn, a pretty lilac-coloured mushroom.

Yours sincerely,

A. Caldwell.

P. S. We are sensible of the kind concern you express for Lord Cloncurry. I got leave to visit him twice: he is perfectly well in health, and his apartment as pleasant as a prison will admit; but two great battle-axes, beef-eaters, in the room night and day, and they are as much prisoners as he is; but they are changed every week. The young lord will never be brought to trial; they have nothing against him, except being warm in opposition, and imprudent. It is not unlikely they may detain him till the war is over. He has a fine fortune, but greatly exaggerated in the newspapers. His three sisters have 30,000*l.* a-piece, which is uncommon here; and very good girls they are. The father did not heap all on the one son. We are very tranquil here and in all parts of the country; yet the newspapers give quite a different account.

From the same.

My dear Friend, Dublin, March 25, 1801.

Our epistolary intercourse is not very frequent, but it must not drop altogether;—that idea I cannot support. I am anxious to know how my amiable friend and sometimes correspondent, Mrs. Smith, does; mention everybody when you find a moment's leisure and disposition to think of the absent.

My botanic study has been much interrupted, and not by more agreeable occupation. I have now begun however to resume. *Flora Britannica* is my constant companion. Verbal description can scarcely be conceived more clear or satisfactory; and I fancy, when the spring is more advanced, it will be a delightful guide through the fields. I perceive, however, that nothing but long and constant practice can render the Grasses intelligible and familiar, so that at the first glance the genus may be known. I doubt what I wish for is not attainable,—that some one, or two at most, distinct characters should be pointed out, that for immediate use would decide the plant, without entering into description and minute examination.

I am heartily glad your travels through Greece are performed at the desk near which I had some pleasing silent excursions before you undertook the task. That climate does not seem to agree with English constitutions. Another traveller of great expectation, Mr. Tweddell, (probably you knew him,) found the effects of that air and country fatal, and is much lamented. I always before this had a favourable prepossession about the healthiness of Greece, and cannot recollect that the ancients took notice of it as being dangerous to the human constitution.

If ever you quit botany for relaxation, you may look into Helen Maria Williams. In the vast profusion of chaff there may be found a few good grains of fact.

Farewell for the present, fellow citizen and countryman! I believe we should rather say *subject:* and to all friends communicate my affectionate tribute from West Britain.

Yours sincerely,

A. CALDWELL.

From Walter Wade, M.D.

Dear Sir, Dublin, Nov. 7, 1801.

Much is due from me on the subject of Irish botany, since the Linnæan Society honoured me so highly by associating me with them. I shall, however, now make a beginning.

A favourable circumstance has lately occurred, which enables me to submit to the Society some account of a plant, which in your very valuable English Botany, vol. xi. p. 733, is noticed as not being found in any part of the world but the Isle of Skye.

In the months of August and September last I undertook a botanical tour through the county of Galway in this kingdom,—my chief intention to examine that part of the county called Cunnamara, a district, I may venture to assert, never examined by any botanist before, although that celebrated and industrious naturalist Llhwyd was in its vicinity about the year 1699.

Cunnamara has its beauties in many particulars, not only in the eyes of the botanist, but perhaps might engage the attention of the poet and painter; —numerous lofty, craggy, heathy mountains, and lakes; meandering extended rivers; hills, bogs, creeks,—and all surged by the Western Ocean; which nothing less than the elegant descriptive pen of the President of the Linnæan Society could do justice to.

Among many other rare plants, I met with the one in question, *Eriocaulon decangulare*, Linn. Sp. Pl. 129. Mantiss. ed. alt. 167–327? Phil. Trans. vol. lix. 243. t. 12. Lightf. Scot. 569. Huds. Angl. ed. alt. 414, 415. *Nasmythia articulata*, Withering, ed. 3. 184. *Eriocaulon septangulare*, Flor. Carolin. 83. Lamarck Encyclop. tom. iii. 276. *Joncinelle décangulaire*, Eng. Bot. vol. xi. 733. *Eriocaulon septangulare*, Willdenow Sp. Pl. vol. i. 486. This rare aquatic decorates the edges of all the loughs, great and small, in Cunnamara, and is to be met with in many places in the county of Galway. The generic and specific characters given in English Botany, with the annexed admirable plate, are highly satisfactory. By Lightfoot's excellent specific description no one can be deceived as to its identity; and the generic character by the classical Hudson pleases me much; his trivial name *articulata* is a good one; perhaps *reticulata* would be better. His reasons for changing the old generic name to *Nasmythia* I cannot trace; may be to honour the memory of Nasmyth, who was a botanist, and surgeon to James I.

Our *Eriocaulon* varies much as to the height of its *scapus*, from a very few inches to nearly two feet; the leaves are pretty uniform, and I think 'grass-leaved' a more descriptive term than *ensiform* or 'sword-shaped'. In all the specimens I have examined, the sheath at the base of the *scapus* is invariable; but the angles undoubtedly vary,—most frequently seven: if my eyes, with the assistance of a watch-maker's glass, do not deceive me, I have counted from six to ten angles on a *scapus;* I am therefore of the late Dr. Hope's opinion, who, although our *Eriocaulon* differs in many respects from the *E.* of Linnæus, yet thinks it the same. Perhaps your valuable Herbarium may clear up the point? I was, on the ground of angles, at one time tempted to enumerate a *few* species, strengthened in some measure by a conversation I had some time since with a Dr. Browne, a very excellent, accurate botanist, who informed me that he had met with *several* species of *Eriocaulon* in Ireland. I would most willingly yield to this gentleman's accuracy and better judgement, on many other points; yet in this instance, for the present at least, I must presume to differ, if the *angles alone* on the *scapus* have tempted him to form a specific difference; which, from our cursory conversation on the subject, I am not authorized to say is the case.

I had almost forgot to inform you that our *Eriocaulon* was in full bloom the latter end of September, when we last parted.

In the forming of my *Flora Hibernica*, which I have the satisfaction of telling you is in much forwardness, I have made it an unalterable rule not to insert the *locus natalis* or modernized *habitat* on any hearsay authority, however respectable, except in addition to my own observation, if I can go or send where a rare plant is to be met with, or unless I receive a living specimen along with the information. The publication of my *Flora* must therefore necessarily meet with some delay.

Wishing to do as I would wish to be done by, I send you some specimens of our *Eriocaulon*, that you and the Society may judge for yourselves.

I am, dear Sir,

With every possible respect,

Your very obedient, humble Servant,

WALTER WADE.

From Mr. Caldwell.

My dear Friend, Dublin, Feb. 3, 1802.

I am so delighted and surprised at the sudden, unexpected blessing of peace, it is always uppermost in my thoughts. Your hint about going to Paris electrifies me. Who knows but that in such a scheme we might coincide? Our enjoyments are doubled by sympathy, and being shared with a friend;— mine, I am sure, would be so. I never have been there; you have tried the ground before. As for me, such a crowd of things to see, to search for, rush into my mind, I am bewildered: I am fearful, if left to myself, it would end in neither seeing nor hearing any thing. You could keep me in order and regulate my proceedings; but I must fairly acknowledge, I feel insatiable and greedy to the last degree.

I believe Dr. Wade, at my suggestion, will transmit to you the catalogue of the plants he found in his excursion through Cunnamara. He set out rather too late in the season, and his stay was too short. I cannot say that my desires are incited to pursue his track; there is not reward enough for the fatigue and the total want of comfortable accommodation. One would endure a great deal for Greece or Italy, and go through with alacrity; but the West of Ireland is not sufficiently tempting.

Being obliged to stay in town the whole summer, my botany was confined to my garden. I was, however, part of May at Lord Besborough's in the county of Kilkenny. I found the little specimen inclosed, on rocks, at a place called Owning; it is, I think, the *Myosotis* β: the flowers were yellow, singularly curled like a crosier. The leaves are hirsute as well as in *a*, but the blossom really yellow. The crosier form is perhaps accidental. The *Rubus Idæus* was also there.

I shall be glad to know if you ever observed that

the *Orchis latifolia* has, occasionally, a very disagreeable hircine scent. This *Orchis* is in great plenty about Besborough; but this peculiar scent I do not find taken notice of by any of you botanic writers. The scent is more frequent when the plant is pulled and put into water. It grows then so strong, it is very unpleasant in a room; we were forced to throw them out. I have also perceived it on the ground, but that seldom. The puzzling circumstance is, that many of the same *Orchises*, scattered close about, have no smell. I have examined numbers, to try whether there was any difference to make out a variety;—never could perceive the least. I have watched them for several days, to see whether it was occasioned by the different stage of the growth, —but to no purpose; that seemed not to produce any change. Linnæus says, *bracteis flore longioribus;* that is not the case in any plants I have ever seen, nor in the figure in Miller's Illustr., nor in Curtis. The figure in Curtis is much better drawn and more faithful than in Miller. The latter gives the root palmated; Curtis's and all the real plants I have found have two bulbs. These disagreements ought to be rectified.

I was at the Bishop of Dromore's, county of Down, in September;—found in that neighbourhood, and particularly on the marshy banks of the river Laggan, the *Bidens* in plenty, but whether β. or not, I cannot determine;—at another friend's place, the *Rubus Idæus* in very wild rocky situations in great abundance. I suppose there is no doubt of this being indigenous?

Quantities of the *Nymphæa* in the Laggan; but being quite out of flower, I could not find whether *lutea* or *alba:* the leaves seem just the same.

The winter here has been severe, but from the accounts not so much so as in England. Vast flights of crossbills, *Loxia curvirostra* I believe, made their appearance the latter end of August, and staid till the beginning of October. They made great havoc in the orchards; they never ate the apple, but cut it to pieces, and picked out the pippins. They came first over to the county of Cork, then proceeded to Waterford, Tipperary, Kilkenny, Wexford, Wicklow and Dublin; but no further north: and yet they say it is a bird of Northern Europe. They are good prognostics of a severe winter. They were observed here, I am told, before the hard frost, and once since; and now this time, when there certainly has been severer frost than usual. I observe in December's Monthly Review a work by R. S. Barton on the natural history of Pennsylvania. He considers the connexion between the seasons and the migration of birds.

My young friend, I expect, may be the better for the lottery prize; therefore the less to be regretted it went to the *rich.* It is truly strange, but I never knew a great prize go to any but the rich. Could I have given a little shove to the wheel of the blind deity, it should have turned in another direction. *N'importe!* others have their riches,—the imagination, the intellectual: I would not exchange.

How kind, how flattering your call on me again to Norwich! Be assured I do not want pressing.

There is a pretty little plant in our stoves, the *Eranthemum pulchellum;* the flowers fine blue in a spike; beautiful *bracteæ:* they say the London gardeners stole it from Kew.

Your affectionate

A. CALDWELL.

From the same.

Harrowgate, Oct. 5, 1802.

* * * * * * *

I have been now at Harrowgate since the latter part of August. The weather at first was rainy, but for three week spast wonderfully fine. These water-places do not suit my disposition; I really feel a great dislike to them. The idle, sauntering way of wasting time is tiresome to the last degree; nor have I had the good fortune to meet with one companion from whom I could derive a new idea. I believe, however, the drinking and bathing has been of benefit; and if the purpose is answered for which I came here, I shall have no right to complain.

A little excursion I undertook some days since has been the most pleasing occurrence that has happened to me here. I was two days at Wentworth House; it is a most magnificent place, both as to house and park and plantations. I next saw Wentworth Castle, now inhabited by Mr. Conolly and Lady Louisa*: it is nearly as fine as the other, and

* Sister of the Dowager Duchess of Leinster, and aunt of Lord Edward Fitzgerald. For the character of these excellent women, and of Lady Sarah Napier, another sister, see Moore's Life and Death of that unfortunate young nobleman.

in prospects and situation much resembles it. I went from thence to York; the minster there everybody has heard of, and all who see it must admire. Castle Howard, Lord Carlisle's, was the next;—all these places are so splendid, one is apt to think the last the finest. This place terminated the excursion, and I returned here.

Yours,

A. Caldwell.

From the same.

My dear Friend, Dublin, Oct. 21, 1803.

It was a great misfortune not to be able to meet you at Liverpool; and when I heard you had left it, I at once abandoned all thoughts of going there this year.

So soon to be deprived of your amiable friend, and your future schemes of intercourse blasted, is melancholy. The frequent losses of those we love is the great evil of life: every other misfortune in my estimation sinks before it. I experienced a severe stroke of this kind but little more than a twelvemonth since; a friendship of more than thirty years was then terminated. What a blank it has occasioned! Such inflexible virtue, so much good sense and information are rarely met with. There was that intimacy that in his society we might each of us be said to be thinking aloud. His name was Mangin, of one of the French Huguenot families. He had been for many years first clerk in the secretary of state's office in the Castle; but from ill health, much increased by unremitting application to business, was obliged to retire seven or eight years ago. His relaxation and favourite amusement was collecting prints and drawings. He had the best taste, and more knowledge on those subjects than any man I ever was acquainted with. He appointed me one of his executors, and left me entirely his vast collection of prints and drawings. This mark of regard, however gratifying, can never give me the same enjoyment as formerly; I am overwhelmed with them; and only trying to look into one portfolio plunged me in melancholy. I am a great deal older than you, and have met with so many deprivations of this sort, that I feel a kind of insensibility creeping on. I consider myself as a spectator only of a fleeting world, and have little interest in any thing that occurs. Nothing you say in these cases avails but religion. It is the best consolation, provided it be pure and benevolent, unmixed with worldly craft,—not such as we have to lament in this country, that sows hatred and animosity, and is the true source of all our mischiefs.

You allude, I suppose, to a future state; the prospect of which religion holds forth. I must confess, however, that I do not feel that horror at the idea of non-existence that Dr. Johnson did: I can conceive a much worse circumstance;—one alone,—with the privilege of immortality, when every thing else has ceased;—that, in my opinion, would be more shocking than any thing that could be conceived.

My taste will not be questioned by my friend Mrs. Smith as to the Ayrshire poet. The four volumes of Burns are lying in my parlour, and I open one or other of them every spare moment. I certainly admire him as much as she can do. The letters surprise me more than the poetry: the refined criticism and good sense in them are astonishing, when his humble birth and education are considered. I feel great concern that he did not preserve Mrs. Dunlop's correspondence; perhaps out of delicacy to her it may have been suppressed. Dr. Currie I am acquainted with; he is a most amiable man, as well as of more than common abilities. Nothing more perfect or complete could be desired; and as he is in very full practice, it is extraordinary he could have performed this task so admirably within one year.

There is another Scotch poet to whom I am extremely partial,—I mean Allan Ramsay. His works are lately published,—the most full and complete edition that has yet appeared. Though not equal in genius to Burns, Ramsay has great merit. His songs and Gentle Shepherd were the first poetry I knew and was delighted with in my earliest years. They bring to me many recollections, many associations that are pleasing, but often melancholy. The editor is not so satisfactory as Dr. Currie; there is a want of illustration and anecdote, that often disappoints our curiosity. Ramsay's day is not so distant but that every thing might be explained.

I dare say you have founded a good sect of botanists at Liverpool. It is a captivating pursuit, and the more known the better liked. Our huge garden here is a most pleasing one for a walk. I saw, the other day, half a dozen sorts of beautiful *Xeranthemums* a lady had got from the ever-marvellous, inexhaustible Cape. One sort, the petals all curled and formed a sort of globe of a straw colour: others of a fine orange: others were flat and open, purple, and at a distance I took them for China Asters. England, I dare say, has had them all, and you probably have seen them.

You mention some noble additions to your books. Though your collection may not be so extensive as Sir Joseph's, I am apt to believe what you have is more choice and perfect than his. I cannot wonder your reputation attracted Madame Bonaparte's attention, and the presents she has sent you are very gratifying. I shall hope to see them with you one day or other. They are not likely soon to reach this part of the world.

My best affectionate compliments to Mrs. Smith.

Yours sincerely,

A. Caldwell.

From the same.

My dear friend, July 11, 1804.

* * * * * * *

All I can say is, that it was a disappointment to me —most grievous, the being prevented from going to Norwich. I was introduced to so many agreeable people at Cambridge, their civilities were such, and their pressing desire to have me stay for the Commencement, so great, I found it scarcely possible to decline. Letters then came from London, that obliged me to come hither without further delay. When I went to the university, my intention was to stay a week at the utmost, and I staid three weeks;—much more kindness I experienced than I had any right to expect. I suppose you have heard of Dr. Clarke: he and Mr. Cripps travelled together for three years. Dr. Clarke visited Pallas in the Crimea, and purchased his herbarium. The Doctor said he was old, and when he died nobody there would think it of value, and it would be thrown out of the window. Clarke has brought home great collections, manuscripts, medals, minerals, drawings consisting of views of interesting places, antique marbles, and the colossal bust of Ceres, discovered above a century ago by Spon and Wheeler, in her temple at Eleusis: it is almost quite destroyed,—not a vestige of the face; something remains of the shoulders and their drapery; she has a great turret or basket on her head, which is called the *calathos*: it is covered with symbols, pretty distinct. There will be a particular account published. What an instance of instability in this world,—the object of such mystery, adoration, and magnificent ceremonies in ancient Greece, has now removed to a college amongst the barbarian Britons! I dare say you will some day or other visit Cambridge, and look over the herbarium of Pallas. You will find in Dr. Clarke one of the most engaging men you ever met with, and he will delight in gratifying you to the utmost extent.

A friend in Dublin mentioned to me, that the

drawing of Mars and Venus, omitted in the prints of the *Elogium Stultitiæ* when you saw the copy, is now engraved in M. Veaux's French translation of the *Elogium*, printed at Basle in 1780, and which perhaps you had not since heard of.

I have had a letter from Mrs. Charlotte Smith. She has had some thoughts of going to Lowestofft for bathing. She wishes it much for the hopes of seeing you there.

I am, dear Sir,

Yours sincerely,

A. CALDWELL.

From Dr. Wade.

My dear Sir, Dublin, Nov. 1, 1805.

I send you nearly 200 species of Mosses, and, with very few exceptions, all in fructification. The great majority of the more rare were gathered by myself, in 1796 and this year, in the county of Kerry, particularly on the mountains, in the woods, glyns, &c., within a few miles of enchanting Killarney, so rich in such uncommon scenes of natural wildness and rural magnificence. I am aware of having sent a few duplicates in separate papers; but at the time I arranged them for you, some trifling difference seemed to occur, which your very discriminating eye will best judge of. Pardon me for sending you some common species which I gathered in this county and the county of Wicklow.

My respected friend may inform Mr. Turner,

that I have specimens of Musci *in quantity* at his service, as he seems to make the muscology of Ireland his favourite object.

You will receive by this opportunity a few specimens of plants which may be considered as rare with us in Ireland,—particularly the *Sium verticillatum* (truly a *Sium*), which I got in marshy ground near Loun Bridge, about four miles from Killarney; *Bartsia viscosa*, whose *habitat* you know (*Pl. Rarior.*). I will now answer for the *habitat* of *Saxifraga Geum* or *hirsuta?* specimens of which I send, having met with it in abundance in that astonishing and bewitching spot to the naturalist, the Gap of Dunloc, about eight miles from Killarney, and which divides the black and craggy sides of those stupendous mountains Mac-Gilly-Cuddy's Reeks from the Tornies Mountain, and is the direct road and best way to the Purple Mountain. As you may suppose, I did not meet with it in flower, as it is a very early flowering plant, and I went there very late in the season. However, I have sown some of the seeds, and planted the roots, by which means I shall be enabled to determine exactly what *Saxifraga* it is; as well as some other species of the same beautiful tribe, particularly some extraordinary varieties of *Saxifraga umbrosa; Pinguicula lusitanica*, common; *Schœnus rufus?* on Purple Mountain in abundance. I can assure you that *Euphorbia hiberna* is uniformly furnished with *umbella quinquefida*, for I have examined above a thousand growing specimens this season, and met with *one only* 6-*fida*. Many measured between thirty and forty inches high; a *caulis simplex* invariable; none branching at the base, and the *axillary ramuli* almost always present. I have a very correct figure of our *E. hib.* nearly finished, and pardon me if I say a correct figure of it is very desirable, as none approaches in likeness to our *E. hib.* but t. 290. *Hort. Elth.*, and, if recollection serves me, in *Rivin. Tr. Mon.* *Serapias latifolia* about the old cobaltic copper-mine holes, very common in the delightful and romantic peninsula of Mucruss.

Orobanche elatior in profusion in the different chambers, &c. of the old abbey at the northern side of Mucruss, together with *Polypodium cambricum*, in a state of great luxuriance. *Aspidium Oreopteris* I consider as a rare fern, as well as *Adiantum Capillus-Veneris;* but as I am forming a collection of them for you, I shall for the present reserve my sentiments on the subject of Irish ferns.

Do you consider *Schœnus mariscus* as rare? Besides the *habitats* in *Pl. Rarior.* I met with it, but rather sparingly, in the fens at the N.W. side of Mucruss. You cannot conceive, my esteemed and respected friend, how abundant *Hymenophyllum tunbridgense* is on all the dripping rocks on the mountains in the county of Kerry; and the uncommon state of luxuriance and magnitude which *Osmunda regalis* assumes on all the islands in the range to the upper lakes at Killarney is truly astonishing, looking more like large highly cultivated shrubs, than a humble fern.

You will be pleased to present to the Linnæan Society the inclosed remarks on *Holcus odoratus* and *Buddlea globosa,* which perhaps are only valuable by being accompanied with excellent coloured figures,—thanks to the artist. Do you know of any coloured figure of *Holcus odoratus?*

I long to see Mr. Turner's prediction in the preface to his *Mus. Hib.* verified, as to a Flora of Ireland by Mr. Templeton: for my part, I fear it will be long before Mr. Turner will have an opportunity of honouring my name with a similar prediction; and I say it with regret, for the reason stated in my humble preface to *Plant. Rarior.*

Our mutual friend Mr. Caldwell desires to be kindly remembered to you, and intends writing shortly to you. I feel I have intruded much on your very precious time, and shall only add that I am, with every possible respect,

Your sincere and obliged humble Servant,

WALTER WADE.

From Mr. Caldwell.

My dearest Friend, Dublin, March 7, 1808.

The grateful feelings for the testimony of your remembrance and friendship that I received yesterday by your letter, surpass all my powers of expression. I read it over and over: it contributes to my happiness, but I am conscious I do not merit such attention; I do not meet with it from many on whom I might have some claim: but your kindness is all from generosity and sincerity, and without any pretension on my part.

I have passed but a melancholy, lonely winter, having been much indisposed since the beginning of November. It is not being alone that is distressing: where one can take a book, and likes reading, amusement and enjoyment run hand in hand; but a heavy sickness renders me incapable of enjoyment: it has been a labour and an exertion at different times to write my name.

It gave me great comfort to find you were well; but though I used to receive or answer a letter from you with delight, I this time find myself disabled. I am, however, growing better, and look forward to the hopes of seeing you in London, where I scheme to be the latter end of next month. It is to be regretted by me that fortune had not placed me within the reach of your society, and some others in England: I should have enjoyed much more constant friendship and attention than here; this place is too dissipated, and there are few whose tastes agree with mine in preferring quiet social intercourse.

I have endured a long truce with botany, but must lose no time in refreshing my memory, or shall forget the very elements. I went to our great garden two days ago,—the first visit since last June. It made no great appearance, and has suffered much from the uncommon severity of the weather: though it is almost ten times as large as the garden at Liverpool, and there is expense to the greatest prodigality, I doubt whether it be half as curious. There is a fine specimen of the New Norfolk Pine, but which will soon be too high for the green-house; a very great Banana, which looks as if it would produce fruit. They have had the Bohea Tea in flower:—and excepting these articles, our garden has nothing to show at present. I believe where we exceed Liverpool is in the large trees and shrubs.

I hear your new work is come over, and I am impatient to see and begin it. If I can make it out so as to be in time for the Institution course, I shall be enchanted. Mr. Roscoe, from his writings, and intimacy with my relations, I think myself well acquainted with: I was introduced to him, and enjoyed his conversation for half an hour when last in Liverpool.

I believe our politics, that is, yours and mine, being founded on strict moral principles, can never differ. I look on the outrage at Copenhagen with as much indignation as if I were a native of the place: I am glad to hear there are some people in England that are sensible of its baseness; but I fear very few, and that is a dreadful symptom of general depravity.

Botany, that in England unites people and classes them in friendship, produces here a contrary effect; they are all at variance: the University has displaced Dr. Scott, an ingenious, lively man with great merit, and a good botanist.

I have read Mr. Johnes's Froissart, and think it a most curious picture of its times. All the little skirmishes and battles that resemble one another so closely, we may skip over in a great measure. How melancholy the consideration, that in all countries and in all ages to fight and to slaughter is the road to glory! I am rather displeased with Mr. Johnes, for not reprobating that mischievous absurdity; and he never produces a sentiment or makes a remark on the occasion.

I am, my dear Friend, yours sincerely,

ANDREW CALDWELL.

This letter, written under the pressure of illness and dejection of spirits, was the concluding one of a correspondence which for many years afforded considerable enjoyment to both parties, and presents a lively image of the conversation and temper of the amiable writer. He died very soon after the date of the above.

CHAPTER X.

Miscellaneous Letters from 1810 *to* 1816:—*Sir Thomas Frankland.—Mr. Walker.—Mr. W. Smith.—Dr. Waterhouse.—Mrs. Cobbold.—Sir J. E. Smith.—Mr. J. Lee.—I. A. Pavon.—Dr. Bigelow.—Sir Joseph Banks.—Dr. Tenore.—Mr. Repton.—Rev. Mr. Walpole.—Baron Humboldt.—Professor DeCandolle.—Duke of Bedford.*

From Sir Thomas Frankland.

Dear Sir, York, March 10, 1810.

THOUGH I have been confined by rheumatism nearly since Christmas, I feel a fancy to write you a gossiping letter. Till last summer I had never crossed the Tweed; but in July my son and I set out to accept the Duke of Gordon's invitation, and to see what we could in a month and a tour of above nine hundred miles.

We went by Newcastle to Dunbar and Edinburgh, where we stayed forty-eight hours, and were enchanted. From thence to Stirling, Crieff, Perth, Brechin, and turned off a few miles to Mr. Alexander Brodie's at the Burn (purchased of Lord Adam Gordon), where we stayed two days: then to Aberdeen, and by Inverurie and Huntley to Gordon Castle. After six days, to James Brodie's for two days. He very kindly went with us to Inverness, where we hired four horses for the Highland road; and in four days they got us to Crieff, 120 miles. Then to Stirling and Glasgow, where we were most highly gratified: but rainy weather began then, and continued throughout the remainder of the journey to Carlisle, Keswick, Ambleside; Lord Lonsdale's, at Lowther, for three nights; and home, seventy-six miles, with ease in a day. It is difficult to say what we saw or did. We were asked how many brace of moor-game we had killed. Answer—We went to gain new ideas of country and inhabitants, and refused all opportunity of sporting. What plants did I find? None, but two or three which the Duke's factotum showed me in the woods; for my son being no botanist, I was determined not to delay him for a moment, and, except his laughing at my being annoyed with *precipices* in the Highland road, we never had a difference; indeed that road is so bad, that, but for an excellent carriage and servant, we should have been lost in difficulties. My son saw the Cumberland Lakes to disadvantage, from an unceasing rain; but I have seen them twice before. Lord Lonsdale's new house is a wonder. It is a gothic palace; and, although only begun two years since, is already inhabited by all the family. Smirke is the architect. Brodie has a botanic garden for British plants, and employs himself almost entirely with them, though I got him to salmon-fishing one day. Where salmon is most plentiful, there is the worst angling; for they are harassed with nets night and day, and have no time to feed. The Duke's fishery lets for 6000 guineas per annum!

and the renters are said to make a fortune. I saw the ice-houses, which are entirely above-ground and have no waste drain, the melted ice sinking into the sand. Two thousand salmon have been taken in one day in that fishery (river Spey). The largest fish I heard of was caught at Aberdeen in May, 1762, and weighed sixty-three pounds, Dutch weight, of seventeen ounces and a half to the pound.

Gordon Castle is five-hundred and sixty-eight feet long, and eighty-four high in one part. Mr. Hoy was endeavouring lately to determine, upon Arrowsmith's map of Scotland, the number of square miles contained in the Duke's estate in the different counties, and found that in Moray, or shire of Elgin, along the side of Spey, there are 13¾; in Banffshire, low country, 24½; in the same, high country, 204; in Aberdeenshire, 107; in Inverness-shire, Badenoch, 325; in Lochaber, 330; —total, 1004¼ miles, and 642,720 acres; all property lands, without including those over which the Duke has only the superiority.

The plants which were shown me in the woods at Gordon Castle are *Satyrium repens* and *Pyrola secunda*. Mr. Brodie gave me *Eriocaulons*.

Satis superque from your most faithful

THOMAS FRANKLAND.

From John Walker, Esq.

My dear Sir, Bedford Square, March 16, 1810.

I beg to offer you my particular acknowledgements for your truly friendly attention in sending me cuttings of Willows. Some of the species, particularly the *Russelliana,* a very valuable one, I did not before possess.

When you come to town, you will have the satisfaction of seeing Sir Joseph Banks looking uncommonly well, and perfectly free from gout, which he attributes to a French quack medicine he has been induced to try, at the pressing solicitation of Lord Spencer. From a single dose of a tea-spoonful and half, he experienced almost immediate relief. Mr. Home left him one afternoon in agony with the gout in eight different places, and with a hard and high-beating pulse at 94. The following day he found him altogether without pain, tenderness alone remaining, and with a pulse perfectly tranquil and reduced to 67. I am aware that a solitary instance is by no means sufficient to convince you or any medical man of its general utility; but to the above instance I may add those of Major Rennell, and near twenty more who have tried it with the same immediately beneficial consequences, however inexplicable they may appear. It has been prepared in France upwards of thirty years, under the name of *Eau medicinale d'Husson.* Chemistry has proved it to be vegetable; but from what plant the very extraordinary powers of this medicine are derived, remains at present unknown.

You will have heard that we have lost Mr. Cavendish,—a man of a wonderful mind, more nearly approaching that of Newton than perhaps any individual in this country since his time. Educated and trained by his father from very early youth to scientific pursuits, they became habitual, and were the sole occupation of a very long life, fruitful in important discoveries, forming the basis of modern chemistry. That his mind remained unaltered and unimpaired to the very last, his paper in the second part of the Transactions for 1809 is a sufficient proof. "*Ingenuas didicisse fideliter artes*" ought to improve the whole man, the heart as well as head. It is to be regretted that science and charity were wholly forgotten in the distribution of his property. He might well have divided 100,000*l.* between them, which I believe would not have been an eighth part of his immense possessions.

Yours most sincerely,

JOHN WALKER.

From William Smith, Esq.

My dear Sir, Parndon, Nov. 20, 1810.

I am really sorry I was not of your party in the so well described library*, with its so well appreciating visitants; but with all due and humble submission, I should have made, I will not say a better, but a more every-hour companion for Lord Spencer than yourself; for when tired, if ever, of the library, I could have gone with him to the field,—from taking down books, to knocking down pheasants, —which I fear you could not.

* Mr. Coke's, at Holkham.

He is an admirable specimen of the pteriplegistic, philosophic statesman,

> "Who rapid turns the keen observant eye
> From men to books, from books to man again;
> Stores and compares; his well-digested store
> Then opes and uses in the walks of life;
> And when the faint o'er-laboured spirit flags,
> Renews its vigour in the sportful field."

I cannot recollect where is your most happy and poetic quotation about the dusty books, if indeed I ever knew it.

Yours truly,

W. Smith.

From Sir Thomas Frankland.

Dear Sir, Thirkleby, York, April 13, 1811.

My son has just returned from Lincolnshire, where he has been hunting these last five weeks. It agrees with him, as far as my eye informs me. In the third week of October last I went to Sir Joseph Banks's in Lincolnshire, where I stayed near a week, and had a most agreeable and interesting visit. The singular fineness of the season and good breeding of game gave me uncommon diversion there; but late in October I perceived a strain in my tendon which was broke in 1792, and, though I favoured it, by a slip (at Duncombe Park) in November I contrived to be crippled. However, I was trout-fishing all this morning, and feel no concern on the subject. *Glaucium fulvum*, of which you gave me some seeds two or three years since, is the most desirable plant of the kind which I know. We found it in full blow on returning from town, and for some months it was in gorgeous flower, renewed daily. I saved much seed, and have distributed it as an acquisition to every garden. I had great pleasure in thinking Sir Joseph particularly well in mind and body, notwithstanding the alarm which many persons have felt respecting the gout medicine which he has latterly taken as an alterative.—Accum tells the Duke of Gordon that it is *Gratiola*; and that a medical friend of his gives that plant with success. In looking into Lewis's *Materia Medica*, I was struck with the effects of *Gratiola* agreeing so exactly with those of *Eau medicinale*.

I found lately a lime-tree branch like a teapot handle, about four feet from extremities; but whether the branch grew upwards or downwards does not appear, the union is so perfect. It is in a row of very fine old trees.

T. F.

From Dr. Waterhouse.

Dear Sir, Cambridge (America), July 24, 1811.

I here send for the library of the Linnæan Society a little work I have just published, in which I have endeavoured to do honour to our great master. I have laboured to make him known to my countrymen during eight-and-twenty years past, and at present Linnæus is no stranger to the inhabitants of these "ends of the earth."

As I was the first who gave lectures on natural history in America, I was obliged to give it the popular dress which "The Botanist" comes forth in. I was compelled to clear the ground to prepare it before I sowed the seed. I found I must first excite a taste, and then try to gratify it. Our elder brethren in Europe know not the difficulties that the first settlers in science have to encounter. I found it particularly difficult when I first gave lectures on mineralogy to persons who had perhaps never seen a book upon that subject. This will account for the popular dress in which botany has been presented to the youth of both sexes in this country, and it will explain why I tried to celebrate certain female characters who had distinguished themselves in natural history. I hope to send you hereafter some paintings of plants and insects, executed by some of our ladies, that Marian herself would not be disgusted at. Natural history will flourish among us, although we may not produce a Linnæus or a Buffon. I observe among us young Swammerdams, young naturalists on whom Fate has written "*Laudatur et alget.*"

My bookseller already talks of a new edition of "The Botanist;" but before I determine on that, I should like to profit by the criticisms of such a great master in the science as Dr. Smith.

My wish is to open such a path in the field of

botany, as would be most useful to such a country as ours, and at the same time be gratifying to the botanists of England.

With a high degree of respect,

I am yours, &c.

BENJAMIN WATERHOUSE.

*From Mrs. Cobbold**.

My dear Sir, Ipswich, March 26, 1812.

The pleasure I received from those lines which you showed me of Mrs. Barbauld's on the King's illness induced me to order her pamphlet† of 1811, which I have read with much attention, and, I must add, much feeling. It is in a high strain of poetry, and possesses a fire of genius and force of language which I should not have expected from her advanced age and what I had seen of her earlier productions; but if I were offered the powers of genius, together with the feelings manifested in that poem, I would reject the combination as a dangerous and deadly gift. It yet remains for the world (I would say a world of false philosophers, were that race not sinking fast into oblivion,) to invent a word so monstrous as to express hatred of one's country, and exaltation of her rivals! Could such a passion have a name, I should say this is the ruling one of

* This accomplished lady is the author of some very beautiful and appropriate verses on the death of Francesco Borone, which were published by Sir James in his volume of Tracts, in 1798.

† Eighteen Hundred and Eleven, a poem.

Mrs. Barbauld's mind. Was it by degradation of his country that Nelson infallibly led her sons to victory? Is it by filling the minds of her sons with gloom and despondence, that we would enable them to support her through the present arduous, unequal, and dreadful conflict? Shall that being dare to assume the name of patriot, who shall call it guilt in Britain to contend with the giant-power of France? Or shall a mortal pretend to read the decrees of the Almighty, and declare that he has doomed the fall of Britain?

Columbia, too, is the land of freedom! the temple of science! Pray tell me how has she improved in science since she lost her Franklin? and how has the pristine vigour of her constitution, and its freedom, been preserved, since she lost her great, her truly exalted Washington? What are her legislators now? One half of them servile knaves, sold, for the most ignoble of all rewards, to the despot of Europe; the other and more virtuous half, distracted by jarring interests, and feeble and disunited in their councils.

America, from her extent of territory, her considerable and increasing population, and the abundance she possesses of all the necessaries of life, can never reasonably be the object of conquest: any attempt to subdue her must increase her moral and physical strength by cementing her civil union. Discord is her great and only danger. She may, and probably will in time, become a great nation; but she is at present only in her youth; strong and agile, I grant, but with a head not sufficiently energetic to direct the limbs in the most efficient manner. Should that head be weakened by fatuity, or perverted by the corruptions of the heart, adieu to all the visions of rising grandeur.

Far, far be from me to say that Britain is all god-like! that she owns none of the human imperfections to which other states are subject! Her rulers have the same corruptions as those of other lands; and perhaps the dreadful examples of anarchy around her, have led her to submit too easily to the encroachments of power. Still there is a vigour and purity in her constitution, which will manifest itself,—not I trust in the rude efforts of brutal strength, but in calm and steady energy. Dearly do I love my country, and I cannot suppress my indignation, when I see any of her sons or daughters, even in imagination, attempting to bow her neck beneath the feet of a nation which I consider so inferior to her in all moral dignity, as America.

It is a most unusual thing for me to enter on any subject connected with politics; but as I know my opinions to be perfectly independent of any party, I thought I might be allowed to express those genuine feelings which arose in my mind on the perusal of a book, which, on the most mature consideration, I cannot find tending to produce any good effect in society; and which is only the more dangerous on account of its poetical excellence. At the present momentous crisis, we ought to be awakened to a sense of our danger; but so much the more should the *amor patriæ* be excited to avert it. Had I been acquainted with Mrs. Barbauld, it is probable that a sense of her private worth might have led my mind to overlook or palliate the gloomy tendency of her poem. As it is, I read without prejudice.

I promised to tell you what I thought of this book, and I have done so without reserve; you would have a right to despise me, if I could sacrifice truth even on the altar of friendship. I would rather lose some little portion of the love of my friends, than one iota of their esteem. Love may be won by flattery; esteem must be the result of good qualities, and the power of appreciating them.

Your ever affectionate Friend,

E. Cobbold.

Sir J. E. Smith to Mrs. Cobbold.

My dear Madam, Norwich, March 30, 1812.

I did not doubt your admiring Mrs. Barbauld's *poetry;* indeed, I think this poem (without any allowance for her age) may take its stand amongst the most lofty productions of any poet, male or female; as I think her hymn, beginning

"Jehovah reigns: let every nation hear!"

the most sublime and poetical hymn that ever was written,—without any allowance for her *youth* or her *sex* in that case either. Indeed she requires no allowance, though ever ready to claim it; for her character and her humility are equal to her talents. Now let me tell you my opinion of the politics of her poem. I conceive the three first pages and four lines of the fourth are a virtuous, *patriotic*, and just view of the evils produced by the usurpation and policy of Buonaparte: what follows is too just, though I hope too gloomy, a picture of what Britain has to fear. As to the 'guilt', *that* involves a question on which very good people differ. Mrs. B. seems to think the war, not being unavoidable, was guilty. Some think it the cause of the tyrant's success, who might never have had an existence, if France had been left to herself. This would lead me too far; you know better than I do what has been said. I only mean that Mrs. B.'s idea is conformable to that of numberless undoubted good Englishmen and good subjects.

From p. 6 to 17 appears to me to express the most enthusiastic *love* and *patriotic* admiration for her country, and all that has ennobled or will immortalize it. It must be a 'false philosopher' indeed, who could pervert this eloquent celebration of British virtue, talent, and distinction into an effusion of '*hatred*'! And where does she exalt the rivals of her country? Does not she express that even America, if it should, *in the lapse of ages*, (for that I know is her meaning,) become what Asia was, and Europe is,—*even America*, I say, will be indebted to England for all her taste and learning, and will fondly dwell on this as classic ground? I surely need not point out to you that the sublime and forcible picture (p. 17 to the end) is a highly poetic personification of the power of civilization. In this she looks forward to ages yet unthought of, taking her stand far above the present events or contemplations; comprehending the whole history of civilized man, all antiquity, in the retrospect, and anticipating what thousands of years may produce. Whatever alludes to Britain in this prospect expresses what has been the fate of all other countries; and it is lamentable for any,—but not less inevitable. Has not the progress of Christianity been from east to west? So we are taught to trust it will embrace the globe. She does not say Columbia *is* the temple of freedom or science, but that it probably *will be*. Far indeed is she from descending to the paltry speculations for which you give her credit about the petty parties of the day, there or elsewhere! Now, my good friend, forget all party, and be (not a *false*, but) a *true* Christian philosopher; take this excellent woman to your heart as a congenial spirit; for if you knew her as I do, I will do you the justice to believe you would love and admire her as much. My wife is copying two productions of hers*, that abundantly evince her

* Mrs. Barbauld's Hymn before mentioned. Long as it is since it first appeared in print, the writer makes no apology for inserting it at length in this place.

HYMN.

Jehovah reigns: let every nation hear,
And at his footstool bow with holy fear;
Let heaven's high arches echo with his name,
And the wide peopled earth his praise proclaim;
Then send it down to hell's deep glooms resounding,
Through all her caves in dreadful murmurs sounding!

He rules with wide and absolute command
O'er the broad ocean and the steadfast land:

patriotism and her piety. How far her precise notions or opinions may agree with yours or mine, I really neither know nor care. I know she is a zealous Christian, but not a zealous Socinian, Arian, or Athanasian; I should suppose most of the second.

Jehovah reigns, unbounded, and alone,
And all creation hangs beneath his throne:
He reigns alone; let no inferior nature
Usurp, or share the throne of the Creator!

He saw the struggling beams of infant light
Shoot through the massy gloom of ancient night;
His spirit hush'd the elemental strife,
And brooded o'er the kindling seeds of life;
Seasons and months began the long procession,
And measured o'er the year in bright succession.

The joyful sun sprung up th' ethereal way,
Strong as a giant, as a bridegroom gay;
And the pale moon diffused her shadowy light,
Superior o'er the dusky brow of night;
Ten thousand glittering lamps the skies adorning,
Numerous as dew-drops from the womb of morning.

Earth's blooming face with rising flowers he drest,
And spread a verdant mantle o'er her breast;
Then from the hollow of his hand he pours
The circling water round her winding shores,
The new-born world in their cool arms embracing,
And with soft murmurs still her banks caressing.

At length she rose complete in finish'd pride,
All fair and spotless, like a virgin bride;
Fresh with untarnish'd lustre as she stood
Her Maker bless'd his work, and call'd it good;
The morning-stars with joyful acclamation
Exulting sung, and hail'd the new creation.

Yet

I can speak positively to her Christian charity, which you and I think the most important point.

Thus I have written as freely as yourself in the true spirit of friendship: and if you find me in the right, it is all I care about; for you are worthy to know, and capable of appreciating, if you did know, my highly honoured friend.

&c. &c.

J. E. Smith.

Yet this fair world, the creature of a day,
Though built by God's right hand, must pass away;
And long oblivion creep o'er mortal things,
The fate of empires, and the pride of kings:
Eternal night shall veil their proudest story,
And drop the curtain o'er all human glory.

The sun himself, with weary clouds opprest,
Shall in his silent dark pavillion rest;
His golden urn shall broke and useless lie,
Amidst the common ruin of the sky;
The stars rush headlong in the wild commotion,
And bathe their glittering foreheads in the ocean.

But fix'd, O God! for ever stands thy throne;
Jehovah reigns, a universe alone;
Th' eternal fire that feeds each vital flame,
Collected, or diffused, is still the same.
He dwells within his own unfathom'd essence,
And fills all space with his unbounded presence.

But Oh! our highest notes the theme debase,
And silence is our least injurious praise;
Cease, cease your songs, the daring flight controul,
Revere him in the stillness of the soul;
With silent duty meekly bend before him,
And deep within your inmost hearts adore him!

From Mr. James Lee.

Dear Sir, Hammersmith, July 9, 1812.

I received your kind letter of the 12th of May with a packet of seeds from Russia, for which accept our best thanks; and also your other of the 22nd June concerning the particulars of the life of Mr. Masson. I mentioned to you that I thought he had been ill paid, considering what he had done for the science of botany. He explored the Cape of Good Hope twice, Madeira, the Canaries, Azores, Spain, Gibraltar, Tangier, Minorca, Majorca, the West Indies, and Canada. Masson was of a mild temper, persevering in his pursuits even to a great enthusiasm, of great industry, which his specimens and drawings of fish, animals, insects, plants and views of the countries he passed through, evince; and though he passed a solitary life in distant countries from society, his love of natural history never forsook him. Characters like him seem for the present dwindling in the world, but I trust they will revive.

I am, dear Sir, with great respect,

Your much obliged Servant,

JAMES LEE.

From J. A. Pavon.

Madrid, the 22 of July, 1813.

To the must celebrated Botanist Sir J. E. Smith.

I conserve in my power and to your order a little offering, which I make to the merit that Sir Smith obtained of the charmer science of the botany. This little present is a *Monography* of eight and twenty new sorts of Laurels that I discovered and drawed in the native places, and under the trees of the South America. The present was prohibited me to publish it, and so I send only to you the engraved patterns; and you will have the goodness to tell me any person to give them in Madrid, along with another exemplar, that I inscribe for your mediation, and in my name to the wise Society of London, asking you and the wise Society will be so good to censure, with kindness, the imitation of the patterns in the botanist part, and tell me the defects which you know in the analitic or anatomic part of Linneo's system.

I send to you the note of collections which I will sail to live; if any wise man lover of the botany will buy some of them, you will have the goodness to tell me by our ambassador the Count of Fernannuñez.

I will send to you the other exemplars of the Laurography so soon as I can publish it. I shall be very glad if you will be so good to answer me, only to have a memory or letter of the wise Sir Smith, honour of the botany for your works that I read every day to learn and to know your erudition.

I was eleven years in the South America, and I experienced great many tempest in the Cape of Hornos and in the South Sea, and by land the revolution of the Indians under the command of Tupac Amaro. In Spain I have been many times to be killed by the bayonets of the famous scholars of

Neron Targuine, and above all them the new Maguiabel Bonaparte.

I wish to be useful to you in any thing, and in the mean time I am,

Sir,

Your most obliged and obedient Friend,

The Spanish Naturalist

JOSEPH ANTONIO PAVON.

The envelope of this letter was thus singularly inscribed:

"To celebrious botanist, Sir Jacob Eduard Smith member of various academys scientificals.— To London.

" Para el celeberrimo botanico el Doctor Smith J. E. presenta esse pequeño fruto de sus trabajos literarios botanicos, su afectisimo consocio el Botanista Español, Autor de las Floras Peruana y de Chile, José Antonio Pavon."

From Dr. Bigelow.

Sir, Boston, Massachusetts, Nov. 18, 1813.

I take the liberty to address you, inclosing the introduction which my excellent friend M. Corrêa has obligingly given me. In the present imperfect state of science in this country, I am every moment sensible of the necessity of an European correspondence, to supply the defect of books and advantages which it is impossible here to command. I cannot but feel highly honoured in having at this time the privilege of epistolary access to the head sources of the science. In the New England states we have a country as yet little explored by European botanists, and having in some degree a botany peculiar to itself. Many of our species resemble those of Europe, but at the same time differ in small particulars, so as to leave it doubtful whether they are species or varieties. For example, our *Salicornia herbacea?* has its joints entire, not emarginate, and its calyx-scales acute. Our *Callitriche aquatica* has very obtuse or cuneiform leaves. Our *Æthusa Cynapium?* of which I inclose seeds, is destitute of a nauseous taste, &c. On these plants it is difficult to decide, without a comparison of actual specimens. The present unhappy war presents a serious obstacle to the conveyance of seeds, &c.; but opportunities must sooner or later occur for making communications or exchanges of this sort. I send you with this the seeds of an Iris, which is new to me, but perhaps is already in your collection. Its character is, *I. imberbis, foliis linearibus, alternis; caule tereti, multifloro* (*7—8-floro*); *germine trigono, lateribus bisulcis.* It has the habit of *I. sibirica,* but differs in its germ, which is smaller; grows in wet lands with *I. virginica,* but is less frequent. I also send seeds of a tetradynamous plant of our salt marshes, which from its habit I think a *Bunias.* I should describe it, *B. silicula articulis binis lævibus monospermis; foliis obovatis, sinuato-dentatis.* The silicle is drupaceous. It answers nearly to *Cakile ægyptiaca* of Willdenow, without the "*dente obtuso*". I wish very much to know whether the *Datura Tatula* of Linnæus be a distinct species. We have here a

Datura with a purple, spotted stem, answering in every way to the description of *D. Tatula,* but which is evidently a variety of *D. Stramonium.* I have sought in vain for any permanent difference.

I have caused to be put in the press in this town an edition of your excellent Introductory work, for the attendants on lectures which I have undertaken in this place. Our botanists are not yet sufficiently numerous to induce the booksellers to publish large works; but as the country grows I hope the taste for science will increase. I have the honour to be,&c.

Your obedient Servant,

JACOB BIGELOW.

From Sir Joseph Banks.

My dear Doctor, Soho Square, Dec. 9, 1814.

We are all here at a loss on the subject of *Hermodactyls.* We have little doubt that the gout medicine is composed of that root, and that the doctor of Yoxford in Suffolk, who sells a medicine similar in taste and smell to the French, uses our *Colchicum,* but have some doubts whether our *Colchicum autumnale* is the same plant as that used in the Greek Pharmacopœia.

Cæsalpinus on the article *Hermodactyle* says, that the flower is white; in other respects the plant resembles our *Colchicum:* but he speaks of two other plants of the same genus, that are poisonous. Have you any thing among Sibthorp's papers that throws any light on the *Hermodactyl* of the Greeks? If you have, I shall be thankful for the communication. If you have not, I intend to write to Constantinople, where I have no doubt the drug will still be found in the apothecaries' shops.

Adieu, my dear Sir James!

Always faithfully yours,

JOSEPH BANKS.

From Dr. Michel Tenore.

Naples, Jan. 6, 1815.

My very respectable Sir and Fellow!

The favourable circumstance of the amiable relations happily established between our countries, and the great desire of enjoying the honour and the advantage that every botanist may obtain by the correspondence with a personage as worthful as the celebrious author of the British Flora, encouraged me to address you this very respectful letter, with a copy of the Royal Botanical Garden's Catalogue.

This establishment, although was born but in these years, is at present notably enriched of every kind of plants, and specially of those collected by my voyages all through the different provinces of this kingdom.

It is several years since I am applied to illustrate the plants of the Neapolitan Flora, as you may see by the prospectus that I have the honour to envoy you; and I pray you to let them circulate amongst the learned men of your country.

If your goodness will not disdain my correspondence, I shall make myself a duty to put every care for showing you the devotion and the high esteem which I am penetrated with for your respectable person. I am the honour to be

Your most obedient Servant,

MICHEL TENORE.

From H. Repton, Esq.

Harestreet, near Romford, July 9, 1815.

My dear Sir Edward,

How mortified have I been ever since I found your card on my study chimney! After nine months' confinement, I at length ventured to spend a week in town, and carried my pains and penalties with me, or had them carried about from place to place in a carriage,—to which the entrance and exit were always felt to my heart. I can now suppose that you intended to pass the day with me; and had I known it, I could have left London a day sooner to enjoy such a meeting.

After having rubbed on for sixty years in this same world, where we all have our rubs, so many old friends are rubbed off, that to me it is a peculiar satisfaction to see a long-known friend; and if you had no other quality than what Dr. Johnson would call sexagenary contemporaneousness, it would be a cause of jubilee; but we have both passed through life without shutting our eyes, like half those whom we remember to have seen, while they saw nothing but vanity and frivolity. You, like your great precursor, have examined from the cedar to the hyssop, and could surprise him by producing your *hortus siccus* to show how little he knew; and yet after all, you still have a list of *desiderata*,—something to add to that mass of wonders which daily inquiry has been collecting almost beyond the power of numbers and of names. If you should ever be at a loss for the latter, and could affix it to some plant of the Ivy tribe, or of any climbing genus, which, like myself, wants to be supported, I should rejoice to have my name recorded by your power of conferring immortality. My great predecessor *Adam* would never have been able to find names for an hundredth part of your vocabulary; but he lived in a garden with one friend, and one enemy, who, like Buonaparte in our days, was the enemy of peace. What wonders have we lived to witness! kingdoms raised and kicked down like a child's house of cards. The mighty empire of the Franks, one week given to the Bourbons, the next to the Buonapartes! then taken away, and now hanging in jeopardy, to be decided by the Russian Cossacs; while a Wellington and a Blucher are supposed to have, according to a favourite French phrase, "covered themselves with glory." Who but must see something beyond the power of man, which has been operating by means unaccountable, to produce these wondrous changes? and while we boast our conquests of kingdoms, and our power and glory of victory, without the most distant taint of Methodism, I am insensibly led to consider whose are the kingdom and the power and the glory. Such thoughts will naturally arise when the body has been so long preparing for that change, which, instead of being an object of terror, is to me a source of joy and hope,—not only as the means of escaping from much torture and some anxiety, but as holding out new scenes of beauty, and a more intimate knowledge of the works of the Creator, which, however numerous, are now become familiar, and excite less of wonder although more of reverence for their Divine contriver. Is it possible for you, who know so much more of the created evidences of the Deity, to feel satiety here, and a wish to enlarge your scene of observation? I have no doubt this will happen to all whose active minds lead them to wish for such enlargement; and then you and I shall meet, and compare our ideas on that and many other subjects:—but I will not believe we shall never meet again somewhere,—till when, God bless you!

H. Repton.

From the Rev. R. Walpole.

My dear Sir, Aylsham, Feb. 12, 1816.

You may remember our speaking once together on the subject of φλομὸς (the modern name both of *Euphorbia Characias* and *Verbascum sinuatum*,) being used for stupifying fish; and I mentioned to you that Sibthorp says, *Mercurialis annua* is used now for catching the *Scarus*. On looking at a volume of the Memoirs of the French Institute, containing a paper on the fishing of the ancients, I find the following passage:—"Ainsi ils attiroient le

Scare avec le *linocotis*, qui paroit être nôtre *mercuriale*." I suppose this is from Oppian, but no reference is made : this, however, I will look to. "Les anciens pêcheurs avoient, ainsi que nous, des recettes pour endormir le poisson, et aussi pour l'empoisonner. Aristote nous apprend qu'ils employoient à cet usage l'ellebore. Suivant Oppien, on faisait des gobbes pour enivrer le poisson avec du cyclamen et de l'argile. Le cyclamen est une plante à raçine tubéreuse que le vulgaire appelle *pain de porçeau*, parceque cet animal en est très friand. Théophraste lui reconnoit entre autres propriétés, celle d'endormir."—*Hist. Plant.* ix. c. 10.

It is remarkable that the French writer should say nothing of that passage in Aristotle which describes fish being caught τῷ πλομῷ; for so it is written.

Will you be kind enough to give me your opinion about the first statement respecting *linocotis ;* and also if you have anything to remark concerning the use of ellebore, as intoxicating? The three plants that Sibthorp mentions with this quality are *Conium maculatum*, *Euphorbia Characias*, and *Verbascum sinuatum.* Will you also be good enough to add the modern name of cyclamen ?

I wrote some months ago to a gentleman who was travelling in Greece on this subject; and he answers me, that one mode of catching fish is "by means of φλομὸς, great Tithymal : they dam up part of the water, and then throw into it some of the herb : this intoxicates the fish, so that they float on the water's surface, and are easily taken by the hand. There are two species of φλομος, one larger than the other. For want of this, *Aconitum* is sometimes used for the same purpose." You see part of this agrees with Sibthorp's statement, and confirms us in the use of φλομος. I will here observe, that Sprengel says *Euphorb. Char.* is χαρακιας of Dioscorides, and τιθυμαλος of Hippocrates.

What can my correspondent mean by *Aconitum?* I am sorry to have intruded thus on your time; but you must blame only yourself, because my troubling you arises from your readiness always to answer what questions I send you.

With our best regards to Lady Smith,

Believe me to remain, dear Sir,

Yours very truly,

R. Walpole.

From Baron Humboldt.

A Paris, 18 Mars 1816.

Je profite de l'occasion du fils de Sir Samuel Romilly, qui part cette nuit pour Londres, pour offrir à Monsieur le Chevalier Smith, en mon nom et celui de mes amis et collaborateurs M. Bonpland et Kunth, le premier demi-volume de nos *Genera et Species Plantarum.* Je désire qu'il veuille bien agréer cet hommage avec cette indulgence qu'il a daigné accorder jadis à mes travaux. Je vous supplie très particulièrement de jeter les yeux sur mes *Prolegomena.* C'est un travail auquel j'ai mis beaucoup de soin. L'ouvrage que j'aurai l'honneur de vous adresser successivement renfermera 4 volumes in-4to. Les planches sont presque terminées. Vous jugerez par l'étendue de cet ouvrage et par notre position sur le continent, combien j'ai dû m'armer de courage pour me jeter dans cette nouvelle entreprise.

J'attends impatiemment le temps où je pourrai vous voir au milieu de vos richesses à Norwich: je tourne mes regards vers la haute chaine des montagnes de l'Inde. J'ignore si je les verrai; je trouve tout naturel que le gouvernement accorde plus de protection à des sujets Britanniques qu'à des étrangers; mais j'ai eu trop à me louer des personnes qui composent le gouvernement, et j'ai trop de confiance dans la libéralité d'idées de votre nation, pour penser que l'on puisse blamer chez vous le désir d'un voyageur d'examiner des roches et des herbes, dont Dieu, je pense, n'a donné le monopole à personne. Vous direz que ceci ressemble à une plainte. Vous jugerez si elle est juste lorsque vous lisez le dernier N° du *Quarterly Review*, l'extrait d'Elphinstone. Pourquoi vouloir décourager les hommes qui parcourent le monde pour casser des pierres. Si le Chevalier Smith n'est pas tout-à-fait mécontent du travail botanique de M. Kunth, je le supplie de l'agréger un jour à la respectable Société Linnéenne. Je demande cet honneur pour mes amis, je saurois l'apprécier pour moi-même.

Mon ami, M. DeCandolle, m'a dit que vous daigniez vous rappeller de moi. Je vous prie d'agréer l'expression de ma vive reconnoissance.

LE BARON DE HUMBOLDT.

From Professor DeCandolle.

Monsieur, Paris, 18 Mars 1816.

J'ai quitté Norwich, ou plutôt votre maison, l'esprit tout occupé des journées intéressantes que j'y avois passées, et le cœur pénétré de reconnoissance de votre aimable accueil : j'ai reparé souvent dans ma mémoire les conversations que j'ai eu le plaisir d'avoir avec vous, et me suis sans cesse félicité d'avoir eu le bonheur de pouvoir maintenant vous connoitre personnellement : je n'ai point oublié l'histoire du *Corchorus*, &c. &c.

Mon premier soin en arrivant à Londres que d'examiner le *Smithia:* je dirais presque si ce n'étoit trop bizarre, que j'ai eu du regret à trouver le genre trop bon dans l'état actuel de nos connoissances pour pouvoir être changé ; le caractère du calice le distingue bien de l'*Æschynomene*, et Mr. Brown m'a dit avoir deux nouvelles espèces qui ont le même caractère, et qui renforceront le genre de Dryander. Obligé de renoncer au plaisir de vous dedier le genre que j'avois envie de vous offrir, j'ai au moins une occasion de me dedommager et de vous temoigner publiquement mes sentimens ; et j'ai redigé sous la forme d'une lettre qui vous est adressée mes observations sur les deux genres à rapporter selon moi à la famille des *Rosacées*. Si vous trouvez cette note digne de l'attention de la Société Linnéenne et la langue où elle est ecrite n'y met pas obstacle, je vous prie d'en donner communication à cette illustre Société. Je vous l'adresse avec d'autant plus de motif, que c'est vous qui avez le premier observé l'un des points fondamentaux de cette petite discussion. Elle etoit prête le jour de mon depart de Londres; mais je n'eus pas le temps de la mettre au mot pour vous l'adresser. Je suis arrivé ici il y a peu de jours, et l'un de mes premiers soins est de vous l'envoyer pour avoir en même temps occasion de me rappeller à votre souvenir. J'ai remis votre lettre à M. Desfontaines, qui a eu grand plaisir à recevoir de vos nouvelles; lorsque je lui ai raconté mes journées de Norwich, je croyois qu'il alloit partir pour vous faire aussi une visite: quant à moi, Monsieur, j'ai beaucoup plus envie d'en faire une seconde que j'en avois de la première, et c'est beaucoup dire. Veuillez, je vous prie, présenter à Lady Smith mes hommages empressés. Je vous prie d'agréer, Monsieur, l'expression de ma considération distinguée et des sentimens avec lesquels je suis

Votre dévoué Serviteur,

A. P. De Candolle.

From the Duke of Bedford.

Dear Sir James, Endsleigh, Sept. 9, 1816.

Accept my very sincere thanks for your kind letter, inclosing one to my gardener Sinclair. I am sure he will feel your goodness to him as it deserves; for your opinion and advice cannot but be highly valuable to him in the pursuit in which he is engaged with so much zeal and earnestness, and which I trust will do him credit, and be of some utility to those who take up agriculture as a science. And surely when every thing tends so much to the depression of agricultural pursuits, those who think them of some importance to the interests of the country cannot too closely investigate every part of this subject, which may in any degree operate to render the earth more productive, and give a stimulus to the industry and skill of the cultivator of the soil.

I was much gratified by your visit to Woburn, and flatter myself that you may be induced to repeat it; and if at any time you should be brought into this western corner of the island, I shall be delighted to show you the humble but picturesque beauties of the scenery surrounding this little cottage, which, though on a more confined scale than the mountains of Wales or Highlands of Scotland, are not without considerable interest.

Believe me to be, with sincere regard,

Your very faithful Servant,

BEDFORD.

CHAPTER XI.

Correspondence of the Abbé Joseph Corrêa de Serra.

AMONG those friends of Sir James Smith, who were created such at first sight by the magic sympathy of natural science, the Abbé Joseph Corrêa de Serra must be placed foremost. He was a botanist of the first rank; and the reader may recollect that his acquaintance with Sir James began at M. de Jussieu's at Paris, and that they afterwards met at Rome.

The Abbé Corrêa was a man of great genius and penetration, of good family and connexions; but, although a priest, appears to have fallen under the suspicion of the Inquisition. The particulars are involved in obscurity, yet some light may be thrown upon them by the letters which immediately follow. The first is from M. Broussonet, to whose acquaintance Sir James was first introduced at Edinburgh.

From M. Broussonet.

Mon cher Smith, Saragosse, 29 Juillet 1794.

Persécuté en France et pret à y périr, je me sauve à travers les plus grands dangers en Espagne, et je desire pouvoir me rendre en Angleterre pour passer ensuite l'Amérique Septentrionale, où j'ai envoyé quelques marchandises, dont le produit me fera vivre jusques à ce que je gagne quelque chose en fesant la médecine, ou en donnant les leçons d'histoire naturelle. Je me rends dans ce moment à Cadiz, où je voudrois trouver une passage sur un vaisseau de guerre Anglais ou sur une frigate.

Peut-être même pourrais-je passer à Gibraltar ou à Lisbonne; mais je crains que les fonds ne viennent à me manquer, n'ayant pas pu en emporter beaucoup de France. Je me suis adressé au Chev. Banks, je le prie de me procurer de passage sur quelque vaisseau de la marine royale: je tremble que ma lettre ne le trouve dans le Lincolnshire. Veuille bien, mon bon ami, l'engager à prendre intérêt à ma situation; elle est réellement digne de pitié.

Envoyez-moi des lettres pour Gibraltar, pour Lisbonne. Aidez-moi à me tirer de la position bien malheureuse où je suis. J'ai été nourri dans le malheur; je puis me contenter de peu; la table même des matelots me conviendra. Si vous pouvez me prêter quelque chose, vous me rendrez un véritable service; il me faut peu, et encore même ce ne sera que pour le passage. Les marchandises que j'ai à Baltimore, les effets que j'ai assurés en France, enfin mon frère et mes sœurs vous repondront de la bonté de ma dette. Je ne connois que vous et le Chev. Banks à qui je puisse m'adresser dans cette occasion. Adieu, mon cher Smith; plaignez-moi, mais n'abandonnez pas

Votre Ami,

AUG. BROUSSONET.

N. Ecrivez-moi aussi sous deux envellopes, la première *Aug. Broussonet*, la seconde *Don Antonio Cavanilles*, Hotel del' Infantado en Madrid.

Ecrivez-moi ; ce sera une grande consolation. Adieu !

From J. T. Koster, Esq.

Sir, Lisbon, Oct. 23, 1794.

By desire of Dr. Withering, to whose good advice on the score of health during the two winters he resided among us here I am much indebted, I send you a specimen of the Creeping Fig with its fruit. Mr. Masson, with whom I had the pleasure of being acquainted here some years ago, I think called it *Ficus pumila;* but my worthy friend the Abbé Corrêa calls it *Ficus repens:* it grows abundantly and luxuriantly in the green-house at the Royal Botanic Garden near this city, covering the trellis on the inside the glass to the very top.

I had the pleasure of seeing the Abbé yesterday afternoon, who called upon me, and brought with him the unfortunate M. Broussonet: this gentleman came in a merchant-ship from Cadiz, under convoy of the America and two frigates on their way to England; but having heard of many French cruizers being out, who have taken another of our packets, he is afraid of proceeding. I have done everything in my power, at the Abbé's earnest request, to get a passage for him in the King's ship; but Captain Rodney's orders are so strict against carrying French passengers, that he could not consent to it. They both begged that I would make this circumstance known to you, and desired to be remembered. M. Broussonet had been at the botanic garden, and seen this plant, which he also specifies by the name of *repens*.

The specimen is packed in a small paper box stuffed with cotton wool, and directed for you. Sir Henry Vane, Bart., who is a passenger in the America, has kindly taken charge of it, and will deliver it to you, if in his power.

I wrote to Dr. Withering some time ago, and told him that the plant would produce its fruit about this time; and in his answer he expresses a wish that the specimen might be sent dried to you; and it is fortunate that so good an opportunity should immediately offer.

After having mentioned to you these four names, I am confident it would be superfluous to make any apology for troubling you with a letter.

I am, with very great respect, Sir,

Your most obedient and most humble Servant,

JOHN THEODORE KOSTER.

From the same.

Sir, Lisbon, Jan. 25, 1795.

I received the honour of your letter of the 16th of December by the packet that arrived the 15th instant, and it was exceedingly unfortunate that she did not come in a few days sooner. M. Broussonet might then have gone in the Boston frigate, and his stay here has been the cause of great mortification and real detriment to our friend the Abbé Corrêa.

I explained to Sir Joseph Banks that his presence here gave umbrage to the two emigrant Dukes of Luxemburg and Coigny, and that they had applied to the court to have him sent away. Since that time still more noise has been made about it, and it has been contrived to supersede the Abbé in a lucrative and honourable place, which had been promised him,—that of public librarian; and he still apprehends other disagreeable consequences: not that I imagine poor M. Broussonet is the real cause, any further than that the Abbé's enemies have eagerly seized upon the opportunity, and alleged his attachment to Mons. B. as a proof of their former slanderous accusations. It is true that our friend the Abbé has been rather too unguarded in expressing his political opinions in such a country as this. Under a weak, irresolute, though arbitrary prince, surrounded by a corrupt, ignorant, bigoted, intriguing court, what else can be expected? It happens too, very unfortunately, that the Duke de Lafoens* is confined in a violent fit of the gout, or he might in some measure have stemmed the torrent.

In consequence of Sir J. Banks's recommendation, I am sure Mr. Walpole would readily undertake it, and could certainly protect Mons. B., and obtain leave for him to stay here as long as convenient; but you may judge of M. Broussonet's feelings on

* John de Braganza, uncle to the Queen of Portugal.

knowing that the Abbé's civilities to him cost him so dear; he has therefore resolved to set out directly for Gibraltar. He carries a passport to our Mr. Pinto, the secretary of state; another from Mr. Walpole, with his letter to General Rainsford, and the admiralty order. I hope he will get safe to that fortress; indeed I shall be anxious to hear of his safe arrival in England; for when once a man's affairs get into a wrong train, it is often so difficult to set them right again, that one cannot help being suspicious and apprehensive about future events. I can assure you that poor M. Broussonet can hardly resolve to set out, and goes with a heavy heart, foreboding a cool reception from General Rainsford, and all the ills that can possibly happen; and I am persuaded that, when you acquaint Sir J. Banks with these circumstances, he will have the goodness to write again to General Rainsford.

The first time you see Sir Joseph, I beg the favour of you to let him know that I have advanced a small sum of money to M. Broussonet, that I have procured a letter of credit for him to Gibraltar, and also to Faro upon his route, of all which I will give Sir Joseph an account in due time.

As soon as I saw your Tour advertised, I sent for it, which the Abbé laid hold of and had the first reading; perhaps this was the work you wished to send him. M. Broussonet has just finished a letter to Sir Joseph Banks at my house, and will set out as soon as ever a man and horse can be procured to conduct him to Faro; from thence our deputy consul will take charge of him, and send him forwards in a proper manner, either through Spain or by sea, as most convenient.

I have the honour to be, most truly,

Your most obedient, &c.,

JOHN THEODORE KOSTER.

Whether the Abbé Corrêa escaped from a real danger or not, there is no doubt that he fled from his country soon after these letters were written to Sir James, and took refuge in his friend's house in Great Marlborough Street, who received him with open arms. This was early in 1795.

The following extraordinary letter explains all that is known of this affair; it is certain his mind was deeply impressed with terror, which has no appearance of being imaginary.

From the Abbé Corrêa.

Dear Sir, Penzance, Cornwall, April 13, 1795.

Perhaps you will be surprised to find me so much nearer you than you believed; but I am sure you will be still more surprised when you will know the motive of it. I have been obliged to leave Portugal by a persecution whose injustice and iniquity is only to be credited when seen, and whose particulars I will explain to you in London; and our friend Broussonet will tell you what part of it he saw himself.

The inquisitor-general bore the greatest part in

that hellish plot, whose final aim was not directed against me, but it was deemed necessary to strike me down before.

As there was no time to lose, and I was not informed of the peculiarity of your alien act, I did not provide myself with the passports which I now find necessary to enter England, and which I own are very just precautions in such a time as this of war and suspicions. Now at Penzance I am detained by contrary winds, and by an embargo laid on all your ships coming from abroad. I cannot without leave from your government take the way of London by land; and even in case the ship will be permitted to proceed to London, according to the master's hopes, I cannot disembark there without such a permission. What I did for our common friend Broussonet at Lisbon in the days of my prosperity, I beg from you now that I am in the same circumstances he was then; and I beg less, because I will not remain in England but just a few days. I will then embark for anywhere else but Portugal and Spain, to be out of reach of my infernal persecutors, who (for some time at least) will, I am sure, be unrelenting, and will not spare any just or unjust means, false or true pretext, to hurt me as much as it will be in their power. Happily I could bring with me money enough to reach whatever part I will choose, and live there for some months; and I could leave in Lisbon, in sure hands, about three hundred pounds to begin my settlement where I will choose to remain. That makes me in some sort tranquil for the future; but that does by no means diminish the embarrassment I now lie under,—the want of this permission of your government, which I entreat you by all the reasons of humanity to endeavour to obtain to me by the interposition of Sir Joseph Banks, who, in his character of general patron of sciences and letters, will, I am sure, not be scandalized, if I apply to his protection for such a narrow permission of staying in London.

I have so full a confidence in your friendship and humane generosity, as well as in his liberal heart, that I entertain no doubt of obtaining it. This favour can make me happy; and what is to me now a time the greatest benefit to receive, is not I confide very difficult to you to bestow. I do not write to the Chevalier Almeyda, our ambassador, because he is the nephew of a great man who is in the number of my persecutors, and from what I have seen, none of the less violent; and this same reason hinders me from thinking of England for the place of my asylum. The ship I am in is the sloop Mary, Captain Anthony Reskruge. The name I have taken is Joseph Porto, which is the family name of a relation of mine, not by any other reason but to get safe out of the Tagus; but now I shall pass by it till I reach London, and the permission I so earnestly pray you to obtain me is to be obtained with that name.

Perhaps a letter for me from Lisbon has reached Sir Joseph Banks's hands, as also now I send a duplicate of this, the more to be assured of your receiving it. I entreat you to beg pardon to him in my name for such importunities; but the friend who probably has written to me does not know your address; but he knows that of Sir Joseph Banks, and also that by your means it would have come to my hands. Did you ever believe that I would have found myself in such circumstances, when you wrote in the third volume of your Travels, "that you wished me long life to reap the fruit of my benevolent aims"? But if you can obtain for me this permission, which is rather of going from England than of staying in it, I would reckon myself very happy; and you will be a great benefactor to your most obedient and affectionate Friend and Servant,

JOSEPH CORREA DE SERRA.

P.S. Do not forget my address here is Mr. Joseph Porto.

From Sir Joseph Banks.

My dear Doctor, April 16, 1795.

A letter will by this evening's post be dispatched to the mayor of Falmouth, directing him to grant a passport to Mr. Joseph Porto to come to London. I inclose you a frank and the Abbé Corrêa's letter, in order that you may apprise him of the circumstance.

Very faithfully yours,

JOSEPH BANKS.

I find it has been reported in Lisbon that he was a little deranged in his faculties; but his letter to you seems to put that matter out of question.

It appears from the following correspondence, that the Abbé continued in England, and remained in tranquillity some years in the neighbourhood of London.

From the Abbé Corrêa.

Sir, London, June 15, 1795.

Till now I have not thought proper to obtrude myself in your way. I know what is to be in the bosom of one's own family, much more on such occasions as this which now brought you to Norwich, and have given you full time to enjoy it undisturbed. But I have my rights also, and will not be negligent about them. I have a right to know how you do, and to be happy in the knowledge that you are so: I have a right to present my best respects to your truly respectable mother; and have a right to tell you what I do here, and what news I receive from my country about my affairs. Did you ever expect such a bill of rights from a man who owes you so many obligations? Don't be afraid; I will not be severe for the moment, because I believe you are nearly about the time of returning to London, where I will know what respects to you, and tell you what concerns me. At Norwich I wish you only to assure your mother of the sentiments of respect and deep esteem I have for her, and how heartfully I wish her every blessing. As for you, I believe you are sure enough how sincerely and heartily

I am, Sir, yours &c.,

JOSEPH CORREA DE SERRA.

P.S. The news from Portugal are far from being bad. *C'est le meilleur des mondes possible,* in Pangloss's language.

From the same.

Dear Sir and Friend, London, July 4, 1795.

Since you wish to have from me a long letter and botanical news, I will readily comply with it, because of all England you are the person with whom I like the better to converse, and botany is now the purest spring of pleasure to me. Your Κυαμος is a pretty name, and I find the ancient botanists very explicit in its signification. They distinguish well enough the *Cyamus ægyptius,* which is the *Nelumbo,* from the *Cyamus hellenicus,* which is the common bean.

It seems that κυαμος was to them a general name of a class of resembling seeds, just as the word *pois* in French, which is given to the *Pisum Ochrus,* to the seed of the *Cnestis,* to the *Cytisus Cajan,* to the seeds of *Guilandina,* &c.; and the English word *bean,* which is given not only to the *Faba,* but to some *Phaseoli,* &c., only distinguishing them by epithets, as the Cyams were.

I wish to ascertain if the beans of the Mareotis and of the Tritonia Palus,—to eat whose (which) was a sin to the Ægyptians, they being under the influence of Typhon, a cruel deity,—were the seeds of the *Nelumbo.*

But there is another reason why the word κυαμος is a happy one for the Tamarà emblem of prolific nature. Its original root κυω is applied in some senses to the act of fecundation and gravidation; and if some of our Greek scholars put any difficulty to the derivation of the word *κυαμος* from *κυω*,—which he can very reasonably do, since there is no forcible analogy in that language to bring the one from the other,—we can stop his mouth, presenting him the verb *κυαμιζω*, a verb by the common consent of grammarians derived from *κυω*. If I had here at hand philological books, as I have of natural history, something should be done to illustrate the archæology of this new genus; but still I will go to the library of the British Musæum, and try to make a classical nosegay for the christening of so classical a genus. So *Cyamus* let it be, and *Cyamus* for ever!

Since your departure three new botanical works have appeared, viz., Thunberg's *Monographia* of the *Hermanniæ*, and Major Velley's and Mr. Stackhouse's works on the *Fuci*. These last I believe to be very good books; but the physiological part is so obscure and vague, that I have dared to do a little memoir myself on the fecundation of submersed plants, which I defer reading till after your return: because I confide in your friendship, you will not treat me in this and other such occasions as Ray treated Sir Hans Sloane. The docility and deference which you always will find in me for your advices, will, I hope, entitle me to your kind patience and inspection in my botanical undertakings.

Days ago I was at Lee's with our fellow Lambert; and, as many species of *Verbascum* unknown to me were in blossom, I tried in all of them the phænomenon of their irritability, with the greatest success in all of them. I remember having spoken to you about it; and as probably the English *Verbasca* are now in flower about Norwich, I will be very glad if you observe it minutely. Strike with a stick, and with some force, the stem of the plant, under the branches, two or three times, and wait patiently for some minutes; you will see the *limbus* of the *corollæ* move towards the centre, and afterwards the basis of the *tubus* slowly detach itself from the receptacle, and fall,—and immediately the *calyx* close itself with impatience, and embrace the seeds; and that will happen to all the open flowers one after another. In Portugal I had observed it in the *Verbascum Thapsi, sinuatum, Blattaria,* now at Lee's in many others, and I wish I could observe all its species and the *Celsiæ*: I know nothing analogous in other plants. Next year this will be followed, and make a memoir for our Society. Mr. Bauer is about to publish the *Digitalis.* Mr. Menzies has written to Sir Joseph from Valparaiso in Chili, and in a few months will be in England.—These are all the botanical news I can send you.

From what you say, I am convinced it is my fate to be obliged to all your family. I will endeavour to justify the good opinion which your mother has given of me. Pray remember me to her; and also where I am not yet acquainted I vow myself, as in the Roman dedications, *notis et ignotis Diis.*

I wish Mr. Voght* was in London, to get acquainted with him; the author of such a benevolent

* Casper Voght, Esq. See p. 71.

pamphlet must be, I am sure, a man of an equally amiable character and sound understanding.

Your compliments have been presented to the botanic circle at Soho Square, and I am charged to send you theirs. To see you again in London, happy and in good health, is the constant wish of your hearty friend,

JOSEPH CORREA DE SERRA.

The writer recollects with much pleasure the Abbé Corrêa's visits to herself and Sir James, when they resided for a short time at Hammersmith in the spring of 1796. She was prepared to admire this engaging foreigner, and was not disappointed. Nothing could exceed the hilarity of his spirits, and the lustre of his dark intelligent eye, while in a sort of poetic rapture he talked of the charms of nature, and of flowers in particular, and especially of the satisfaction it would afford his mind to have a plant named after him. This he declared was the lasting honour he desired. He had witnessed the events of the French Revolution, had seen rank and fortune overthrown, escutcheons effaced, and names obliterated;—while in this mode of perpetuating an honourable fame, nothing human could destroy it; the remembrance of him would be renewed every season with the revolving year, as long as the world endured!

His friend had the pleasure of dedicating to him a beautiful genus, native of New Holland, of which three species only are known: that called *Corræa virens* is figured in Exotic Botany. How much soever the Abbé might pride himself in this possession, he found he was in danger of innovation where he thought himself most secure. "Do you know," says he, "that La Billardière has transformed the *Corræa* you granted me into *Mazeutoxeron?* What a barbarous name! It seems to me as if a Vandal or a Longobard had been put in possession of my territory. God help me; is it not misfortune enough to be stripped of my property, but I must be robbed by such a monster as Mazeutoxeron!"

From the Abbé Corrêa.

London, Jan. 30, 1797.

Do not judge of my readiness in answering your letters by the long time I have waited to answer yours of the 3rd instant. I hope I will seldom be interdicted the use of my right-hand to write, as I have been in the present occasion, by an abscess in the second finger of it, which is at present perfectly healed, after a tedious and they say dangerous suppuration in this cold season. By this time I am sure your work is advanced: I wish I could say as much of mine; but I begin to feel the necessity of drawing by myself, otherwise it will be impossible to advance as fast as I wish. Willemet's *Herbarium Mauritianum* I have seen in Usteri's Botanic Annals, and is, as you say, a poor performance. The Conway of which the preface speaks is the French general commanding in India, who perhaps did persecute Willemet rather from difference of party in politics, than from any dislike to natural knowledge.

They are now in France planning an expedition to the interior of Africa, according to the report of the secretary of the class of natural sciences in the public meeting of the Institute in the 4th of January, where also an historical eulogium of Professor Sibthorp was read, composed by M. de Guys.

The only work of natural history worth mentioning comes from foreign countries. Since you are at Norwich is the *Musæum Ichthyolithologicum Veronense,* which is a splendid publication, and the written part of it deserving commendation. As for English publications, I am sure you have seen Mr. Salisbury's *Prodromus,* and found me turned into a star,—a nebulous one indeed; others will rather say meteor.

Afzelius has brought very extraordinary plants from Africa, and truly paradoxical. His *Flora Guineensis* will be a great step advanced towards the improvement of botany. Few remain to be examined, and I believe in a week more they will be over.

I have lately received a very friendly letter from Don Rodrigo de Sousa, and he sends compliments to you. I believe you will do well if you write to him.

My respects to Mrs. Smith and your venerable mother. Since Miss Fanny deigns to recollect and claim my old acquaintance, you will be so good as to give her my most grateful thanks.

As for you, believe me ever entirely yours,

JOSEPH CORREA DE SERRA.

From the same.

My dear Friend, London, March 23, 1797.

What have you thought of my silence ?– Be sure it was not want of friendship or of gratitude. When you are acquainted with what it was, you will laugh perhaps.

I have in my hand your diploma from the Lisbon Academy, which I do not send to Norwich, because you are to visit London in the next month. The Academy sends to you a present of all its works, and they are at Lisbon (ready to be embarked) in Koster's hands, who has written to me to inquire if you would think them worth paying the duties on them, which he says are very high on books in England. Sir Joseph Banks, to whom the Academy also made a present of its works, tells me that he is very glad of having them all. I wish to know what you intend to do.

In your last letter to Lambert, I saw with the greatest pleasure your division of genera of the *Protea* family. It happened that I had just then examined the Cape *Proteas* in Masson's collection, which has occupied us in these last weeks, and will occupy for some more; the variety of seeds and seed-vessels, and the different ways of aggregation in the flowers had struck me. When your letter to Lambert came to my hands, I was enchanted at the reading of it, and must give the due praises to your discriminating powers, which I always have told you will constitute your proper character of a botanic writer. All your genera are, I suppose, taken mostly from New Holland plants; some of them are not familiar to me, but I can see the justness of the characters you establish. I wish you would also examine the Cape *Proteas*, because the two Floras are much more alike than it is supposed they are,—not only parallel, but are both fragments of a whole. *Styphelias* are found at the Cape. Masson has brought two species, and God knows if in the different genera of the *Protea* family many are not common to both Floras, and to be reciprocally illustrated by each other. Will you allow me the liberty of a friendly observation? The general tendency of the family is to be *tetrapetalous*, notwithstanding the insertion of the *stamina* in the *corolla;* so I wish you would try those which appear *monopetalous*, before a definitive determination, because many of them may be only slightly connated petals in a tubular appearance, as I will be able to show you in the *Lambertia*, which is perfectly *tetrapetala*, and the petals of which are only *alæ*, which accompany the filaments to the insertion under the *germen*, extremely alike to the *ala* which accompanies the *pedunculus* of the flowers in the *Tilia*, and which is commonly reputed a *bractea* *.

I beg pardon, my dear friend; but if the caution is not a good one, it comes at least from a friendly heart.

Be so good as to present my compliments to Mrs. Smith, and to your respectable mother.

I am most heartily yours,

JOSEPH CORREA DE SERRA.

* I don't say these petals are *bracteæ*, but only the form of them is like the *bractea* of the *Tilia* flowers.

J. E. Smith à son Altesse Royale Monseigneur le Duc de Lafoens, à Lisbonne.

Monseigneur, Août 12, 1797.

Lorsque j'ai été à Londres pendant quelques mois du printems de cette année, mon très cher et estimable ami M. l'Abbé Corrêa m'a mis entre les mains un riche paquet de livres, que votre Altesse Royale a daigné m'envoyer de la part de l'Académie des Sciences de Lisbonne, dont les recommendations du même ami m'ont rendu, depuis quelque tems, membre. Je suis très sensible, Monseigneur, à l'honneur que votre Altesse Royale et l'Académie m'ont fait, et je vous supplie d'agréer mes remercimens les plus respectueux. Je serai charmé d'avoir quelque chose touchant l'histoire naturelle digne d'être offerte à l'Académie, et de lui témoigner ma reconnoissance.

Mais je saisis avec encore plus d'empressement l'occasion qui se présente d'offrir mon hommage le plus sincère à son illustre President, dont le caractère et les lumières relèvent même son rang élevé, et doivent le rendre cher à sa patrie et au genre humain.

Je suis charmé que M. Corrêa se trouve si bien en Angleterre. Il s'est devoué à la botanique, surtout ce qui regarde l'économie des plantes, et il me semble qu'il ne se ressent nullement de ses malheurs passés lorsqu'il se trouve dans un jardin de botanique.

Je suis, avec les sentiments les plus respectueux, de votre Altesse Royale le très humble et très obéissant Serviteur,

JAMES EDWARD SMITH.

From Don Rodrigo de Sousa Coutinho.

Monsieur Smith, Lisbonne, ce 22 Nov. 1797.

Je viens de recevoir par mon ami M. l'Abbé Corrêa, votre très estimable lettre en date du 16 Août, et ce n'est que trop vrai que je me suis empressé de demander de vos nouvelles à peine arrivé ici.

Le souvenir que vous daignez garder d'un ancien ami, que vous avez même voulu honorer avec le titre de votre confrère dans la Société Linnéenne, a été pour moi très flatteur, malgré que mes occupations m'eussent depuis long tems privé de l'honneur de vous écrire. J'attens avec impatience les deux ouvrages de votre Tour sur le Continent, et des Lectures Botaniques, que vous avez la bonté de m'adresser, et que je n'ai point reçu. Je donne l'ordre à mon libraire à Londres pour votre magnifique ouvrage sur les Insectes de Georgia, que je n'avois point vu annoncé, ou que peutêtre j'aurais oublié, car le regne animal a toujours eu moins d'interêt pour moi, que les deux autres. J'attens aussi le livre de Mr. Staunton, votre ami, dont j'ai lu quelques passages sur Madeira, Praya, et Rio de Janeiro, qui changeront maintenant en mieux par les soins éclairés et par les ordres de mon auguste maître, dont les vertus sont supérieures à toute expression. Votre amitié est la cause de l'illusion que vous vous faites pour mes foibles talens; mais si je n'ai point des droits aux qualités que vous me supposez, au moins j'en ai à votre estime, car je préférerai à bien d'autres choses le doux plaisir d'aller vous voir à

Norwich avec M. l'Abbé Corrêa, et d'y parler sur les sciences, dont les progrès devroient intéresser bien plus le genre humain, que d'autres objets dont il ne rétire aucun profit essentiel.

Les progrès de la Société Linnéenne me font le plus grand plaisir, et je viens aussi de donner des ordres pour qu'on m'adresse tout ce qu'elle a publié.

Je n'ai point reçu des nouvelles du Dr. Bellardi, qui pourroit bien venir ici s'il n'avoit point une si nombreuse famille, mais qui poursuit toujours ses travaux botaniques avec passion.

Daignez agréer les sentimens d'estime que je vous voue, et permettez que je fasse des vœux, pourque vous veniez un hiver herboriser en Portugal, et que je puisse de vive voix vous renouveller les assurances des sentimens d'estime et considération avec lesquels j'ai l'honneur d'être

Votre très humble et obéissant Serviteur,

D. Rodrigo de Sousa Coutinho.

From the Abbé Corrêa.

My dear Friend, London, Sept. 3, 1800.

Since you are determined to do our Academy the honour of writing a paper for her, I thank you for it, and will take proper care of forwarding it by a safe conveyance.

You are so good as to take notice of my horror of the *Mazeutoxeron;* but I can tell you, that not only is your publication much anterior to that of La Billardière, but that Jussieu is decided to support the old name. Jussieu has written me a long friendly letter, and entreats me to make not only his compliments, but his excuses also to you for not writing to you as he wished.

Yours most sincerely,

JOSEPH CORREA DE SERRA.

Thus was his desire established; and still may each returning season perpetuate the memoryof the friend who conferred and the friend who accepted so beautiful a memorial! In the world he had tribulation; and whether real or ideal, the Abbé's sufferings excite our tenderest compassion, as he again appears under the influence of a power armed with most direful means of inflicting terror and despair.

From the same.

My dear Friend, London, Sept. 16, 1801.

When your letter of the 8th instant reached London I was in Birmingham, and this is my good excuse for not sending a prompt answer. The description I had made of the *Doryanthes* I had long been at a loss to find again; but to satisfy you, I have written it again in fewer words, and you will find it here inclosed, together with the two plates already engraved for it.

About my Excellency and my diplomacy I cannot tell you; but there was no unluckier fatality than, after being six years out of Portugal in peace, to fall again in the clutches of the same family who had driven me from my country, after having brought me to madness and lunacy. All the ancient hatred, sanctified in their eyes by the motives you know, without any scruple about the means, for the same reason are and have been exerted by this monster of ambassador to effect my ruin, either by sending me back to Portugal, where the Duke of Lafoens is now in disgrace, or in England if he can. I have been for many and many weeks almost driven to madness, living the greater part of the time with him, and have afterwards left him and given my resignation, which he did not choose to accept, but I have sent it directly to Portugal. He is the son of the Marquis of Ponte de Lima, cousin to the Grand Inquisitor. I am, my dear friend, in the situation of Laocoön when encircled and torn by the serpent,—and this is a poisonous one.

Believe me, I am most sincerely yours,

JOSEPH CORREA DE SERRA.

The Abbé's next letter, written twelve years after, is dated from America, whither he went in a high official capacity, as Portuguese minister plenipotentiary at Washington.

Dear Sir and Friend, Boston, Nov. 18, 1813.

From what a distance do I address you at present! but it is to send you not only the assurances of my unalterable friendship and respect for you, but also a compensation, and a thorough one, for my long silence, as time I am perfectly confident will show to you.

Botany is still in her infancy in America: some

amateurs are scattered through this continent; and some European rather travelling gardeners than real botanists are collecting plants and shrubs and trees for the gardens of European rich men,—or for the nurseries. In the two years that I have lived and travelled in the United States, I have found only a man*, and a young one too, a physician of this town, who shows a true botanic genius and great zeal for science; he has already acquired a great scientific knowledge, only aided by his industry and the limited number of botanic books which the country affords. He will be I am confident an illustrious botanist, if he is put in correspondence with the chiefs of the science, and helped by their knowledge in the doubts that he may encounter.

I am persuaded I do a great service to botany as well as to him by introducing him to you, our venerable patriarch. By fostering his efforts, and by resolving his doubts, you will I am sure in a few years bring forth more thorough knowledge of North American plants than we have hitherto had. He will write to you and send this little memorandum. His name is Dr. Bigelow, and his fixed residence Boston.

If I live some years, I am confident you will thank me, in the name of the botanic church, for having been instrumental to open this correspondence.

From Philadelphia I intend to write to you and send you some few things. Farewell till then!

Most sincerely yours,

JOSEPH CORREA DE SERRA.

* Dr. Muhlenburg excepted.

The Abbé's latest letters have the date of 1821, when he was again in England, and then preparing to take a voyage to the Brazils, whither he was appointed by the King of Portugal to superintend the sciences and arts, and to a very honourable and lucrative situation.

Dear Sir and Friend, London, July 23, 1821.

I promised you a letter before my leaving England, and I keep my word. The day after to-morrow I go to Falmouth to embark in the packet destined to sail the 29th or 30th of this month, and hope to be in Lisbon by the 10th of August, if not before. It will be a new sight for me, after twenty-six years and five months absence. All is changed: God grant it may be for the better! My health is very precarious still, but on the whole is ameliorated, and my impatience raises my spirits. I beg you to present my compliments to Lady Smith, and for you my best wishes for your happiness, peace, and long life. You see by the erratas of this letter, the state of weariness which my mind is in this moment, by the little teasing numberless details concomitant to a *déménagement*, that may be called the horrors of removal. At all times and in every country be sure of finding in me an obliged and grateful friend. and an obedient faithful servant.

JOSEPH CORREA DE SERRA.

Soon after this period the venerable Abbé returned in very ill health to his native country, and died at Lisbon in 1823.

In reference to a foregoing letter, dated July 4, 1795, the two following ones are added for a further illustration of the same subject. They are from the hand of the learned and elegant scholar Dr. Frank Sayers, of Norwich.

Dear Sir, Sept. 12, 1808.

The discussion into which you were so obliging as to enter yesterday evening, induced me to look a little more for the earliest meaning of the word κυαμος. I do not find that it is used more than once by Homer in the Iliad or Odyssey; the passage which I noticed to you yesterday in the Iliad (v. 589) is thus translated, I find, by Damm (in his celebrated *Lexic. Homeric.*): "A ventilabro in area saliunt fabæ fuscæ et pisa—nam color harum fabarum est fuscus et rufus." The same writer translates κυαμος, "faba maxime ea species quam Germani *walsche bonen*, vulgo *saubonen* vocant, et quas Græci υοσκυαμους appellant." In the Batrachomyomachia (of which the æra however, as you well know, is somewhat uncertain,) the word κυαμος again occurs:—in the arming of the Mice it is said, l. 123,

Κνημιδας μεν πρωτα περι κνημησιν εθηκαν
'Ρηξαντες κυαμους χλωρους, ευτ' ασκησοντες.

The κυαμος, thus ingeniously used for greaves by the mice, must of course have been of the shape of some of our ordinary beans. The *common* meaning of κυαμος, then, before the time of Pythagoras is thus sufficiently plain. It appears, too, from the use of the word κυαμος in the writings of Plato, Aristotle, and others of about the same æra, that this ancient sense of κυαμος was retained in the time of Theophrastus (as in fact it was *always*), and possibly he may have been the person who *first* extended the name of κυαμος to the *Nymphæa Nelumbo*, and may thus have thrown the light which you so skilfully discovered upon the precept of Pythagoras.

In Suidas and Julius Pollux I find no hint of the *Nelumbo* under the head of κυαμος. Suidas has κυαμος, ειδος οσπριου. Pollux also enumerates the κυαμος among the οσπρια or *legumina*, and says κυαμοι οἱ και πυαμοι. Hesychius makes it an οσπριον, but in a note to him (from Eustathius) is added καλουμενον κολοκασιον.

In Athenæus, however, I find mention made not only of ordinary κυαμοι eaten at feasts, but also of the Κυαμος Αιγυπτιος, of which he gives a description extracted from Theophrastus (viz. that of the *Nymphæa Nelumbo*, not of the *Lotus*); he also quotes Nicander, who recommends the sowing of this Κυαμος Αιγυπτιος, saying that garlands may be made of its *flowers*, and that its *fruit* and *roots* may be eaten. This Nicander (who was certainly no Pythagorean) lived about 137 B.C. In his time then, and probably *before*, (but how long before I know not,) the *Nymphæa Nelumbo* was known as a plant *fit for food*, and the name κυαμος was *commonly affixed to it*, perhaps taken from Theophrastus.

The only apology which I have to make for sending you these crudities, is the desire which I felt to make you some slight return for the amusement and instruction which I yesterday received from you.

I am, dear Sir,

Yours most truly,

F. Sayers.

I must leave our learned friend to settle the etymology of κυαμος, as I am not fluent either in the Coptic or Cushite. Herodotus, when he mentions that the Egyptians did not eat the κυαμος, and that the priests thought it unclean (ου καθαρος), uses the word οσπριον in the same paragraph as including the κυαμος.

From the same.

Dear Sir, Close, Oct. 3, 1808.

Since I had the pleasure of sending you a few passages which I had hastily collected respecting the κυαμος of the ancients, a supposition has occurred to me, by which I think some of the difficulties arising on that subject might possibly be removed. I have therefore thought proper to trouble you with it. As we have found no Greek writer, *prior to Theophrastus*, who has used the word κυαμος with any any other meaning than that of the ordinary *legumen* so called,—does it not appear possible that the Hindu κυαμος (or *Nymphæa Nelumbo*) may have been *first imported from the East, at the time of the conquests of Alexander?* Might not even the king himself have ordered so celebrated a plant (and probably other curious natural productions) to have been sent to his preceptor Aristotle, from whom a knowledge of it would have been readily obtained by Theophrastus? But without supposing any interference of the conqueror himself, specimens of such a striking vegetable as the *Nymphæa Nelumbo* could hardly fail of finding their way to Greece from the East. Its introduction into *Egypt* may, I think, be similarly accounted for, and may be reasonably fixed at about the same period. The *κυαμος* mentioned by Herodotus as being held in abomination by the Egyptians, is certainly *not* the *Nymphæa Nelumbo*, I conceive; he expressly calls it an *οσπριον*, and the circumstance of its being held in *abomination*, of its being deemed *ου καθαρος*, sufficiently points out that it could never have been the *holy*, *adorable Nymphæa* of India. Herodotus, then, knew nothing of any other kind of *κυαμος* in Egypt than of the ordinary bean. But the *Nymphæa Nelumbo* might very probably have been introduced into Egypt about the time of the first Ptolemy: to *Nicander*, who lived at Alexandria under the seventh Ptolemy, it would of course be known; but it might still be so little cultivated as to induce him to insist upon its excellence in his Georgics. What effect this exportation might have, I know not; but the cultivation of the *Nymphæa Nelumbo* appears to have continued in Egypt in the time of Pliny, who mentions two genera of Egyptian *fabæ*, one of which he calls "*rotundius et nigrius*;" this I conceive to be the *Nelumbo*: the other kind I presume was the ordinary *κυαμος* of Herodotus. If the above hypothesis be true, it is certainly somewhat unfavourable to your supposition of Pythagoras having borrowed his precept from Egypt,—supposing it, I mean, to apply to the *Nymphæa Nelumbo:* in case of its *so* applying, Pythagoras must I think have taken it direct from India. But it appears to me more probable that he *did* borrow it from Egypt, and applied it (as the Egyptians themselves seem to have done,) to the ordinary κυαμος (of the time of Herodotus). Upon this supposition the precept itself would have been very intelligible to the followers of Pythagoras (although the reasons for it were not understood); but if it contained any allusion to the *Nymphæa Nelumbo,* such allusion must have been totally obscure and *unavailing* to the inhabitants of a country where that plant appears to have been unknown.

I am, dear Sir,

Yours very truly,

F. Sayers.

The beautiful plant which has led the botanist and scholar into this elaborate disquisition, was figured in Exotic Botany under the name of *Cyamus Nelumbo,* the Sacred Bean of India, tab. 31, 32. It is denominated κυαμος by Theophrastus; *Nymphæa Nelumbo,* by Linnæus and Aiton; *Nymphæa indica major,* by Rumphius; *Nelumbo nucifera,* by Gærtner; *Nelumbium speciosum,* by Willdenow; and *Tamará,* by Rheede. Those who do not possess the Exotic Botany may not be displeased to hear Sir James Smith's reasoning concerning the name and character of this celebrated plant.

"Germen superior, green, smooth, inversely conical, its upper broad flat surface perforated with several holes, opening into as many cells. Each cell contains the rudiment of a seed, protruding through the orifice, and crowned with an oblong, obtuse, perforated, yellow, permanent stigma. The whole germen becomes a coriaceous entire capsule, which in process of time separates from the stalk, laden with ripe oval nuts, and floats down the water. The nuts vegetating, it becomes a cornucopia of young sprouting plants, which at length break loose from their confinement, and take root in the mud. This peculiar mode of propagation has evidently occasioned the plant, in conjunction with water, to be adopted as the symbol of fertility, in which point of view it has, from the remotest antiquity, been considered with religious veneration in India, and makes a conspicuous figure in the mythology of that ancient country. It is most generally known to the learned of Europe under the name of *Lotus;* the natives of Hindustan call it *Tamará;* the people of Ceylon *Nelumbo*. It has been confounded by very able writers, even lately, with the *Lotus* of Egypt, *Nymphæa Lotus* of Linnæus; see *Andr. Repos.* t. 391, and *Curt. Mag.* t. 797. We presume the latter to have become important in the Egyptian mythology, only as a substitute for the former. The *Lotus* of Egypt is a real *Nymphæa,* bearing its seeds much in the manner of a Poppy, and scattering them in the mud. There is nothing peculiar in its appearance or mode of growth which could have caused it to be chosen for an emblem of fertility, were it not from the general resemblance of its leaves and flowers to our plant, the original *Lotus* of India. Hence I have for some time presumed to deduce an argument in support of the doctrine, now prevalent on other grounds, that the religion of the Egyptians was adopted from the East.

" Innumerable illustrations respecting the *Tamará*, *Lotus*, or *Nelumbo*, as connected with the poetry or religion of the Hindus, may be found in the learned works of Sir W. Jones, Mr. Knight, and others. In the fourth volume of the *Amœnitates Academicæ*, p. 234, a carved horn of a rhinoceros, sent to Linnæus from China, is described. This is now before me, and is an exquisite specimen of Oriental sculpture, evidently alluding to the mythology of India. The whole inverted base of the horn is carved into an elegant leaf of *Nelumbo*, rising from the water amid a group of perforated Chinese rocks. It is encompassed with various plants of a more diminutive proportion; a peach tree and a medlar (or rather perhaps the mangostan), with *Sagittaria*, *Pothos*, and the *Nelumbo* itself in flower and seed, cover the outer surface. Some fantastic lizards, with bunches of grapes and the Lit-che fruit in their mouths, are crawling over the whole.

" We have to add some remarks on the botanical characters and name of this plant.

" Adanson, Gærtner, Jussieu and Willdenow are most unquestionably justified in separating it from *Nymphæa*, with which Linnæus and other writers have confounded it. The very peculiar fruit, unlike anything else in the vegetable kingdom, and the stigmas so totally different from those of *Nym-*

phæa, sufficiently distinguish it. The chief question in dispute is the name. *Nelumbium* is formed from *nelumbo*, a Ceylon word of very confined use. If it must have a barbarous appellation, *Tamará* would be preferable, as being in general use among the learned and the vulgar throughout Hindustan. Happily we have no occasion to adopt either, for the plant has already a classical Greek name of primary authority and antiquity, being the real κυαμος of Theophrastus; and therefore the word *Cyamus* is what by every right and title belongs to it. *Nelumbo* may be retained as a specific name, rather out of deference to Linnæus and Gærtner than for any good reason; for *Tamará*, being more universal, would be more proper; and *speciosus*, given by Willdenow, more expressive. We wish however to respect the right of priority, and to avoid all needless changes.

"We claim no merit in the restoration of this ancient generic name. Bodæus A Stapel in his Commentary on Theophrastus, p. 446, and Hermann in his *Paradisus Batavus* have amply illustrated the subject; and others, as Plukenet, have alluded to it. But it is remarkable that no recent writer on the mythological history of the *Nelumbo* should have been aware of its being the celebrated κυαμος or Pythagorean bean, which is so evident from the description in Theophrastus. The 'cellular head like a round wasp's nest, with a bean in each cell projecting a little beyond its orifice; the rose-coloured flower twice as large as a poppy,' as well as all the rest of his account, are strikingly characteristic. By this discovery many things, hitherto difficult of explanation, are elucidated. We can no longer wonder at the prohibition of these beans to the Egyptian priests or the disciples of Pythagoras. A plant consecrated to religious veneration as an emblem of reproduction and fertility, would be very improper for the food, or even the consideration, of persons dedicated to peculiar purity. The Egyptian priests were not allowed even to look upon it. Authors scarcely explain sufficiently whether Pythagoras avoided it from respect or abhorrence. However that might be, we need not, in order to ascertain his motives, have recourse to any of the five reasons supposed by Aristotle, nor to the conjectures of Cicero. Neither can there be any doubt that the prohibition given by Pythagoras was literal; and not merely allegorical, as forbidding his followers to eat this kind of pulse because the magistrates in some places were chosen by a ballot with black and white beans, thereby giving them to understand that they should not meddle with public affairs. Such far-fetched explanations show the ingenuity of commentators rather than their knowledge. As the Pythagorean prohibitions are now obsolete, perhaps these beans, imported from India, might not be unwelcome at our tables. The root of the *Cyamus* is also used as food, but we have many vegetables preferable to it."

In some unfinished notes upon the subject of monumental inscriptions, which Sir James Smith left, he speaks of the elegant memorial of Pope Urban in the Augustine church at Rome.

URBANO VII. PONT. MAX.
brevis imperii principi,
sed memoriæ diuturnæ.
Illud Fortunæ fuerit;
hoc erit Virtutis.

Which he has thus rendered, observing, "It is not easy to retain its conciseness in a translation:"

TO POPE URBAN VII.
a prince whose reign was short,
but whose memory will be lasting.
The former Fortune might controul;
but the latter his Virtues will command.

"The intelligent friend*," he continues, "who first pointed out to me this inscription, is himself the author of one not inferior to it in simplicity.

Humanitati Sacrum.

Cineribus HENRICI FIELDING, Angli,
quæ heic absque honore jacebant,
Johannes de Braganza
monumentum hoc ponendi curavit,
ne Musis inhospita
hæc tellus videretur.

Sacred to Humanity.

The ashes of HENRY FIELDING, an Englishman,
having been here for some time buried in obscurity,
John de Braganza erected this monument,
that the land might not seem altogether
inhospitable to the Muses.

This intended tribute to our admirable historian of human nature proved abortive. The virtues, the

* The Abbé Corrêa.

taste, the liberality, and even the illustrious rank of the Duke of Lafoens (John de Braganza), uncle to the Queen of Portugal, had no weight against monkish fanaticism, which would by no means connive at such a compliment to a heretic; and the monument was never executed."

Sir James was himself the author of several inscriptions prompted by affection for departed worth; —one only shall appear in this place. Its excellence consists in its being perfectly just and appropriate.

In memory of
Mr. James Dickson,
one of the original Fellows of
the Linnæan Society,
and a Vice-President of
the Horticultural Society of London;
a man of a powerful mind,
and of spotless integrity;
whose singular acuteness and accuracy,
in the most difficult departments of botany,
have rendered his name celebrated
wherever that science is known.
He departed this life at Croydon,
August 14, 1822,
in the 85th year of his age.

CHAPTER XII.

Miscellaneous Letters to and from Sir J. E. Smith, from 1817 *to* 1827.—*Sir Thomas Gage.—Hon. DeWitt Clinton.—Rev. R. Walpole.—Sir J. E. Smith.—Dr. Wallich.—Sir T. Gage.—Sir T. G. Cullum.—Sir T. Frankland.—Panzer.—Rev. J. Yates.—D. Turner, Esq.—Mr. Sinclair.—Professor Hooker.—Bergsma.—Mr. Lambert.—Mr. Talbot.*

From Sir Thomas Gage.*

Florence, Casa Dini Borgo Santa Croce,
June 11, 1817.

My dear Sir James,

I HAVE for a very long time had in my possession the books from Mr. Targioni, which I have at last

* Of Hengrave Hall in the county of Suffolk, Bart. Sir Thomas Gage died at Rome on the 27th of December, 1820, in the 40th year of his age, and was buried there in the *chièsa del Gesù.* "Enthusiasm and delicacy distinguished his character, and were blended in a manner as happy as unusual. Had these been supported by strong health, there was no perfection in art or science to which he would not have been capable of attaining. His tastes and pursuits were all elegant. Whatever he said or did was eminently marked by gentlemanly feelings. It was both from nature and from cultivation, and scarcely less from cultivation than from nature, that he possessed a tact, which, while it was essential to the pursuit of botany, his favourite science, rendered him tremblingly alive to the beauties of art and the more sublime charms of creation. In the more abstruse parts of the vegetable world he had laboured hard, by the lamp as well as the sun; studying the works of his predecessors in his closet, and exploring the objects themselves in the fields."—*History and Antiquities of Hengrave.*

found the opportunity of sending by a friend;—chances of this sort are rare, and I would not trust them to the merchants, having myself lost a packet by this means. I ought also long since to have written to you; but the account I could have given of myself, till within these few months, was such that it would neither have given pleasure to my friends nor myself. I am happy now to be able to say, that I have escaped from a very serious state of ill health; and am far better than when I last had the pleasure of seeing you.

I have not neglected my botanical pursuits, and have made great additions, particularly to my *Cryptogamia.* I have collected everything of that kind I could meet with here, and have purchased a very fine collection, particularly of *Lichens,* from Schleicher of Bex, whose list I send you. His price is so moderate and his specimens so good, that, should you or any botanical friend desire to have anything from him, I would readily undertake to procure it. He sends me constantly specimens as well to purchase as to determine; and I am much advanced in my drawing to illustrate Acharius. The present state of botany at Florence is respectable, but not much encouraged. The Grand Duke is fond of plants, and has lately purchased a copy of English Botany; and I believe some other works, of which I furnished a list. He has built a fine range of stoves and houses in the Boboli garden, where many good plants are cultivated, though it cannot be considered a regular botanical garden, the best plants having been removed a few years since to the new *Orto Fisico* under the care of Mr. Targioni, which is a liberal establishment and an extensive ground well laid out, a part being destined for agricultural experiments. There is a want of glass, but improvements are making, and the collection of plants increases. Mr. Targioni is a man of much acuteness, and has great love for science, and of a most amiable disposition, and highly liberal in communicating. I have not only had the advantage of selecting specimens from the garden, but also of examining all the *Lichens,* to the number of several hundred, in the rich herbarium of Micheli, and of taking what specimens I pleased; I have therefore verified all the synonyms of that author, and made out a list for Mr. Targioni of the Acharian names. Agricultural and botanical lectures in the Linnæan system are given at stated periods, both at the garden and the academy of the Georgofili. Mr. Targioni's son is professor of chemistry, and is a good entomologist.

At the head of the Royal Museum, which is a magnificent establishment, especially in minerals, is Mr. Radi, a young man of the first talents, indefatigable as a botanist, and in every branch of natural history, of which he has more practical knowledge than any person I have ever known. We are constant companions, and make excursions into the mountains occasionally. I have been able to assist him in the *Lichens*, so that now he has become well acquainted with all the *Cryptogamia* of the country. He had already discovered new Mosses, particularly the *Fabronia pusilla*, of which genus

Swartz has found another specimen. He is now engaged in publishing an account of the *Jungermanniæ* of this country on a new system, which will appear shortly, and of which I hope to send you a copy. I wish to engage him in a pocket *Flora Toscana*, in hopes to encourage the same spirit which has so contributed to botanical knowledge at home. Here there is little taste of that kind; for the nobles are not rich enough to purchase science, nor wise enough to esteem it. There are some exceptions: the Princess Corsini has a good collection of plants, as also the Marquisses Feroni and Pucci. Nothing is published here in the way of natural history. At Pisa Mr. Sair has begun a *Flora Etrusca*, which has great merit, but advances slowly. It is a rich country in every branch, and deserves to be well known.

Though I was very ill when I crossed Mount Cenis, I collected several rare plants, and I thought of you when I found *Cetraria cucullata* near the Hospice, as mentioned in your Tour.

My residence in this city has been pleasant enough, particularly since my health is improved, and it is not my intention to leave it before the month of October, when we talk of going to Rome.

The quantity of English who have been here during the winter has been excessive, though I by no means think it a good place for a delicate constitution at that season. The air is very damp, and I should think bad; most of the houses in the lower parts of the town having stagnant water under them, and the changes from heat to cold are rapid and severe. Violent inflammations are very frequent, and we have had a bad typhus fever, which has caused great mortality among the lower classes, chiefly owing to the badness of the crops of last year, and the want of food in the mountains. It is now subsiding, and the *mal aria* fortunately does not extend to Florence. Lady Mary Anne and our children have enjoyed constant good health; the latter are grown stronger since they left England.

I ought not to forget to thank you for the consequences of the letters you were so kind as to give me. At Geneva I got acquainted with several persons of science, as Mr. Pictet, MM. Jurine and Morricand; at Turin by chance with Professor Biroli of the garden; and at Milan I delivered your letter to Oriani, and was made known to Mr. Herman of the garden and Mr. Breislach, the mineralogist, who gave me a letter for Mr. Targioni. I was obliged to send yours to the Marquis Ippolito Durazzo, whose sister is here, married to the Sardinian minister Brignoli,—a very amiable family.

After proving myself so bad a correspondent, it would be very unreasonable for me to hope to trespass on your time, which is so much better employed; but should you find a leisure moment, I should feel very great pleasure in knowing you and Lady Smith are well, as also in hearing anything concerning Mr. Turner, Borrer, or any of our F.L.S. who may ever think of me.

I am, my dear Sir,

Yours very sincerely,

THOMAS GAGE.

From His Excellency De Witt Clinton, LL.D. Governor of the State of New York, and President of the Literary and Philosophical Society there.

Sir, Utica, July 18, 1817.

A few days ago a farmer stopt with his waggon at a house in the village of Rome, about fifteen miles from this place. A respectable gentleman, who was conversing with him on business, observed among some hay, lying in the waggon, a few stalks of a strange plant, inquired what they were; and on being told that they were wild wheat, and were cut with common grass in a beaver meadow and on a wet soil in the town of Western in this county, he took out a few grains, and gave them to an honest and industrious farmer in his vicinity, who planted them in his garden. The second crop produced about a peck of grain, which yielded upwards of twenty bushels the third year. Wheat of the same species has also been found in a wild state in a swamp covered with trees near Rome. It is said to differ from the common wheat in a variety of respects,—in the compactness of the stalk, in the largeness of its leaves, in the peculiar position of beards at the apex of the head, which is in all other respects bald, and in its superior height, being considerably taller. Since the comparative scarcity of snow, which formerly served as a protection against the attacks of frost, our wheat has suffered severely by what the farmers denominate winter-killing. Our ground freezes during the winter about a foot and half in depth. When the sun resumes its vernal power, a partial thaw of two inches takes place in the course of the day; and owing to the porosity and hollowness of the common wheat, the water is absorbed in it. On the return of night, the ground is again frozen, and then the expansive power of frost produces the destruction of the plant, by eradicating it, or breaking the roots to pieces, and bursting the stalk where the water has penetrated. Rye is not affected in this way, because it is not so permeable by water, and because its roots are stronger, more elastic, and strike deeper into the earth. In like manner the wild wheat of Oneida County is said to resist the power of frost; and this is imputed to the same causes which protect the rye.

As I am persuaded that the history which I have given of this wheat is substantially correct, it presents a very interesting subject for investigation. Is it indigenous, or was it originally imported wheat, and accidentally conveyed to the places where it was found? If the former, it is the very grain which nature created for our soil and climate. If the latter, it has been evidently improved by its wild state and spontaneous growth;—a circumstance of an anomalous character, and contrary to the usual course of nature.

Although I am not prepared to give a decided opinion on this subject, yet I may be permitted to observe, that there are cogent arguments against the latter hypothesis. The plant was found in a swamp and in a meadow, and appears to delight in a wet soil, which is not congenial with the common wheat. It presents not only a different aspect, but seems to have peculiar and characteristic qualities.

Linnæus, if I remember rightly, made six species of *Triticum*. Sixteen species are now enumerated, besides varieties; and these are found in the most diversified climates: the Murwaary Wheat of Barbary; the Spring Wheat of Siberia; the Spelter of Germany; the Wheat of Egypt, of Switzerland, of Poland, and of Sicily, cannot be derived from the same country. Ceres, who, according to the heathen mythology, discovered corn, was said to have had her principal seat in Sicily; but this granary of the ancient world has no exclusive claims to the most important of the *Cerealia*. The *Froment tremais*, which arrives at maturity in these marshes, is as different from the other kinds of *Triticum*, as it is possible for different species to be; and it unquestionably could not have had an identity of local origin with them.

I have been a long time of opinion, that many of our native plants have been improperly considered as naturalized; and as I am anxious to claim the most important culmiferous plant as an indigenous production, I have no hesitation in denominating this wheat, discovered near Rome, *Triticum americanum*.

I also transmit by this opportunity specimens of a plant called the Wild Rye, which grows spontaneously and in considerable quantities in the country bordering on the upper parts of the Mohawk River; and I believe it might be cultivated to advantage.

The opinion on this subject of such an eminent botanist as yourself will be very acceptable.

Yours most respectfully,

De Witt Clinton.

P.S. To hear from you will always be acceptable. Nothing would give me more gratification than to see the distinguished proprietor of Holkham, whose name, as well as that of Sir J. E. Smith, is highly appreciated in this country. But I am unfortunately denied this pleasure at present.

From the Rev. R. Walpole.

My dear Sir, Aylsham, Dec. 8, 1817.

Can you give me any information from your own or Sibthorp's sources respecting Greek melons? I find in a work I have been lately reading, that our melons were known to the Greeks and Romans: Σικυος, it appears, was the genus, the species being *Melo, Melopepon,* and our cucumber. The *Melopepon* is what the French call *Sucrin:* what is our word corresponding to *Sucrin*? There is a passage in Galen which clearly distinguishes the *Melopepon* and the *Pepon;* after citing it, my author says, "*ex hoc loco patet pepones Galeni esse nostros melones, &c.*" Aristotle, he adds, knew them well; for he asks (Prol. l. 20), διατὶ οἱ σίκυοι πέπονες ἄριστοι γίγνονται ἐν τοῖς ἑλώδεσι πεδίοις οὖσιν ἐνύγροις, οἷον περὶ

'Ορχόμενον καὶ ἐν Αἰγύπτῳ,—Why the best melons are produced in marshy and moist places, as about Orchomenus (in Bœotia) and in Egypt.

Will you be kind enough to favour me with a line at your leisure;—and believe me to be, dear Sir,

Yours very faithfully,

ROB. WALPOLE.

Sir J. E. Smith to Dr. Wallich.

Dear Sir, Liverpool, Sept. 16, 1818.

Your most polite and obliging letter of the 12th of January last (accompanied with those of Colonel Hardwicke and Mr. Loring,) did not reach me till a few days before I left home for this place. I find letters are readily sent to India from hence, and therefore I will not delay thanking you for your liberal offers of communication, in which I fear I shall be the chief gainer, unless you have any botanical doubts or queries which you think I may have a chance of solving. Nothing could be more desirable to me than to receive from you dried specimens for my herbarium, especially of any of your new or rare plants. I am particularly desirous of *Scitamineæ, Orchideæ, Liliaceæ,* which experienced botanists now learn to dry well, but which, being rather difficult, have formerly been given up in despair by collectors, and are rare in herbariums:—I find a weak solution of corrosive sublimate, with a little camphor dissolved in it, preserves dried plants from the attacks of all insects. My whole Linnæan herbarium is thus poisoned, by means of a camel-hair pencil; but the specimens must first be *dried thoroughly*. I long much for specimens of Dr. Roxburgh's *Aeginetia* and any *Orobanche*, or plant of the parasitic kind, which cannot be cultivated here. You see, my good Sir, how ready I am to grasp at your kind offers. I trust you will not find me less willing to fulfill my part of the treaty, and I beg you will send me anything that is doubtful or difficult. You may depend on my not *telling more than I know;* for if I find difficulties, I will freely confess my ignorance. The Linnæan Society will be very happy to receive any communication from you, and also to have you enrolled on our list of Fellows, whenever you may be pleased to permit it: Colonel Hardwicke will inform you of the laws and obligations. I take the liberty of inclosing a letter for this worthy friend of mine, and one for Archdeacon Loring.

I am extremely happy to find you are in correspondence with my valued friend Mr. Roscoe, and also with Mr. Shepherd of the botanic garden here, no less excellent in his way. The latter has a nephew who is very expert in raising ferns from seed, and you will render a great service to botany in sending him fragments of any of that tribe with fructification in a ripe state.

I received one of your rich packets of seeds, and after letting Mr. Shepherd take his choice, I have divided the rest among several friends, particularly the worthy Bishop of Winchester, who propagates the Cinnamon-tree by seeds ripened in his own hot-houses! The other packet, containing the *Rho-*

dodendrum arboreum, and the new white-flowered species, &c., has not yet reached me.

I remain, with much respect, &c.

J. E. Smith.

From Dr. Wallich.

Botanic Garden, Calcutta, July 24, 1819.

Dear Sir,

Although it is only a short time since I had the honour of answering your favour of the 16th of September, and transmitting a parcel of specimens, I feel confident that you will please to pardon my thus early intruding on your time, especially in so voluminous a shape as this sheet exhibits, which Mrs. Archdeacon Loring has kindly supplied from the stores which she has received from her father, for the express purpose of writing on their *single-sheet letters*, without any risk of surcharge from the post-office.

My long expected regular supplies of Napal specimens have at last arrived from Katmandu, after I had almost despaired of their reaching me in safety, if at all; and in so beautiful a condition, and such vast numbers, as to enable me to have the honour of presenting you, Sir J. Banks, Mr. Lambert, and Mr. Rudge with abundance of plants from that highly interesting country. This large collection, which has been made entirely by my faithful and deserving men, stationed under the protection of the resident at Katmandu, during the latter end of 1817 and the whole of 1818, arrived in boats from a place near the frontiers Segowly; and so anxious was I to avert the ruin which would inevitably result, were the specimens allowed to remain in Bengal during the rains, which have been pouring down latterly in excessive quantities, that I have been at work since the arrival of the boxes (on the 4th inst.), in order that the dispatches might be in time prepared for embarkation on the Blenheim. The circumstance of my having used the shortest possible time for arranging the specimens to avoid the dangers which such extensive collections would be exposed to from the extreme wetness of the present season, will, I flatter myself with the hope, at once account and apologize for the fewness of the labels which have been attached to the individual species; at the same time it will evince to you how gratefully sensible I am of the high honour you have conferred on me, by granting me the privilege of corresponding with the first and most illustrious botanist of the age. Accept, dear Sir, again and again, my warmest and most respectful thanks for so flattering a distinction,—a distinction which I have so long felt ambitious of enjoying; and do me the justice to believe, that I shall incessantly endeavour to show myself not wholly unworthy of your kindness. The tribute of respect and gratitude which I have now the satisfaction of presenting to you, I trust may not prove unacceptable, and I am not without hopes even of its containing some specimens *not* included in that splendid collection which my most esteemed friend and predecessor Dr. Hamilton (Buchanan) presented to you from the same quarter.

At the close of this year another invoice will be sent down to me; partly collected by my own people, and partly by those of my valuable correspondent and friend Mr. Robert Stuart, at present acting British resident at the court of the Rajah of Napal, of which, of course, a large proportion shall be forwarded to you in due time. Besides the specimens which I am daily receiving from thence, the most astonishing treasures are frequently sent to me from Mr. R. Colquhoun*, from the province of Kamaon; from Dr. G. Govan, the intelligent superintendent of the botanic garden at Sahrunpore; from Dr. Gerard at Nahn, and other elevated tracts; from Captain Webb at Srinaghur, with whom also is stationed a collecting party, which has within these few days transmitted to me some very fine packets, containing many of the beauties recorded in the sixth volume of the Asiatic Researches, by my inestimable benefactor and friend Colonel Hardwicke.

From all these stupendously high countries bordering on the snowy mountains, you may rest assured that supplies shall be sent to you; so that I may indulge a hope that a number of new plants may in time be added to your splendid herbarium, "haud sine veneratione summâ adeundum," to use the words of DeCandolle, in the preface to his matchless work *Regni Veget. Syst. Nat.*, which I had the happiness to receive yesterday from Mr. Lambert.

I ought not to omit mentioning two other sources; namely, the expedition of my excellent friend Dr. Moorcroft, who has set out on an arduous tour to the

* Now Sir R. Colquhoun, Bart.

north and west, and has taken two of my plant collectors with him, and Singapore in Sumatra, from whence I expect extensive collections from my highly valued friend Mr. W. Jack, the personal surgeon to Sir Stamford Raffles, who has already favoured me with ample harvests of rare and interesting plants gathered by him at Penang. A letter from Sir Stamford which I got yesterday has already prepared me for the arrival of some highly curious things, amongst them a new *Nepenthes*, of exquisite beauty and gigantic dimensions, parallel in its enormous size to the marvellous flower, measuring in circumference six feet! * found last year in a tour through the island of Sumatra, and of which an account has, I learn, been sent to Sir Joseph Banks. Lastly, I expect large additions from an expedition sent up to Gossain Than (*alias* Neel-Khaunt), by Mr. Robert Stuart, from Katmandu, for the express purpose of investigating the natural productions of that part of the Himalaya or snowy range of mountains, and of collecting specimens and seeds for me.

I beg you will pardon the length of this digression, which has entirely originated from my anxious wish of becoming a not altogether unprofitable correspondent and contributor.

A good number of the plants in your chest (which I beg leave to observe are all natives of and collected in Napal, at and about Katmandu,) I have named in my dispatches to England; so that you will meet with abundant, and I fear but too numerous occasions for correcting my errors and mistaken

* *Rafflesia Arnoldi.*

notions; a favour which your generous offer induces me to solicit, and which will lay me under infinite obligations. I entreat and beseech you to grant this my earnest request, thereby enabling me to judge with precision of the nature of the specimens I may have the satisfaction of presenting to you.

Among the contents of the chest you will find a number of *Scitamineæ*, of which you will be glad to hear. I have this very day the following species in the fullest blossom in the garden:—*Hedychium coccineum*, Buch., id. *var. carneum* mihi, both of them inconceivably grand; *H. flavum*, Roxb.; *H. ellipticum*, Buch., a most superb species; *H. spicatum*, Buch., the flowers of which are most delightfully perfumed, like the carnation and *Echites caryophyllata*;—all these are in profuse blossom with me, together with *H. coronarium* and *angustifolium*, R. The name of the last is not so well adapted, because there is a more distinct character in the leaves,—their *distichous* position, which would have at once afforded a name and a specific difference.

This species (*angustifolium*) and *coccineum* are very much alike in stature and inflorescence; the former differs chiefly in its more rigid and perfectly *distichous* leaves. *Roscoea*, so beautifully represented in your Exotic Botany, is also in flower, which, however, are of a very pale blueish colour, precisely like those of *Kaempferia rotunda*. I have a vast number of plants of every individual species above mentioned, of which, immediately after the rains, I shall do myself the pleasure to send you a share. There are likewise in your chest of specimens a great variety of *Orchideæ,* among which your splendid *O. gigantea, pectinata; Epidendrum præcox;* your *Neottia acaulis,* and a very large species which I call *procera:* of these two tribes you will receive from time to time large supplies, according to the wish you have expressed, as also of *Orobanche* and *Aeginetia.* Of the former of these I have the pleasure to present you on this occasion with a species, which I cannot distinguish from *indica* of Buchanan; of the latter, there are also specimens in the chest.

I presented Mr. Roscoe with a good many *Scitamineæ,* which were forwarded in the Cornwallis. That illustrious man, as also the Botanic Institution at Liverpool (which in 1816 conferred the flattering distinction, of electing me their first and only honorary member), have placed me under personal obligations for some excellent collections just received from the ship Bengal. What a clever and experienced man Mr. Shepherd must be!—out of one hundred and one sorts of plants packed up in a light chest, according to his judicious mode, not less than forty-seven were in the highest vigour; eight were in a state of dangerous weakness, but still holding out hopes of recovery! If it is considered that the chest was embarked on the 22nd of January; that more than half a year had elapsed since the roots and plants were packed up; and that the chest remained on board the ship after its arrival here, for some time at least, on account of its being placed so far down in the ship's hold as not to be come at without taking out part of the cargo;—if all this is considered, too great encomium cannot, in my humble opinion, be bestowed on the judgement and success of the curator of the Liverpool Botanic Institution. I have resolved to give this method ample and frequent trials, and I shall begin by returning the compliment to that excellent garden. I shall not forget sending, also, ferns.—Lastly, among your specimens you will find numerous *Melostomeæ*; *Labiatæ*; *Scrophularineæ*, among them my *Scr. urticifolia*; several *Ternstroemiaceæ*, *Polygoneæ*, *Liliaceæ*, *Melanthaceæ*, viz. *Anguillaria indica* of Brown; *Filices*; *Gramineæ*; *Cyperaceæ*; several species of *Hydrangea*, *Viburnum*; *Ranunculaceæ*, *Umbellatæ*, *Corymbiferæ*, many *Querci*. There is a *Rhus*, which I take to be the genuine *Sitz* of Kaempfer, and which I call *R. juglandifolium*. A seemingly *dioecious* species of *Cynomorium*, two species of *Myrica*, one of which is Colonel Hardwicke's *Kaephul* in Asiatic Researches, vol. vi.; the other is a doubtful species, which answers to Rumphii *Lignum Emanum*; its fruit and female flowers I shall do myself the pleasure to present to you on a future occasion. Several species of a genus which has a close affinity to *Incarvillæa*, and with that forming, according to Mr. Brown's suggestion, a distinct section of *Bignoniaceæ*. I call the genus in question *Didymocarpus*. It has three stamina, one of which is abortive. The capsule *linearis pseudo-4-locularis, 2-valvis, dissepimento contrario 2-lobo: lobis valvulis parallelis margine involuto seminiferis*, seminibus nudis! *Herbæ humiles, incano-pubescentes venoso-punctatæ, aromaticæ.*

One of the species, my *D. aromatica*, is in high

estimation at Napal, its young leaves being employed in a dried state as a perfume and for sacerdotal offerings, under the name of *Koom-Koom* (which is also that of Turmeric), and *Ranee-Govindha.*

A plant intermediate between that genus and *Incarvillæa* I retain under the latter, and call it *I. dubia.*

But it is time that I should conclude this long, digressing letter. Happy, most happy shall I be, if you are pleased to pardon its faults, and if the collection of specimens to which it alludes chiefly, prove in any manner acceptable. My venerated friend Colonel Hardwicke enjoys a better state of health than has been his portion for several years past; he wrote to you by the Isabella, and he requests me to offer you his kindest regards.

I have the honour to remain, with sentiments of the most profound respect and esteem,

Dear Sir,

Your highly obliged and devoted Servant,

N. WALLICH.

Sir J. E. Smith to Dr. Wallich.

Norwich, March 6, 1820.

[Observations on Dr. Wallich's plants, and his letter of May 27, 1819.]

Rubia Munjistha of Roxburgh is precisely *R. cordifolia*, Herb. Linn, and we therefore avoid a most uncouth name. Such names I think are seldom eligible.

R. alata, Wallich, a new species, and an excellent name.

Galium elegans, Wall.; seems new, though I have one very near it, if not the same, from Madeira, which appears not to be described.

G. Aparine I have carefully compared with English specimens, and find no difference.

G. asperifolium is new to me. The name surely cannot be found fault with.

I cannot undertake your unsettled parcel of *Spermacoces* at present. It would be better for you to determine their distinctions in a fresh state. At least I will now proceed to more certain and urgent matters; nor is it necessary to follow the order of your letter, or any other.

Your "*Stellaria? triandra*" is a very interesting plant. It is a *Holosteum*, *very* like the *cordatum* in general appearance, but the calyx affords a clear distinction. *H. cordatum*, native of the West Indies, (where it is used in bleaching linen and taking out iron-moulds, being called *Moron* by the French in Cayenne,) has *folia calycina uninervia, glabra, planiuscula*. Yours, which I would call *striatum*, has *fol. cal. trinervia, sulcata, scabra*, vel *pubescentia*. Seed would be welcome.

Aegiceras majus—good.—"*Ajuga Hyoscyami*," —why so called? I see no resemblance, and I cannot commend such names in the genitive case. See Introd. to Botany, ed. 3. 290. If I am wrong in my principles or opinions, I beg with all my heart to be corrected; but if I help to improve science, I wish to be supported by classical botanists like you.

I have not Brown's *A. sinuata*, but your plant seems to answer to his character. If new, *cana* would be a good name.

Berberis pinnata, Roxb. I have this fine plant from Dr. F. Hamilton, by the names of *Leontice fruticosa*, and *Berberis Miccia*. It may be *Ilex japonica* of Thunberg, as you suspect; but how could he overlook the pinnate leaves? What is most remarkable is, that I have the very same plant from California, gathered by Mr. Menzies, who likewise called it *B. pinnata*, an excellent name.

Saxifraga ligulata, Wall.—You have, I doubt not, a reason for your name, though it does not appear; but I have an idea of there being somewhere a *S. ligulata*. On searching, I find there is a *lin*gulata of Bellardi, which however is my *callosa*, published in Dickson's Dried Plants. Your plant is next akin to *S. crassifolia*, and might have been called *fimbriata*, from its essential distinction.

Geranium from Kalmouda, called *Tissoo soah*, is next akin to *G. bohemicum*, but the flowers are little more than half so large, and the *bohemicum* has the capsule corrugated transversely, which in yours is even (an excellent mark) and rather less hairy. Your leaves too have rounder and more deep lobes. The various pubescence of the stem is singular; in some parts very dense; then suddenly smooth. It would be charity to help you to a specific name in such a great genus. Perhaps *heterotrichum* might do, if the above character be constant.

Pavetta tomentosa, precisely mine.

Ixora lanceolaTA, Colebr.—new to me. I presume the name alludes to the form of the leaf; *lan-*

ceolaria, if it *were latin*, would mean that the tree or wood served for little spears.

Paris polyphylla.—Thank you for this fine supply.

Colebrookea oppositifolia—good.—*Ixora cuneifolia*, Roxb., and its variety,—new to me.

Triumfetta oblonga, Wall. Very good. I have it from (Buchanan) Hamilton, with a question whether it may not be *T. annua* of Linnæus; but this I have no means of determining, having no specimen. He also refers to *T. glandulosa* of Willdenow, as probably like it. This I have not.

"*Callicarpa arborea*, Roxb." This seems what I have from Dr. Roxburgh by the name of *C. villosa*. Are they the same?

C. macrophylla, Vahl.—seems right. I have it not.

"*Dalrympela pomifera*, Roxb.?" With this and its history I am quite unacquainted. Whom is it named after? Is the name Dalrym*pel* or *ple*?

Holmskioldia sanguinea—very fine and acceptable.

Loranthus bicolor, Roxb. I have one very near this, from Tranquebar, by the name of *L. lonice-roides*, which it may be.

Anguillaria , Brown, precisely *Melanthium indicum*, Herb. Linn. I think, as you say, it ought to be an *Ornithoglossum*.

Teucrium macrostachyum, Wall., is my *Leucosceptrum canum*, Exot. Bot. t. 116.

Commersonia echinata. Yours looks different from Forster's original Otaheite specimen, whose leaves are only 2 or 3 inches long, and 1 or 1½ broad. The pubescence of their under side is very fine and dense, scarcely visibly stellated.

Thank you for your fine specimens of *Rhododen-*

dron arboreum, and the white-flowered one, supposed a variety. I think a specific difference may be founded on the back of the leaf; but of this you must judge. Both are growing in England, from your seeds.

Daphne Gardneri, Wall., I have from Dr. Hamilton, by the name of *Dais Bamutis.* I know not why he made it a *Dais,* unless it be decandrous, which is of no consequence. I retain your name of course. May not this be (*ob capitulum*) *Daphne indica* of Loureiro?

Daphne cannabina of Loureiro is probably right, for I pay little regard to the opposite leaves he attributes to it. I take under this name your plant with glaucous leaves, yellowish beneath. This I think is Hamilton's *D. papyrifera* in his MSS., though the leaves of his flowering specimens are less glaucous. They are glaucous, like yours, in his fruit-bearing specimens. But I presume to think you have herewith mixed two other species. One is *D. Botlua* of Hamilton, (I would call it *D. saligna* if there be none already), whose leaves are much narrower, not glaucous; flowers larger, without the involucral leaves of the *cannabina.* The branches are long and *vimineous.* The other is a smaller, more densely branched plant, with smaller leaves, not glaucous, of a broader figure; flowers smaller, with fringed deciduous involucral leaves.

Ormosia dasycarpa, Jackson Tr. L. Soc. v. 10, *right*—a native of the West Indies. I hope it will afford you its beautiful seeds in due time. I thank you much for the flowers, as my specimen is in fruit.

Hedyotis gracilis—new to me. It seems a pretty plant, and I hope this name, which you have now chosen, will suit it well.

Clematis smilacifolia—a very curious new species, so I must not complain of the badness of the specimen.

Mespilus from the mountains near Katmandu—seems to me very distinct from *japonica* (which has now borne a *very* hard winter in my garden without any shelter). Yours has elliptical, nearly entire leaves, not decurrent at the base as in *M. japonica,* nor sharply serrated as in that *. *M. elliptica* would be a good name.—I find a few teeth here and there near the points of some leaves.

Buddlea paniculata, a fine species, new to me, though probably among Dr. Hamilton's plants, which I have not yet all arranged.

Cornus aggregata. I must rely on you for the genus, as the specimens are not in a state to determine it.—*C. oblongifolia* is new to me. Both these indeed may be among Hamilton's plants.

Cupressus from Almora. May not this be *C. pendula* of Thunberg and Willdenow, though perhaps not *Fi Moro* of Kæmpfer, which seems by his short account to be a *Juniperus* ?

Ligustrum napalense. I think this only a downy variety of *L. japonicum.* Thunberg's original specimen, though in fruit, retains some pubescence (exactly like yours) on the flower-stalks. I find no difference at all in other respects.

Andromeda ovalifolia. This seems what Dr. Ha-

* I see you remark these particulars.

milton first named *A. latifolia,* and then *A. capricida.* The latter, if true, would be preferable.

Acrostichum flagelliferum. A very fine fern, new to me; *frondes apice radicantes, seu gemmiparæ,* occur in several species.

Bauhinia corymbosa, Roxb. Beautiful! new to me.

Botrychium zeylanicum.—Good and neat specimens of this would be one of the greatest treasures you could send European botanists.

Bauhinia racemosa, with a great downy legume, I had long ago from Roxburgh by the name of *scandens.*

Juglans pterococca. Of this I know nothing, nor of *Ulmus virgata* of Roxburgh.

I know not whether this fine *Inga* in flower, "*Mimosa Djiring* of Roxb.?" be supposed different from other specimens with young fruit. They seem to me the same. Is *Bua Jiring* the Malay name?

Your pretty *Primula,* with short radical flower-stalks, was called by Dr. Hamilton *P. Cushia.* He gathered it near the source of the Bagmutty. I am not partial to such specific names, but they are not the worst of all.

Uvaria dioica, Roxb. New to me.—*Gualtheria fragrantissima,* Wall., I have from Dr. Hamilton as an *Arbutus.* It comes near *G. erecta* of Ventenat, *Jard. de Cels,* t. 5, but has smaller flowers, and more dense, downy (not hairy) clusters. Leaves also more glandular beneath. *Gaul*theria is certainly right.

Heynea trijuga, Roxb. Very acceptable, though I want the flowers.

Combretum costatum, Roxb. A very noble plant. —*Epidendr. moschatum,* very precious. I wish you could send me the leaves.

An Orchideous plant, without a ticket, is a *Satyrium*, called by Hamilton in his MSS. *Orchis bicornuta.*—A *Viola* also without a ticket, is called by our said friend *longifolia,* not very happily, but it is difficult to find a good name for it.

Hoya viridiflora, Br.—right, of course. I do not find that I have it.

Fraxinus floribunda, Wall., is, as you say, very near *Ornus,* but in yours the corolla and stamens are 3 or 4 times shorter than in *F. Ornus;* the leaflets more stalked, more pointed, and sharply serrated. They are certainly distinct.

Callicarpa, with long leaves, sent from Sirinagur by Sir R. Colquhoun;—how does it differ from Vahl's *macrophylla?*—*Vitis parvifolia,* Roxb., well named; new to me.—" *Connarus* from Penang," I know nothing about. It is a magnificent plant.—A few that remain, I am obliged to pass over, either for want of time, or because I have nothing to say.

Thus my dear Sir, I have endeavoured to show, as well as I could, my gratitude for your bountiful communications. A vast parcel of very fine things which came last year, before the present parcel, I have not yet been able to study; but I have devoted many days to the present, and to your very kind letter of May 27, 1819. Pardon me if I make remarks freely, especially about names. I wish those who, like you, are competent to take the lead, to consider and to study to keep the science as classical, elegant, and correct as possible. Do you support me in this, against illiterate botanists, &c. &c. You may see by my articles in Rees's Cyclopædia that I keep these things in view, and defend the Linnæan fortress as stoutly as I can; but I am thankful for, and concur in, all sound correction and improvement. Mr. Brown and I are sworn friends, though he uses the natural arrangement. He is one of the most amiable, acute, and worthy of men. Pray tell my dear cousin Mrs. Loring, with my kindest regards, that I have just buried my venerable mother, who died at 88, without any previous infirmity, the delight of all, old and young.

Great part of your parcel was half devoured by insects. I shall distribute the duplicates as you order. I wish however you would send me, in future, what you destine to me only, compact well pressed specimens. This sheet, or rather half of it, shows the size of my herbarium, being the paper I use for it. I have nothing at hand to send you at present, but the 3rd and last edition of my *Compendium Flora Britannicæ*, perhaps the best thing I have done, but I wish to correct its defects. Great numbers of the seeds you sent me have succeeded very well, and some of the roots; but most of the latter were eaten by rats in the ship. Seeds are always acceptable. Pray send *Ormosia* if perfected.

Believe me, dear Sir,

With every sentiment of regard and respect,

Your obliged and faithful Servant,

J. E. Smith.

From Dr. Wallich.

Dear Sir, Botanic Garden, Calcutta, May 10, 1820.

I have the satisfaction to inform you that a chest of specimens to your address, directed to the care of Mr. Kindersley, has been embarked on board the Henry Porcher. Its chief contents consist in Napal plants, among which I hope there may be found some not formerly sent, and worthy of a place in your museum. They have been collected by my people about Katmandu, and by some who were dispatched by my late friend Mr. R. Stuart to Gossain Than, at the foot of the Himaleya, or rather within its first range. The other specimens are from this garden, and I have to make an apology for several of these which belong to exotics in this country. They were preserved by my people among other plants of this garden. I have reason to fear they are not of any value to you.

There are likewise specimens of Sumatra and Penang plants which were left with me by my ever-valued friend Mr. Jack, surgeon to Sir Stamford Raffles, for the purpose of their being presented to you in his name. Some more shall follow in my next dispatch. Of the Scitamineous plants, I request you will do me the favour to communicate some to Mr. Roscoe, with my best compliments. I have lately had some most heart-cheering letters from that great and good man, to which I shall reply shortly. From them I learn that a package I had the satisfaction of forwarding, of Scitamineous plants, had proved acceptable to yourself and to him;—among them were what I rejoice to find that you consider as *several new species* of *Roscoea*. He also kindly communicated a highly interesting synoptical view of all the *Hedychia* I had sent. I anticipate the most glorious harvest in that tribe from my Napal journey, for which I set out in July next. I hope that I need not assure you of my cordial desire to transmit to you such collections of every description from thence as I may anticipate will be favourably received. Yourself, Sir Joseph Banks, Mr. Lambert, and Mr. Colebrooke, shall receive ample supplies, and I propose continuing to send them until I learn that you want no more of them!

It was my full intention to have presented you with a chest of specimens, with a request that you would kindly offer them to the Linnæan Society as a feeble tribute of my respect and esteem; but the necessary leisure was wanting. I may however still be able to effect this intention before my departure, because I am amply prepared with treasures for that purpose.

Mr. H. Graves, proceeding in the Henry Porcher, will present you with a copy of the first volume of Roxburgh's *Flora Indica*, which I request you will do me the favour to accept, as a feeble testimony of the high esteem and respect which I entertain towards you. He will likewise deliver into your hands another copy, which I request you will do me the honour to present to the Linnæan Society, for their library.

I ought to have mentioned that almost all the

Primulas and *Androsaces* are from the Himalaya, and all the *Lysimachiæ* are from the neighbourhood of Katmandu.

Mr. and Mrs. Loring are in excellent health: they are at present living at Barrackpore, and if they knew I was addressing you, would certainly have desired me to present you with their compliments.

Believe me, with the greatest possible respect and esteem,

Dear Sir,

Your faithful and humble Servant,

N. Wallich.

From Sir Thomas Gage.

Castel a Mare, July 30, 1820.

My dear Sir James,

The object of my troubling you with a letter from this distant place, is to state, that having spoken of our Linnæan Society to the Prince de Butera, a Sicilian nobleman of the first consequence in that country, he expressed very strongly a wish to become an honorary member. The Prince is extremely fond of botany, and cultivates with great success a number of very rare and curious plants.

The Prince informs me there is a good garden at Palermo, and well kept. I cannot say as much for that of Naples, which is completely in a state of disorder. Though introduced by letter to Tenore, I have never been able to make his acquaintance.

He is publishing, on an expensive plan, a *Crambe recocta* of indigenous and foreign plants, cultivated in his garden. Petagna, Stellati, and Briganti, are the only naturalists I have heard of here: the latter I am assured is a man of considerable talents. Nothing is published on the subject in this country.

My health, which still continues very uncertain, prevents me from taking the advantage I should have among these fine mountains of Castel a Mare. I have been obliged to give up my favourite pursuits, as I walk with great difficulty.

The *Ophrys Speculum* and *Distoma* of Bivona are the best things I have got of late. The country about Rome affords a much greater variety of rare productions, and there is now a good Flora published by Sebastiani and Mauri. The former I am sorry to learn has fallen into an unfortunate state of health, which affects his mind; and the latter, a young man of great promise, disgusted by want of encouragement, told me he had been obliged to give up his studies, to endeavour to live by teaching Italian. A part of the Farnesina garden has been given for a botanic garden, but not the means of putting it into order. I see by our papers, that even at Bury St. Edmunds we are further advanced, and that a garden has been made even there. I have applied myself a good deal to the study of insects, and have made a collection of about two thousand. How very little have we advanced in this branch of natural history! Kirby, Spence, Marsham, and MacLeay, must help us to something better than what we have got yet. I saw with great regret in our late papers, that we have lost our venerable and excellent Sir Joseph Banks, to whom I at least owe the greatest obligations for his assistance when a young man.

Having written so much on my favourite subject, you will see at all events that we enjoy perfect security and feel no apprehension for our safety, though living in the midst of a revolution. Indeed, unless a foreign power interferes, I think this place as likely to be quiet as any in Europe. Nothing can exceed the attention paid to public security and respect to the persons and property of every individual. In short, we see a whole nation fixed and united for the attainment of a grand object, and despising all the common and little means which in general are adopted, and sacrificing even the natural violence of its character in the general cause.

I shall remain here for the benefit of the waters, if possible, till October; and I think I shall again pass a winter at Rome.

Lady Mary Anne and my children are well. I trust Lady Smith is so, and beg you will accept our united compliments.

Believe me, dear Sir,
Very sincerely yours,
THOMAS GAGE.

From Sir Thomas G. Cullum.

Bury, August 27, 1820.

My dear Friend Sir James,

My sincerest thanks for your affectionate congratulations on Lady Cullum's recovery of her sight.

The operation on one eye did not exceed two minutes, nor more than three minutes in the other; and I am happy to tell you that the operation has succeeded so well, that, with the help of spectacles, she thinks she sees as well as ever she did in her life; she never suffered five minutes pain after the operation:—but how fortunate has Lady Cullum been! She was couched on Monday morning the 19th of June, and on Saturday the 24th the intense heat came on: had this intense heat come on the first or second day after the operation, the loss of her eyes would probably have been the consequence. The operators do not regard the advanced age of their patients, nor the season of the year, except in extreme hot weather. If I was forty or fifty years younger than I am, I should never forget the 19th of June, it being the day on which Lady Cullum underwent the operation, and the day of the death of Sir Joseph Banks; and the last time I saw him he particularly inquired how she did. I saw Sir Joseph at his Sunday's levée about three weeks before; and I said to several friends I should never see him more! I have been over most of the places you mention in your Tour, particularly Haddon and Bakewell.

In my last I mentioned to you a *Dianthus*, the seed of which was collected on the ruins of Catullus's villa at Sermione, on the Lago di Garda in Italy: it grows upwards of three feet high, with a profusion of blossoms. I take it to be the *Dianthus virgineus*. There is a figure of it in Bot. Mag. tab. 1740. In one of the plants, most of the flowers have only four petals: in some flowers the *stamina* are included in the *faux corollæ:* in others, one, two, or three, or more, are elongated.

The evening before Lady Cullum was couched I was desired to put one drop of the infusion of the leaves of *Atropa Belladonna* into each eye; and the next morning, an hour before the operation, the same;—the effect, to make the pupil of the eye dilate, is wonderful. Some of the oculists use for the same purpose the *Datura Stramonium* (and, if I am not mistaken, the *Hyoscyamus*).

Ray seems to have confounded the *Belladonna* and *Datura,* as to their effects in dilating the pupil of the eye (see *Syn.* p. 226.). It is not clear by his notes, which of the plants the lady used,—it may be one or the other; but it is now ascertained that both have the same effect. Ray has given the hint, and the modern oculists have profited by it.

I have many questions to ask you,—more than I can put on paper. How much pleasure it would give me and Lady Cullum, if you and Lady Smith would come and see us in October! Our gaieties begin about the 9th: come on that day, and stay as long as you can, and fix no time for your leaving us. Thanking you again and again most heartily and sincerely for your kind and interesting letter, I remain, with the truest regard,

Yours affectionately,

T. G. Cullum.

Sir J. E. Smith to Sir T. G. Cullum.

My dear Sir Thomas, Norwich, Aug. 31, 1820.

I must thank you for your very kind letter, and the good news it contained, which is beyond our hopes. Pray tell Lady Cullum her little postscript is worth more than its weight in gold, and we value it accordingly.

The confusion about *Belladonna* exists in Ray's own second edition of his *Synopsis*, where three lines relating to *Solanum pomo spinoso* &c. *Datura* stand (in p. 150) between other paragraphs, which all evidently belong to *Atropa Belladonna.* There are no berries (*baccæ*) for children to eat in the *Datura.* Dillenius has not corrected this error in the third edition, p. 266; though he has there printed what regards *Datura* in larger characters, and marked it as a species with the §. But this only increases the error, as he leaves the remarks as they were.

"*Somniferum est et malignum,*" though it may be true of *Datura,* proves, by the word "*baccas*" and all that follows, to belong to the *Atropa.*

I gathered *Dianthus virgineus* on Mount Cenis, —exactly like Bot. Mag. t. 1740.

I was at Sir Joseph Banks's on Sunday evening, the 21st of May, and left the room with the same melancholy forebodings that you did a fortnight after. I was with him alone a few days before, and he told me he was quite easy about the event, which he knew could not be far distant, considering the state of his stomach.

I received a letter from Sir Thomas Gage the other day from Castel a Mare, near Naples, giving a pretty good account of himself and his family.

Once more, my dear Sir, our most affectionate regards attend you and Lady Cullum.

J. E. SMITH.

From Sir Thomas Frankland.

Dear Sir, Thirkleby, Nov. 5, 1820.

I would sooner have acknowledged your kind letter from Holkham, but that I had no matter worth writing in return. A letter to a sportsman from Mr. Coke's in the high season was a curiosity; but I cannot wonder that your mind was more occupied by the business which has so long engaged the house of lords. My politics on the subject agree very much with Lord Harewood's speech in this day's paper.

My son and daughter have been amusing themselves for some weeks in placing single trees and shrubs on the lawn; and so zealously that the lady carries various articles, and even digs. All this of course delights me more than the renounced fox-hunting did; and the pretty intelligent children I cannot keep my eyes from. My son takes a little shooting; but I seldom go with him, preferring the having his own way, and taking care that he is well attended. For my own part, though my vitals are good, debility and rheumatism make me so helpless at a hedge, that I am generally obliged to be pushed over by two stout men, and sometimes a third lifting my foot up! Still I cling to my "ruling passion," and shoot well till over-fatigued. That my garden thrives, I will only say that we had asparagus on the 3rd, and have pease this day. The former is raised in one of the pigeon heated frames invented by M'Phail; and it seems that as no vapour of manure can affect the bed, what is raised in it must be materially sweeter than by the common process. I had a letter lately from a Radnorshire man who has been visiting Mr. A. Knight, and reports a fig-tree which has had nine crops in fifteen months, and has now a tenth on it! This is effected by drenching the plant with a mixture of pigeon's dung and water many times every day.

We have so few woodcocks that I have a bad opportunity for ascertaining whether my suggestion that the *males* only come over in the first flight, is correct. We have had four, and I inclose the exterior quill-feathers of one bird as a perfect example of the male,—the female having a white line running most of the way from the quill to the extremity.

As for tearing out likenesses in paper, there is much chance in it; for if you get wrong in the simple outline, it is almost impossible to correct. I have sometimes succeeded with strongly marked features, at a concert or an assembly,—usually tearing up the thin paper of the bill of fare. The inclosed is the only one which I possess, but will give you a perfect notion of the thing. It represents an old Mrs. Warburton, well known in the society of York 30 or 40 years ago. Her daughter was married to Mr. Burgh, who had formerly been distinguished in the Irish parliament. All the three parties died some time since at York. I should wish to have the *portrait* again at your leisure; for it is considered as strong a likeness as Lawrence's art could have produced.

I have just written into Scotland for a specimen of *Menziesia cærulea**, and shall attempt the *polyfolia*, through my sister Lady Roche, who resides in Dublin. My daughter harrowed up these *desiderata* in my herbarium.

I am, dear Sir,

Yours most faithfully,

T. F.

Viro perillustri atque generosissimo J. Edw. Smith, Præsidi Societ. Linnean. reliq.

S. D. P.

G. W. F. Panzer.

Ex quo literas tuas, calend. Januar. anni 1818, humanissime ad me datas accepi, quibus me tibi

* "We have more than usual satisfaction in announcing this as a British plant, on account of its rarity and beauty, and of the opportunity it affords of adding to our Flora a new genus, dedicated, long ago, by the writer of this, to one of the worthiest men that the native country of the plant ever produced. The original *Menziesia* was gathered by Mr. Archibald Menzies on the west coast of North America. The present species was discovered at Aviemore in Strathspey, and in the western isles of Shiant."—*Engl. Bot.* fig. 2469.

summme obstrictum reddidisti, in dies ardor, denuo humanitatem tuam literis adeundi, increbuit, adeo, ut nisi veritus fuerim modestiæ et verecundiæ legibus adversari, vix me jam pridem temperare potuissem, quin a tuo et in me ingenuo et liberali animo, benevolentiæ atque benignitatis tuæ, iteratum efflagitâssem documentum. Verum vicit cupiditas, et ut aperte fateor, desiderium, super nonnullas mihi dubias stirpes Linnæanas, abs te principe, omnium consensu botanicorum, et a tuo, universi orbis botanici oraculo, de meliori, informari. Atque nunc eo usque processit audacia mea, ut tandem apud me constituerim, collectionem variarum stirpium, tum regionibus nostris tum aliis familiarium, ad te transmittere, simul ac summa animi observantia, abs te petere atque contendere, ut eas benevole accipere, et quas dubias inter illas deprehenderis, distinctione curatiore benignissime velis illustrare.

Obstrictissimum animum, pro tanto in me et singulari favoris et benignitatis documento, per omne vitæ spatium spondeo, ac simul me, ad quævis obsequii et observantiæ summæ officia, nunquam non, paratissimum. Quod etiam alias, ob fugam vacui, præter mihi dubias, stirpes, addidi, ignoscas velim; quum facile fieri poterit, ut eæ, plurimæ, nisi omnes, jam pridem abs te collectæ fuerint; tunc rejicias, vel nihili habeas quæso.

Cryptogamicarum stirpium, et inter has, muscorum frondosorum specimina, me nulla addidisse, de industria factum est; quippe mecum reputans, me iis, vix ac ne vix, ad collectionem tuam jam absolutam augendam, conferre potuisse: attamen indicem muscorum frondosorum, præsertim in Germania obviorum, et a nostro Funckio editum, eo fine addidi, ut singulas mihi indicare velis species quas forte desideras, quasque ad te lubentissime pervenire jubeam. Felicissimum me habebo, si vel in quacunque re, in Germania nostra, tibi officia mea præstare possim, ad quæ peragenda, me semper offenderis paratissimum.

Rob. Brown *Prodrom. Plantar. Flor. Nov. Hollandiæ*, frustra hactenus, quamvis anxie, in Germania, tum et in Anglia quæsivi; semper responsum tuli, hunc egregium librum auctorem ipsum, fatali modo, suppressisse.

Mihi et Dr. Trinio, auctori *Fundam. Agrostograph.* Viennæ, 1820, jam sæpius, *Alopecurus alpinus* tuus fuerat species exoptatissima; quare, si tibi adhuc specimina aliquot superflua restant, unum ex illis, nobiscum benevolentissime communices quæso.

Ratione mercedis cursus publici, opportunitate nundinarum Lipsiensium, in præsentia, usus sum, ut iste fasciculus plantarum ad te perveniat, id quod etiam aliquando, nisi recuses, faciam, et me tuis jussis facile submittam.

Vale, vir generosissime, et me tuo favori, aliquando habeas commendatissimum.

Dabam Hersbrucci prope Norimbergam,
in Bavaria, d. Cal. Octobr. 1821.

From the Rev. James Yates.

Dear Sir James, Birmingham, Feb. 26, 1822.

I have been long wishing for an opportunity of expressing to you and Lady Smith the pleasure and obligation which I feel whenever I call to mind my visit to you last autumn; and at the same time to inform you that all I expected to find at Holkham was even exceeded, both in the magnificence of the place itself, and its richness in the various objects of taste and curiosity, and also in the hospitality and goodness of its noble-minded possessor,—and of his daughter, of whom it is the highest praise to say that she is worthy of her father. From Mr. Odell, also, I received the most obliging attentions; and the kindness of Archdeacon Bathurst was quite overwhelming. Indeed, I never took any journey which afforded me so much of the pleasures and advantages (and these I reckon among those which are of the highest order,) of intercourse with the truly great and good. Among these I number not only your venerable Bishop, but others of the inhabitants of your classic and intellectual city, whom I met in your company, and who, though in humbler stations of life, are the real ornaments of society,—the patterns and promoters of whatever is most estimable in human nature,—the enlightened and consistent friends of liberty, truth, and virtue. I would beg you to give them my kind and respectful remembrances, if it were not that, from the number of the kind and good friends with whom you made me acquainted, the commission would be too extensive to trouble you with. I am sorry that a fondness I have both for carrying and sending letters by the hand of a friend, rather than by the post, has occasioned so long a delay in the expressions of my thanks to Lady Smith and yourself. But my impressions of gratitude and of admiration for what I saw and experienced in Norfolk, are still as lively as if it were a thing of yesterday.

Your Friend and Servant,

JAMES YATES.

Viro præclaro doctissimoque
G. W. F. Panzer, M.D.
S. P. D.
J. E. Smith.

Gratias summas, vir optime, tibi ago, ob literas tuas benevolas, mihi gratissimas, plantarum speciminibus comitatas, quarum dubia, ut potui, in chartula hic annexa, solvere conatus sum.

Doleo maxime quod *Alopecurus alpinus* mihi non amplius in duplo est.

Amicissimus R. Brown exemplaria omnia *Prodromi Novæ Hollandiæ*, non adhuc emissa, donec opus absolutum erit, arcte retinet.

Desiderata mea e muscorum catalogo ditissimo Funckiano alia vice, te favente, mittam. Plurima sane ibidem mihi nova sunt.

Cl. Reichenbach tibi proculdubio notus est. Literas meas, hoc fasciculo inclusas, mittas quæso, ut tuis auspiciis me condonet vir optimus.

Salute minus valida adhuc in patria mea detentus, a Londino et Societate Linnæana diutius quam solet abfui. Hebdomada proxima ineunte, ut spero, urbem et amicos salutabo.

Mitto tibi, vir amicissime, libellos binos Linnæanos, raritatis ergo forsitan non ingratos. Paucis omnino innotuere. Accipias etiam Reliquias Rudbeckianas meas, jam olim curiositatis causa, in amicorum usu impressas.

Vale, vir doctissime, mihique semper faveas!

Dabam Nordovici die 8 Maii 1822.

Sir J. E. Smith to Sir T. G. Cullum.

My dear Sir Thomas, Norwich, June 29, 1822.

I cannot neglect this opportunity of recalling myself to your remembrance.

Mr. Denson has sent me 30 specimens to determine, and I have told him all I know about them.

I feel much improved in health and strength since I came home. We had a pleasant walk to Mr. Crowe's, and back again, after tea on Thursday last. The willows are all growing very finely.

I have placed all Lady Mary Ann Gage's insects in my cabinet, and return her box, which I beg the favour of you to let her ladyship have at some convenient opportunity.

I know not whether I ever gave you a copy of Linnæus's *Orbis eruditi Judicium*, a little work in justification of himself, and the only thing of the kind he ever printed. I fancy he distributed it but little; for few of his foreign biographers have ever seen it, and it is generally reckoned his rarest work. Such, also, is the case with his *Observationes in Regnum Lapideum*, printed for his pupils, and which I also send, as you may like to add it to the many curiosities of your library.

I must next week return to my English Flora. Edward Forster, who has been here a few days, looked at it critically, and approved what he saw. He is a good English botanist, and a sincere friend. I understood you were to meet him at Mrs. Gould's.

With my kindest and most grateful remembrances to your excellent lady, whose benevolence is not wasted on an insensible heart,

I remain, my dear Sir,

Most faithfully yours,

J. E. SMITH.

From Dawson Turner, Esq.

My dear Friend, Yarmouth, Nov. 1, 1823.

Mr. Roscoe is kind enough to promise to come to us the latter end of the approaching week, and if you and Lady Smith can be prevailed upon to favour us at the same time with your company,—not, however, bounding your stay by his,—you will make Mrs. Turner and me extremely happy. Under other circumstances I should scarcely have ventured to propose to you a journey to Yarmouth at this cold and dreary season of the year; but, happily, I can offer you a warm room in a warm house, and you are by no means in the number of those who depend for their comfort or amusement upon things without; and our opportunities of meeting are now so rare, that I am but too glad to catch you in any time or place. In the course of my downward journey through life, there are but few things have vexed me more than the interruption to those annual visits which used to pass between us, and which were periods on whose recurrence, like a schoolboy for his holidays, I used to reckon from year to year. If possible, do let them be renewed: not many more summers, in the course of nature, can be allotted to either of us; and our time of life is not such when we can with prudence expect, or wish, to form many new friends; still less can we be willing to part with those we have: every year, as it goes, necessarily diminishes their number. Since 1823 more than ten of the persons with whom I was then in correspondence are dead. I discovered the fact accidentally in turning over my letters, and I do not know when I have felt so painful a chill. But the effect will be good, if it leads me the more to value those who are left, and among these there is assuredly no one to whom I owe more, or for whom I feel a greater regard, than yourself. Do, my excellent friend, let us endeavour, as far as in us lies, to recall the times that are past, when I used to mark the years by your visits, and counted upon their recurrence from August to August. As one advances in life, and sees the friends of our youth gradually drop about us, the number decreases of those to whom it is possible fully and freely to open our minds: it is painful, therefore, to lose any opportunity, when

> "tibi, dulcis amice,
> Excutienda darem præcordia, quantaque nostræ
> Pars tua sit, mi care, animæ, tibi
> Ostendisse queam."

Ever affectionately yours,
D. TURNER.

From Mr. G. Sinclair.

Sir, Woburn Abbey, Jan. 7, 1824.

I have the honour of your letter of the 23rd December, and humbly beg leave to say, that I have no words by which I can convey to you the deep sense I have of your great condescension and kindness. The approbation of your illustrious name is an honour and reward to me which I feel I cannot too highly prize. Encouraged by your approbation of my labours, I applied to the Duke of Bedford for the honour of his signature to my certificate, which His Grace, in the kindest manner, instantly granted, and accompanied his consent with a mark of his munificence, to render me a free member of the Linnæan Society for life. The condescending and most kind assurance of your support has alone induced me to take steps for such a purpose. I humbly beg leave to say, that my object in intruding on your valuable time, at this moment, is solely to offer you my grateful acknowledgements; for without the permission to do this, I should have felt unhappy.

The Duke of Bedford has of late devoted more time than usual to botany and gardening. Within the last twelve months he has formed a fine collection of Heaths,—the most extensive, I believe, that is anywhere to be found: a compartment for the growth of the more hardy species of *Erica* adjoins the Heath-house. As soon as I have completed my work on Grasses, I purpose to enter on an examination of this beautiful and interesting family of plants,—in which attempt I shall hope for your assistance. The Duke is very anxious to see your new Flora. The important and novel improvements in the generic characters of the *Umbelliferæ* will be found productive of the greatest utility in practice; and in this I feel certain that I speak the sentiments of every practical botanist. The generic characters founded on the general and partial *involucra* I have ever found most difficult to profit by, or use in determining with certainty the groups of this natural and numerous order of plants; and such a lessening of labour and clearing away of difficulties will, I feel certain, be received with gratitude by every student in English botany.

I have the honour to be, &c.

G. SINCLAIR.

From Sir Thomas Cullum.

Bury, Feb. 26, 1824.

My dear Friend Sir James,

I had a letter about a fortnight ago from Mr. Lambert, of whom I had not received a line since I left Fonthill. His letter, I may say, cost me 1*l*. 5*s*., as he expressed himself all in raptures that De Candolle had published Part I. of a new work, entitled *Prodromus Syst. Nat. Regni Vegetabilis*. I sent for it immediately; and though I cannot but admire the pains he has taken, yet I lament that he did not go on with his *Systema Vegetabile,* which probably now I shall never see, as, if I understand him right, he intends to publish another volume, first, of his *Prodromus*.

The chemical names of medicines are so much altered, that I dare not send for a dose of Epsom salts, or some calomel, without looking into the *Pharmacopœia Lond.* to see by what name I am to send a note to the druggist. The new London Pharmacopœia has just been put into my hands. I could not help observing that they took the hint that if that barbarous word *Elettaria* should be abolished, that you would recommend *Matonia* to be substituted for the *Cardamom*.

In the last Dispensatory, the Cantharides were called *Lyttæ,* but now the old name is restored; how far that is proper I must not pretend to question. Three species of *Cinchona*, the *cordifolia*, *lancifolia*, and *oblongifolia*, are still continued. Camphor is still said to be the produce of *Laurus Camphora,* and "concretum sui generis sublimatione paratum." Mr. Miller always told me that all the best camphor was from very large timber trees at Sumatra, sometimes (according to the age of the trees) either in a fluid or concrete state. The whole trade of making camphor is in the hands of the Dutch. Do they make it from the *Laurus,* or from the undescribed trees growing in Sumatra or Borneo?

A few weeks ago Lord Stradbroke sent me a couple of insects, wishing to know their name, as he was apprehensive he should lose all his Scotch firs, as they were sickly and dying in great numbers every year. I had no difficulty in acquainting him that the insect was the *Sirex Juvencus,* which puts me in mind, that many years ago I observed in Mr. Port's garden plantations at Ilam, in Derbyshire, several fine young oaks were lying on the ground; and inquiring of the gardener the reason of their being hewn down, he told me the hornets had bored into them and destroyed them. I now conclude that they were not hornets, but the *Sirex Gigas,* which is more like a hornet than the *Juvencus.*

In Mr. Lambert's splendid book I do not recollect that the *Sirexes* are mentioned as the destroyers of the *Pinus sylvestris,*—but Lord Stradbroke having sent me some pieces of a pine-tree, I will send them to the Linnæan Society if you think it would be acceptable to them.

Yours affectionately,

T. G. CULLUM.

From Professor Hooker.

Glasgow, November 30, 1824.

My dear Sir James,

I received only the latter end of last week your valued letter from Edward Rigby; and I take the very earliest opportunity to return you my most sincere thanks for it. You have given me a great deal of botanical information, and some useful hints, by which it will be my own fault if I do not profit: but what I prize far more than them, is the kind, friendly, and affectionate manner in which you have throughout expressed yourself. I hope I shall continue to deserve your good-will and your friendship, with which I assure you I cannot but feel myself honoured.

I shall now go on to answer your kind inquiries about my proposed *Species Plantarum*. It will neither come under the denomination of a translation of De Candolle's *Prodromus*, nor can the greater portion of it pretend to the title of originality: the very nature of the undertaking would forbid this.

I have De Candolle's full permission to translate and copy whatever I choose from him; and his work (because it will be the most complete,) is what I shall follow in the *general arrangement and character*, especially of the orders and genera. Here, however, I shall make alterations when I may see fit; and with regard to species, I shall, whenever I can have access to a good specimen, draw up the character myself; when I cannot, I shall copy (but *never* without acknowledgement,) from others. Here, again, I shall often be obliged to copy De Candolle's characters, because his are drawn up with reference to *all* the other species of the genus. The species (presumed ones, I mean,) I shall reduce whenever I have materials to enable me to do so; and this is really no light task, for I quite believe, in some extensive genera he has given twice the number of species he ought to do: witness *Thalictrum, Aconitum, Helianthemum,* &c. I have this moment received a good many of his supposed species of the latter genus from Montpellier, and these will enable me to make some corrections. When I may *think* him wrong without sufficient materials to prove him so, I shall copy his characters and state my opinion. I can make nothing of his *Drabæ*;—indeed I am greatly puzzled with our Scotch species. What varieties of *hirta* we have from the Arctic regions! These I think will throw some light on De Candolle's doubtful ones.

As to terminology, you cannot differ in opinion from De Candolle upon this subject more than I do. Much harm has, I think, been done to science by the numerous newly invented terms, many of which are worse than useless. I shall adhere as closely as I can to your language, and I assure you, that *one* great cause of my rejoicing at the appearance of your English Flora, before my going to press with the System of Plants, is, that I shall thus be able to profit by your latest ideas upon the subject. I think in general I have followed you pretty closely. I would prefer, however, *cordate* to *heartshaped,* perhaps from custom; as I would *lanceolate* to *lanceshaped*: neither is, perhaps, generally received English. The word *even* does not express to me what *smooth* does—which I have been in the habit of applying to a surface free from roughness; whilst *glabrous* I have used to signify a surface free from hairs. There appears to me an advantage in using Latin words with an English termination, where they are not otherwise objectionable; that is, in being able to unite with them, the little words *ob* and *sub*.—You can say *OBcordate*, but you must say *inversely heartshaped*;—you can say *SUBglabrous*, but you must say *somewhat smooth*. As to synonyms, I must curtail them as much as possible. There will be 50,000 species of phænogamous plants to be described, and these the bookseller is most anxious should be comprised in ten volumes. I shall give the *first author* who named the plant; De Candolle's as the most complete Flora; and refer to one or more good figures, preferring always the most popular of these: I shall also include those authors who have particularly studied any genus or species. This may perhaps give you some idea of the nature of my book (about which I cannot help feeling very anxious); and the specimen I inclose will give you an idea of the *type*. I need hardly say how thankful I shall be for anything you may suggest for the improvement of my work;—as yet, I have consulted nobody. For any specimens, too, that might come into the earlier or later volumes, that you can well spare, I should be very grateful. I set a great store by those you have already given me, and find them very useful. Your names of new species I should of course hold as sacred, and be most proud to publish them, as well as any remarks you might think proper to make. I never had Nepal plants from Wallich, except our collection five years ago. I have now understood that the India Company require that *they* should have the distribution *of all the specimens*. Mosses, however, I continue to receive from him, and some very good ones too.

I rejoice that the Company have sent you so fine a parcel of Ferns, and I thank you for offering to lay by some for me. The little *Darea* I am quite delighted with; I had previously picked a wretched bit of a barren frond from among some mosses, and admired its beauty. Your specimen I should like to figure in my Exotic Flora, if you have no objection. I wish a fern-like plant I inclose for you may please you half as much as the *Darea* has done me; and pray give me your opinion of it: it is an aquatic from Guiana, and is found in no other state than what you see. It cannot, I think, according to your character of the Order *Filices*, really belong to it: with the general aspect of the fructification of a *Pteris*, there is no ring to the capsules. It cannot be one of the *Hydropterides* of Willd., because they have the fructification among the roots. At any rate, I think I cannot be wrong in forming a new *genus* of it, which I am anxious to dedicate to its discoverer Mr. Parker. Can the East Indian *Pteris thalictroides* (which I have never seen,) belong to the same genus? or Beauvois' *Pteris cornuta*? which you observe comes near the *Pteris thalictroides*; but which again, if Beauvois has represented the fruit correctly, must be widely different.

It would make me very happy if I could think that I had other plants that would be worth your acceptance. I am increasing my foreign correspondence, and have one most excellent contributor in

Demarara, and another in the island of St. Vincent's. I have more mosses too, which I think you might like to have. I wish I could consult you about *Orchideæ* : some fine specimens every now and then flower in our garden ; but according to the present characters *every* new *species* forms a *new genus*. I am glad you think favourably of my Exotic Flora. I have now by me some rare and interesting subjects ; such as the *Marcgravia umbellata, Cassytha filiformis, Cytinus hypocistus,* and some remarkable *Orchideæ,* besides two new *Tillandsiæ*. I have had two plates engraved of the Nutmeg, and am about to figure the *Artocarpus incisa* and *integrifolia,* from very fine drawings made in the West Indies, aided by specimens in spirits.

I have been lately much engaged with exotic plants. I wish you would look at an original specimen of *Lepraria iolithos,* and see if it be not the same as the *red snow* of the arctic people. I have some from Captain Parry, and I have the same plant from Captain Carmichael, found in Argyleshire ; but it does not lose its deep red colour by keeping, nor give out an agreeable odour. I have some reason to think that it is only that filamentous plant (young *Conferva aurea?*) which turns pale by keeping, and smells like violets, and which has perhaps by Linnæus himself been confounded with the granulated kind. Agardh assures me the red snow is common in Sweden.

We may now expect great things from the hitherto unexplored parts of North America. I have been the means of sending out two very zealous and able botanists to the N.W. coast,—Menzies's country. One of them will join Captain Franklin, and return overland with him. The other will explore the vicinity of the Columbia. I have got an appointment for a third to go out with Captain Franklin and Dr. Richardson on their expedition. He will go as far as the Saskatchawan with them, and there remain two entire years, botanizing. He will be within reach of the declivities of the Rocky Mountains, and explore the head of that immense basin or plain which opens towards Mexico, and where he will, it is expected, meet with many of the plants of Nuttall, James, and Bradbury. Franklin and the officers who accompany him will have instructions to collect plants from the mouth of the M'Kenzie river, during their journey to Behring's Straits. Douglas will pass one season on the west coast, and then crossing the Rocky Mountains in lat. 55° (after passing through ten degrees of lat.), will fall in with Captain Franklin at Isle de la Crosse. Richardson will confine himself principally to the country between the Coppermine river and the M'Kenzie river. Their various collections, together with what is now doing in Canada, Labrador, and by Captain Parry's party, will form valuable materials for a Flora of the British possessions in North America. Dr. Hamilton of Leny often speaks of you, and so did Dr. Boott, when he was in Scotland. And now, my dear friend, I have reason to think that long before you come to this part of my epistle you will cry for mercy; but be assured that I never shall, let your letters be ever so long. Do write to me when you have time, and never think of postage.

I am, my dear Sir James,
Your very grateful and affectionate Friend,
WILLIAM JACKSON HOOKER.

Viro Clarissimo J. E. Smith
C. A. Bergsma S. P. D.

Per aliquod tempus, in conscribenda dissertatione de *Thea* occupato, multa obvenerunt adhuc incerta et minus rite explicata. Inter quæ imprimis illud est: quod nondum intelligo, quare clar. Linnæus duas species descripserit generis *Theæ*. Qua de re, suadente viro clar. Jos. Aug. Schultes, audeo te, vir clarissime! rogare, ut nota tua eruditione me edoceas.

Linnæus in prima editione *Specierum Plantarum* (Holmiæ 1753, tom. i. p. 515.) et in *Systemate Naturæ* (Holmiæ 1759, tom. ii. p. 1076.) unam tantum speciem assumsit, nempe *Theam sinensem*. Doct. vero Joh. Hill primus fuisse videtur qui duas species constituerit; sed ipse fatetur, tum, se diu incertum fuisse num quidem constituendæ essent, tum, species quas agnoverat ab aliis confirmandas esse. (*Exotic Botany*, London 1759, tab. 21. & 22.) Eum secutus esse videtur Linnæus. Hill autem forsitan facilius ad duarum specierum constitutionem conclusit, quia Linnæus in *Specierum Plantarum* prima editione animadverterat, se flores in aliis hexapetalos, in aliis enneapetalos vidisse.

In dissertatione Tillæi, Linnæo præside defensa, (*Amœn. Acad.* vol. vii.) una tantum species citatur, et in Linnæi *Materia Medica* etiam, quantum scio, de una tantum specie loquitur, quod et fecit doct. Thunberg in *Flora Japonica.*

Plurimi recentiorum botanicorum, inter quos clar. viri Loureiro, Aiton, Sims et De Candolle, *Theam Boheam* et *viridem* tanquam varietates considerant. Equidem non intelligo, quare duo postremi, *Theam Boheam* varietatem *viridis* esse putant: potius vero contrariam sententiam defenderem; primo, quia *Thea* a doct. Kaempfero delineata, magis cum *Bohea* convenit; deinde quia doct. du Halde (*Déscription de l'Empire de la Chine*) refert *Theam viridem* (Songlo) majori cum cura coli quam *Theam Boheam* (vouy Tsia); imprimis autem quia *Thea* in sua patria fere semper est hexapetala, quod mihi affirmaverunt qui plantam sæpe ibi crescentem conspexerunt.

De hoc argumento litteris conscripseram viro ill. Schultes, cum quo consuetudinem jungendi occasio mihi fuit hac ætate, quum per meam patriam, ad tuas oras accederet; is suam sententiam mecum communicavit, judicans definitionem a Linnæo, in *Philosophia Botanica,* ad speciem a varietate distinguendam, traditam, minime hodie nobis sufficere; verum enim vero mihi suadet, ut tandiu defenderem species Linnæi quandiu hac de re dubius hæream; imprimis autem hortatus est ut te rogarem ut hæc dubia ex herbario in botanicis summi viri Linnæi explanares.

Tu ad me de istis rebus omnibus scribas velim

quam diligentissime, quod ut perficias vehementer te etiam atque etiam rogo.

Vale.

Dabam Trajecti ad Rhenum, kal. Febr. 1825.

Viro clarissimo C. A. Bergsma
J. E. Smith.

Vir Clarissime,

Dubia certe et obscura in *Theæ* historia botanica, nec non œconomica, ad nauseam usque invenies.

Theæ Boheæ exemplaria duo, floribus hexapetalis, manu ipsius Linnæi nominata, in herbario Linnæano conservantur; ut et folia tantum duo *T. viridis*, ex hortulano celebri Gordon missa, sed nullæ auctoritatis.

Forsitan plures, et inter se diversæ species, sive *Theæ* generis, sive *Camelliarum* botanicis adhuc ignotarum, in Europam mittunt Chinenses vaferrimi, sub nomine Thearum diversarum.

Omnia hæc vero, de industria, obscura et confusa reddunt.

De differentia specifica inter *T. viridem* et *Boheam* nihil certum unquam invenire potui—nec in hoc negotio ullo modo prodest herbarium Linnæi.

Cultus tuus devotiss.

J. E. S.

Dabam Norvici, 3 id. Febr. 1825.

Festino calamo hæc scribo, pluribus negotiis occupatus, quæ inquisitionem ulteriorem vetant.

From A. B. Lambert, Esq.

My dear Friend, Boyton House, July 22, 1825.

Perhaps you will hardly believe it when I tell you, the whole of the drawings of natural history of the celebrated Bruce are now all safe at Boyton ! ! !—arrived about a fortnight ago direct from Kinnaird: Mr. Cummings Bruce, who is now in possession of the estates and museum, marrying the heiress of Bruce, and whom I never saw. His brother, Sir W. G. Cumming, married my cousin, Lady Charlotte Bury's eldest daughter.

The drawings, which are about three hundred, consist of birds, fish, quadrupeds, and plants,—the latter most interesting,—all drawn and described on the spot where he found them, in Abyssinia, with dissections of the parts of fructification, and which would do honour to many of the botanists of the present day, and most truly prove him to be a most accurate, diligent observer, and perfectly do away all those malicious aspersions which envy has brought against him.

Yours sincerely,

A. B. LAMBERT.

*From H. F. Talbot, Esq.**

Dear Sir, Island of Corfu, March 31, 1826.

I imagine it will be agreeable to you to have some

* Of Laycock Abbey, Wilts., nephew to the Marquis of Lansdowne. This gentleman was introduced to Sir James in the autumn of 1825 by A. B. Lambert, Esq. of Boyton House.

account of the botany of this island, which you may compare with that of Sibthorp and others who have visited it, probably at different seasons of the year. The aspect of it is very mountainous, and clothed with olive woods: near the city a fertile plain. I have not yet extended my ramble to any distance from the city. The orange-trees are magnificent, and are now loaded with fruit. The olives are covered with their black berries. Gigantic *Arundo Donax*, and *Cactus Opuntia* of considerable size, diversify the prospect. The extensive cultivation of the Artichoke is curious; it serves for hedges, and very pretty ones. There are hedges, too, of scarlet Cape Geraniums, more I think for ornament than protection. In fact, it is an open country,—a garden.

Here is a sort of calendar of Flora, which you may compare with Norfolk on the same day, if you have made similar notes:—

Corfu, March 31.

Elder-tree in flower; Brambles, ditto; Pomegranate in young leaf, having the appearance of a red bush; Fig-tree in young leaf; *Scrophularia peregrina* in flower; along with the following plants in flower:—

Chrysanthemum segetum; Common Scarlet Poppy; *Lathyrus alatus; Fumaria capreolata; Geranium,* unknown to me,—like *molle,* but five or six times larger,—not *pyrenaicum; Thlaspi Bursa pastoris,* of a size little inferior to *Cardamine amara; Asphodelus ramosus* and *Phlomis fruticosa,* very showy plants; *Anemone hortensis; Hyacinthus racemosus; Ornithogalum exscapum* (of Tenore), *nanum*

(Sibth.?) which seems a mere variety of *umbellatum; Helleborus viridis*—this is not in *Prodr. Fl. Græc.; Lycopsis variegata*, a charming plant, forming here extensive fields of blue;—Sibthorp says it is rare in Peloponnesus and the Archipelago, so that I suppose this island is its head-quarters; *Ajuga reptans*, extremely large; *Symphytum tuberosum; Linum usitatissimum; Euphorbia dendroides; Chrysanthemum coronarium; Erodium malacoides* and *moschatum; Lotus ornithopodioides; Echium calycinum* (not in *Prodromus*); *Tordylium officinale*,—this is the commonest umbelliferous plant on the island. At Ancona, in Italy, I saw the other day a scarlet corn-field;—arrived at the spot, I found it owing to a profusion of *Tulipa Oculus Solis*.—The weeds of the South of Europe are very handsome.

April 2.

I can add to the list two *Ophryses* which are not in the *Prodromus*,—*lutea* and *aranifera*, besides *tenthredinifera*. I observed today, in flower, *Anagallis arvensis* and *cœrulea; Hippocrepis unisiliquosa; Saponaria ocymoides* (not in *Prodromus*); *Lunaria rediviva; Astragalus monspessulanus; Galium cruciatum; Spartium villosum; Cistus salvifolius; Anthyllis vulneraria* (*flore rubro*), and *Vicia lutea? Veronica Buxbaumii; Bellis annua; Scorpiurus subvillosa; Phalaris utriculata*, very abundant (not in *Prodr.*); *Trifol. resupinatum; Sherardia arvensis; Geranium dissectum; Lithosp. purpureo-cœruleum;* and a noble *Echium*.

I am convinced the beautiful Geranium of this island is distinct from *pyrenaicum*: the flowers are a fine rose colour, of the size of a sixpence, ornamenting all the road sides.

Hedysarum coronarium (not in *Prodr.*). I wonder this showy flower did not fall in Sibthorp's way.—I have found a *Senecio* that may be new: it is like *vulgaris*, wanting the ray florets, but much larger, and pale sulphur yellow, with long slender florets that are protruded and make a sort of bush.

April 4.

There are now in flower some plants not in the *Prodromus*, viz.

Lotus maritimus; *Vicia grandiflora* (*sordida*, Willd.). *Cerastium manticum* (a curious species, hitherto little understood, to which I would give the following specific character: "*Cerastium erectum glaberrimum pedunculis longissimis erectis, calyce membranaceo, petalis fere integris.*" A curious little *Euphorbia*, (*nova species?*) resembling *E. Peplus* a little, but has a 5-rayed umbel, and *echinate* capsules: the petals also are entire, not lunate or horned. A species of *Muscari* with blue flowers of a large size.—Besides these, which seem additions to the *Flora Græca*, I see a beautiful Malvaceous plant,—a *Malope*, with rosy flowers three inches in diameter; blue Lupins; *Bartsia latifolia*; *Serapias Lingua*; *Delphinium Staphisagria* (not in flower); *Coronilla cretica*; *Medicago circinata*; *Ornithopus compressus*; *Trifol. subterraneum*; *Picridium vulgare*; *Thelygonum Cynocrambe*.

I remain, dear Sir, with the greatest regard,

H. F. Talbot.

From the same.

Dear Sir, Corfu, April 10, 1826.

Trusting you have received a letter I sent you some days ago, I will now give you some further account of the vegetation of this island. I have, however, only examined the environs of the town. In the remoter parts, if I may credit the director of the botanic garden here, there are some very extraordinary productions; as, *Echium giganteum ;* two species of *Othonna*, the *pectinata* and the other; two species of *Osteospermum*, *pisiferum* and *moniliferum ; Nolana prostrata ; Chrysocoma Coma aurea ;* and a beautiful yellow climber, genus unknown to me,—whether *Dolichos* or *Crotalaria* I cannot tell; but as I have not myself observed these plants growing wild, I hesitate about them.

Now to come to my own observations.—I mentioned in my last that I had found a new? *Euphorbia*, resembling *E. Peplus :*—on revisiting the spot, I found (which is singular) another *Euphorbia*, growing with it, and very like it, but admirably distinct in its characters, especially in its *sharply hexagonal* capsule, which I have not seen in any species of this genus. The two plants are thus characterized: (they both resemble *E. Peplus* or *E. exigua.*)

No. 1.	No. 2.
Umbellæ radii 5,	Umbellæ radii 5.
Planta glaberrima,	glaberrima.
Folia semiamplexicaulia,	sessilia.
obtusa,	acutiora.
apice obscure serrulata,	ubique eleganter serrulata.
Petala minima, integra,	idem.
Capsulæ triquetræ angulis echinatæ.	acute hexagonæ.

I should be glad to learn whether these plants are described anywhere. I have found a large white-flowered species of *Lithospermum*, with trailing stems ; *Stachys spinulosa ?* with white flowers,—a large plant; *Veronica syriaca*, an elegant little plant with large flowers; *Phleum felinum ? Convolvulus tenuissimus; Crambe Corvini* (not in *Prodr.*); *Lotus tetragonolobus ; Lathyrus setifolius ; Bunias Erucago ; Ornithopus scorpioides ; Andropogon distachyon ; Aristolochia rotunda ; Allium subhirsutum ; Urospermum picroides ; Hesperis verna ; Rhagadiolus stellatus ; Polypodium leptophyllum ? Anthyllis tetraphylla ; Euphorbia amygdaloides*, abundant (not in *Prodr.*); *Geranium*, with very large flesh-coloured flowers,—petals entire ; *Vicia bithynica ; Coronilla securidaca ;* &c. &c. &c.

I am about to visit Santa Maura (the ancient Leucadia), and hope to give you a good account from thence. The season is backward: there are a vast number of plants springing up everywhere.

I remain, dear Sir,

Yours faithfully,

H. F. Talbot.

P.S. The *Umbelliferæ* of the Ionian Isles are numerous and extraordinary. They seem little known, and I cannot refer them even to their *genera* in some instances.

A copy of the *Flora Græca*, which I met with in the Grand Duke's library at Florence, was of great service to me.

From Sir Thomas G. Cullum.

My dear Sir James, Bury, Aug. 22, 1827.

We have lost from our Society the Bishop of Carlisle. I preserve one of his letters to me, bearing date October 31, 1796, which I value much. At one of the Linnæan dinners in May we were comparing our ages, and it appeared that the Bishop was nearly two years below me; he being born in 1743, and I in 1741. When I looked at him at the Linnæan Club at the Thatched House Tavern, I was much concerned, thinking what an alteration illness makes in a man whom I remember as strong as Hercules. When I look at the private list of the Linnæan Society in 1788, and the printed one in 1789, I will not express how few are alive!

I feel the infirmities of age creep upon me; yet I am thankful for the comforts I enjoy. That we may soon meet in cheerful spirits and health, is the sincere wish of your affectionate

T. G. Cullum.*

J. E. Smith to Sir T. G. Cullum.

My dear Sir Thomas, Norwich, Sept. 4, 1827.

Your letters are always peculiarly welcome, and I had long been wishing to hear from you. I need not say how much I lament the good Bishop of Carlisle. Dr. Latham and I are now the only sur-

* The venerable baronet himself paid the debt of nature on the 8th of September 1831, aged 90 years.

viving original members of the Linnæan Society. He writes to me that he is two years older than the Bishop. I was prevented by the death of my wife's brother, and my own illness, from going to London this year. I suffer severely from rheumatism, which has for three weeks past affected my eye,—the same which suffered from erysipelas twenty-five years ago, and frequently since, becomes cloudy, weak, and painful from cold: it is now very troublesome, so as to hinder my usual pursuits.

I have lately received from Kent the true Swiss *Ophrys arachnitis* figured in Haller,—a great acquisition to our Flora.

Believe me, dear Sir, &c.

J. E. Smith.

CHAPTER XIII.

Correspondence of Mr. Roscoe and Sir J. E. Smith.

A few letters of Mr. Coke of Norfolk are also inserted in this chapter,—a name which it is impossible to mention without a wish to do some justice to the feelings it inspires; but this might seem like adulation, yet fall short of the truth: it is not therefore sufficient, but it may be expedient, only to remark, that Mr. Coke's friendship for Sir James began from the year in which the latter became an inhabitant of the same county, and was perpetuated in annual and perennial acts of kindness during the remainder of Sir James's life.

Of a character so well known as Mr. Roscoe, it is unnecessary to say more than that Sir James's first acquaintance with this distinguished person had its origin in a request imparted to the latter to give a course of botanical lectures at Liverpool in 1803, upon which occasion a mutual esteem was formed: it cannot be said to have *grown* between them, for it arrived at its full strength and stature in so short a period, that time was unnecessary to the development of a friendship which proved as durable as it was decided at the first acquaintance.

The following letter to Dawson Turner, Esq., describes the impression which this visit made upon Sir James:—

My dear Friend, Allerton Hall, July 16, 1803.

At length I sit down to write you a letter,—*literally*, but not, I fear, *metaphorically*, with the pen of a Roscoe,—that very pen which has just been correcting his manuscript Life of Leo X.

I am here at his charming villa, six miles from Liverpool, looking over Cheshire and the Mersey to the Welsh hills.

Our friend Hugh Davies travelled with Drake and me in the mail to Chester; our ride and voyage thence were delightful*.

My lectures are numerously and brilliantly attended, and seem to stir up a great taste and ardour for botany. The botanic garden promises well, though in its infancy, except the stove, which is well filled, and in the first order. The curator, Mr. Shepherd, is the properest man I ever saw for the purpose. I hope to procure him some useful correspondents, one of which shall be our friend Watts of Ashill.

You are acquainted with Mr. Roscoe's taste and genius;—his manners, temper, and character are equal to them. I am surprised to find him so good

* In a letter written about the same time to his sister, Mrs Martin, Sir James informs her "that he had been to dine with Mr. Roscoe at his country house, quite retired, in a most beautiful situation, with fine views over

'*Cheshire and Lancashire both,*'

(Do you remember the old ballad of Childe Waters ?) Wales, the Mersey, &c. I felt as if I were with Lorenzo de' Medici at his villa; for of all the men I ever knew, Mr. Roscoe most surpasses my expectations."

a practical botanist. His library is rich in botany, and especially in Italian history and poetry. I fancy myself at Lorenzo's own villa. I expect my friend Caldwell from Dublin every day, and have some hopes of Mr. Griffiths coming to see me. Two Miss Gleggs, very fine girls, his neighbours, have lately been here : one of them is a botanist, and we have had some rambles together.

The most interesting place I have seen, in itself, is Mr. Blundell's, of Ince,—rich in a profusion of antique sculptures, pictures, and marbles. We had much entertaining talk about Italy, as he has often been there.

I just saw my old friend Broussonet for a few days : he touched at London in his way from Teneriffe to France. He is going to be professor of botany at Montpellier.

Yours most truly,

J. E. Smith.

As a botanist, Mr. Roscoe is distinguished for having thrown the best light upon the *Scitamineæ*, a splendid tribe of plants, previously but little understood,—" a very natural and important order, the eighth among the *Fragmenta* of Linnæus, and equivalent to the *Cannæ* of Jussieu. Its name alludes to the aromatic qualities of most of the species, and particularly to the use made of some of them in cookery or in medicine,—*scitamentum* being expressive of anything rendered grateful to the palate by seasoning or other preparation. Accordingly the Ginger, Turmeric, Zedoary, and various sorts of

Cardamoms, belong to this order. It stands between the *Orchideæ* and *Spathaceæ*.

"The genera referred to this order by Linnæus are *Musa*, *Heliconia*, *Thalia*, *Maranta*, *Globba*, *Costus*, *Alpinia*, *Amomum*, *Curcuma*, *Kæmpferia*, *Canna*, *Renealmia*, and *Myrosma*.

"The order of *Scitamineæ*, including the *Canneæ*, coming at the very threshold of the Linnæan artificial system, and being in themselves very attractive, curious, and rare, have particularly engaged the attention of several distinguished botanists, but with very unequal success. No plants have been less understood by Linnæus and his immediate followers, with regard to their genera, and the principles upon which they ought to be founded.

"Professor Swartz, who has so well illustrated the *Orchideæ*, and whose attention was called to the *Scitamineæ* by their near affinity to that tribe, has not thrown any light on their generic distribution. The French botanists have done absolutely nothing to clear up this family, but have adopted the ideas and all the mistakes of Linnæus.

"The most unfortunate attempt relative to the genera of *Scitamineæ* was made by Giseke, in his edition of the lectures of Linnæus upon the natural orders of plants, printed at Hamburgh in 1792. This writer, working with other people's materials, and destitute of practical experience, boldly undertook to new-model the whole order. But as Gulliver's mathematical tailor of Laputa, having made a mistake in the beginning of his calculation, brought him home a whole waggon-load of clothes, so Giseke, setting out on erroneous principles, has presented us with a rumbling waggon-load of new hard-named genera, dismembering the old ones, not only by insufficient characters, but by characters that do not exist, and establishing new ones with as little scruple or success.

"Mr. Roscoe first suggested a method of reducing the genera of the *Scitamineæ* to regular order, by essential characters derived from the structure of the stamen, particularly its filament.

"This principle is found to be the only one which, while it is clear and precise in defining technical essential characters, leads to the establishment of natural genera.

"It must be observed, that the learned author of the *Prodr. Nov. Holl.* (Mr. R. Brown,) follows the hypothesis of Jussieu in not allowing a corolla to these plants, inasmuch as they are monocotyledonous.

"Mr. Roscoe's ideas of a *Scitaminean* flower are exactly consonant with Mr. Brown's, except that he considers the inner perianth of the *Thalia* as a corolla with a double limb, which is the most obviously natural mode of considering it, and in which I without hesitation concur.

"The *Scitamineæ* as well as the *Canneæ* are properly placed in the *Monandria Monogynia* of the Linnæan sexual system." (See Sir J. E. Smith's article *Scitamineæ* in Rees's Cyclopædia.)

In one of the latest letters addressed to Mr. Roscoe by Sir James, in 1827, he congratulates him on the approaching conclusion of this "highly valuable and honourable work on *Scitamineæ*, one of the few really original valuable and learned works in botany that this age has seen."

The following imitation of the style of Ossian, by Sir James, may not be considered misplaced here: it is recalled to recollection by its title of "Allerton," though it was written many years before the author had any knowledge of Mr. Roscoe, or of the spot where their friendship was begun and so happily matured; and the scenery it describes applies, almost as aptly as the name, to the residence of him who, in every place and situation, was the individual round whom Sir James's pride and his affection equally rallied, and who, to express his esteem for the subject of this memoir, had called him by the flattering appellation "the friend of my early days but lately found."

"Fair art thou, O Allerton! Lovely are thy green places and thy pleasant groves. The forests of Gleddow are spread out as a carpet before thee, and thou numberest the cattle upon a thousand hills. The great and the wealthy daily offer up incense at thy feet.

"The north wind maketh unto himself bowers of thy myrtles, and slumbereth on the velvet of thy lawn. Thy fair plants bow before him at his awakening, and thy saplings are nursed in the storm.

"Terrible art thou, O Allerton, when the tempests of heaven surround thee, when the sweet-smelling herb is swept from the face of the field, and the torrent stayeth not; when the Thunderer establisheth his throne on thy mountain's brow, and the forests shiver before him. Or, when the north wind rusheth forth in his fury, and fixeth his seal on the streams of the valley, then thou dealest out the snowy whirlwind, and the fruitful field becomes a trackless waste.

"Yet even then thy tender mercies are over thine adopted children; and while thy more hardy offspring sport in the whirlwind, and brave the unrelenting storm, thou shelterest in thy warm bosom full many a tender nursling, and sufferest not even thy summer zephyrs to visit their face too roughly: they are grateful for thy kindness, and offer their flowery tribute to thee in peace.

"Yea, truly glorious art thou, O Allerton, in thy summer attire, when thou, kind stepmother! sendest forth the children of Flora to sport on the lawn, where they drink health and beauty from the nectar of the evening, and give their fragrance to the morning gale: then exultest thou in their well-being, as they open their pure bosoms to the sun. He smileth on them, and they are glad.

"Yet even most lovely art thou, when the lengthened shadows of the mountains obscure the valley; when the declining sun gildeth the tufted forests of Gleddow, and empurpleth the little hills on every side; when thy clear eye seeth new mountains afar off that the noontide splendour had eclipsed. Then a little while, and the moon cometh forth in the east the shades of Gleddow acknowledge her silver footsteps. Is the fair moon thine handmaid, O Allerton! that she watcheth thy children as they sleep? Or doth she delight in the music of the nymphs of Gleddow, as their light hands touch the harp of many silver strings? Sweet are the wood-notes of the nymphs of Gleddow. Methinks their imperfect sounds now vibrate in mine ear; or do I hear the drowsy tinkling of the flocks, as they wind slowly along the distant mountain's brow? Now the rising zephyr swells the lengthened notes,—even now—Hark!—they are still!"

J. E. Smith to Mr. Roscoe.

My dear Sir, Norwich, Aug. 25, 1803.

Finding myself at length settled at home, and a little composed, I sit down to converse a few minutes with you, and to transport myself in thought at least into your society. How many times shall I wish that I could do so in reality! I could not at parting say anything that I wished to say. My young companion, as well as myself, thought everything during our ride the rest of the day uninteresting, and Manchester the most comfortless place we had ever seen. We stopped there merely to take my favourite refreshment, tea, and then proceeded to Disley. Here we found ourselves in quiet, and could talk of Allerton and you. The next morning at seven we mounted the top of a coach, and had a delightful ride over the highest parts of Derbyshire, through Buxton and Ashbourne,—a beautiful spot. We stepped into the church, while the organ played the Hundredth Psalm, to see the tomb of poor Sir B. Boothby's Penelope. The statue was covered, but I had seen it in London. You know the lines from Petrarch which are on the base:

> "Le crespe chiome, l' angelico riso, &c.
> Poco polvere son che nulla sente." *

There is also a passage,—I think from Rousseau, but am not certain,—something to this effect: "*Son cercueil ne la contient pas toute entière. Il attend le reste de sa proye; mais il ne l'attendra pas long tems.*"

We arrived at Matlock by dinner-time on Sunday, and spent the two following days in that sweet spot. We saw Burleigh House in our way. I had seen it twice before, but did not remember that it contained so many very fine pictures.

My heart sinks as I now see before me plants and books, the relics of my beloved friend Davall, who languished to see me in Switzerland,—as I do to have you here,—and never had his wishes gratified: yet I will hope; as Allerton is nearer than Orbe,

* The lines from Petrarch quoted in this letter were great favourites of the writer of it; and he attempted to give them an English dress, retaining their peculiar expression:

> "The crisped locks of pure refulgent gold,
> The lambent lightning of the angelic smile,
> That made on earth a paradise e'erwhile,
> Are now but senseless dust, beneath this marble cold."

and you have promised. In the mean time, I am eager to welcome Lorenzo to my library.

Allow me to subscribe myself

Your most obliged and affectionate Friend,

J. E. Smith.

From Mr. Roscoe.

My dear Sir, Allerton, Aug. 28, 1803.

Amidst our frequent recollections of you at Allerton, we had begun to feel some anxiety on your account, which would by no means have been diminished, had we known that you had been careering over the hills of Derbyshire on the top of a coach. Your letter has arrived just in time to alleviate our apprehensions, and to add to the cheerfulness of our Sunday's dinner, where you have as many friends as we number individuals. In rejoicing with you, as I most truly do, on your restoration to domestic happiness, I feel however a selfish hope that it may encourage you at no distant period to pay another visit to Liverpool, and that you will prevail on Mrs. Smith to accompany you. I had almost begun to suspect that the cares of the world or the lapse of years had blunted in me those feelings and diminished that capacity of attachment which in youth is so ardently experienced: but the fortunate incident which introduced me to your acquaintance has restored me to a better opinion of myself; and however I may regret that we did not meet sooner, I gratify myself in regarding you as a friend of my early days but lately found; if indeed

I can be said lately to have found one whom I have known so long in his writings, and to whom I have been indebted for much pleasure and some improvement.

Towards the beginning of October, our curator Mr. Shepherd will, I presume, make his excursion to look through the nurseries of London, and is highly pleased with the thoughts of seeing you at Norwich on his return. I believe no person living will regard the sacred relics of Linnæus with greater veneration than this uneducated son of the art. We have lately devoted a day to the examination of the *Pancratiums*, *Crinums*, and *Amaryllises*, and have found it a subject not without difficulty. I shall only observe, that the plant which we saw at Lord Derby's, which is now in flower with me, and which I called the *Crinum americanum*, is, I believe, the *C. latifolium* of Linnæus,—the *Amaryllis latifolia* of L'Héritier,—which I mention only for the purpose of correcting my own mistake.

Whilst debarred of your society, I shall enjoy the prospect and possibility of seeing you for a few days at Norwich, and extending still further those attachments of which I find myself yet susceptible, among those connexions and friends whom you have already taught me to esteem.

And now, my dear Sir, I have only at present to say, that if I thought our epistolary intercourse were to close with the interchange of a letter, I should be highly mortified. I well know the multiplicity of your avocations, and should be truly sorry to intrude on them: nor am I in general a very vigilant correspondent; yet I shall occasionally take up my pen, if for no other purpose, to give you some account of our proceedings here, and to arrange with you many things which we talked about but left unfinished.

Believe me at all times

Your very affectionate and faithful Friend,

W. Roscoe.

Sir J. E. Smith to Mr. Roscoe.

My dear Sir, Norwich, September 23, 1803.

I am afraid you will begin to think me unworthy of your wish that "our correspondence should not be limited to two letters." If, however, you imagine that this wish or any other kind expression in your truly acceptable letter is not deeply and gratefully felt by me, you do me injustice. Let me explain what has prevented my writing sooner. Early in this month we went to Yarmouth, and spent five days with my friend Dawson Turner; he and I had enough to do with our Mosses and Lichens, and I had much to learn from him about German correspondence; for he is so good as to relieve me from the burthen of dealing with those indefatigable writers and questioners, the German naturalists. He is about printing a neat little book on the Mosses of Ireland. We then visited my wife's relations at Lowestofft and Saxmundham, twenty-four miles further in a very pleasant country, tame compared with your noble scenes, but rich, woody, highly cultivated, and in most seasons verdant,—though the present summer has almost burnt up this whole eastern country. Here was scarcely any rain from April till the middle of September.

I could like, one day, to show you this very tour we took; it would be new at least to you, and I think you would find objects of amusement and curiosity. The printer's devil summoned me home, and I am not yet out of his clutches with regard to *Flora Britannica*, a sheet or two of which stands still till I have made up my mind about some knotty points respecting Mosses. You can hardly be aware how much labour these little plants cost me, nor what obscurity there is among them; yet, when carefully studied, they amply repay our care by their beauty, the precision of most of their characters, and the admirable exercise they afford to a systematic mind. The difficulties are chiefly caused by the inaccuracy of preceding writers. The *Jungermanniæ* and Lichens, which I must next take in hand, are in some instances more variable and uncertain; but I must get through them,—and then I think all other botanical labours will comparatively be repose. The sulphur-coloured Lichen I found on your house is, as I thought, *L. orostheus* of Acharius, never before noticed amongst us. I am particularly pleased at having to quote Allerton for a new plant.

I shall soon begin eagerly to expect my friend Shepherd. How welcome will a Liverpool face be to me! Of *Salices* you know I promise him a harvest. Is not your plant *Crinum erubescens? C. latifolium* as figured in Rudbeck (copied, I believe, from *Hort. Malabar.*) seems broader in leaves and petals.

Yours seems *C. erubescens* of Redouté's *Plantes Liliacées*, fasc. 5. Are not your flowers tinged externally with purple? and sessile?

How gladly shall I cherish the hope of seeing you and Mrs. Roscoe here! Drake is much flattered by your mention of him. He is now reading your Lorenzo, having just read Tenhove. He enters fully into the subject, and reads with as much profit as most young people: I perceive your character has made a deep impression on his mind. You did not perhaps observe that he was in tears almost all the way to Prescot. I know not whether to regret that age when *tears of the mind* so easily find vent at the eyes; but I love to contemplate it in others. I am anxious to hear how Mr. W. Roscoe's health is: we have just been admiring his "lily by moonlight." I think his youthful spirit aims at removing by exertion what would most easily yield to repose and care. Inflammatory disorders are never to be rubbed off by neglect; they are the reverse of nervous or low diseases, and much more rare.

Your ever affectionate Friend,

J. E. Smith.

Sir J. E. Smith to Mr. Roscoe.

My dear Friend, Norwich, October 6, 1803.

I did not think of writing to you just now; but a melancholy event has just reached my knowledge which you no doubt have heard already; and on the subject of which my first feelings seem inclined to relieve themselves by having recourse to you.

My good and kind host, my oldest friend out of my own family, is no more!* We were children together; and, in all our childish amusements, he, being a year older than I, was always my instructor and example. When he went to boarding-school, it was the first real trial of my fortitude; and he was my first correspondent. I used to dream night after night that he was come home for the holidays; and then a more impressive dream would tell me that though the former were delusions, he was then certainly come. He then went to Liverpool;—time, new connexions and pursuits, made us less interested in each other; but we had no other cause of estrangement; we met occasionally in London with mutual pleasure, though we had not so many ideas in common as heretofore. At length my fortunate visit to Liverpool,—in the planning of which I found all the original friendship and benevolence of his heart exerted themselves as warmly for me as ever,—promised the revival of our old attachment during our lives, and I little expected that period would be so very short. He was literally to me the "recovered friend of youth," which you in so flattering and engaging a manner imagine me to be to you. Is it not evident how warmly I accede to the adoption, when in my present anxiety I recur to you for comfort, and trust to you to bear with my tedious complainings, as is the duty and pleasure of an *old* friend? We parted for the last time at your door; you had wit-

* Mr. Thomas Taylor of Liverpool, brother to Mr. John Taylor of Norwich (see above, page 100), and youngest grandson of Dr. John Taylor, author of the Paraphrase on the Epistle to the Romans, &c.

nessed his eagerness and anxiety to serve me; nor can I ever forget his benevolent, unobtrusive, and unostentatious hospitality,—his hearty welcome, the freedom with which he told me all his concerns, and the pleasure he took in recalling the events of our youth. Even on subjects of mere amusement or taste we had much to say, and he had a peculiarly obliging attention to pursuits that were more exclusively my own, because they were mine. His heart was in no degree worse for living in the world, but in some points I have thought it improved; this might be owing to a slight cloud of ill success which had but too often accompanied his path. He was always more successful in serving others than himself. But I sincerely believe he had nothing to regret on his death-bed. He never gave his friends a pang while he lived; but many a tear of the purest, most disinterested affection will be shed over his grave. Excuse me, my good friend, for writing you so unentertaining a letter; it is really for my own pleasure rather than yours, and yet I will not conclude it without a piece of business, about which I had it in contemplation to give you a line in a few days. I forgot in my last to mention some directions for Mr. Shepherd your curator, when he gets to London. If he will call at Sir Abraham Hume's in Hill Street, Berkeley Square, I will previously give Lady Hume a letter, and I doubt not she will show him her plants at Wormleybury, fifteen miles north of London. He will find her extremely ready to give or exchange plants. He ought also to see Mr. George Hibbert's garden at Clapham, and may go and introduce himself there as my friend. He is very rich in Cape plants, and you could help him to West Indian ones. Mr. Woodford's garden at Vauxhall he ought by all means to see, and he will need no introduction except telling who he is :—he may make any use of my name.

Adieu, my dear friend! I shall think it long till I see your handwriting. Remember me most kindly to Mrs. Roscoe and all your family.

Believe me ever most truly yours,

J. E. SMITH.

Mr. Roscoe to Sir J. E. Smith.

Primo giorno dell' anno 1804.

Buon capo d' anno, Signore mio stimatissimo, ed a voi e alla vostra carissima Signora Sposa!

How can I begin the present year better than by addressing myself to you, from whose society and friendship I have derived so much pleasure during the past! The day is fine with us; a great part of my young family are visiting their aunt in Liverpool. I have thrown aside my ponderous volumes about the ambition of princes and the intrigues of popes, to take a momentary retrospect of the past year, in the most pleasant part of which I find you occupy the principal station: you will therefore regard this letter as *un pegno d' amicizia,* to be annually renewed. To promise for our whole lives at once would be improvident; as it would deprive us of that repetition which the ordinary course of Providence may yet allow, and which I trust we shall many times have occasion to employ.

A most violent effort to free myself from the heavy task in which I am engaged, and the continual pressure of business, with my journeys between Allerton and Liverpool, have so devoured every moment of my time, that day by day has passed on till the conclusion of the year, without my being able to fulfill my wishes. I am now, however, determined to be somewhat more my own master. Since you left Liverpool, I have copied and prepared for the press as much as will compose my two first volumes. The remainder is in great forwardness, and if I enjoy my health for a few months, will I hope be completed. M'Creery begins to print with the new year, and promises to proceed with great rapidity. My arrangements are all made to my satisfaction, and some time in the ensuing year I expect to make my appearance before the public in the pompous shape of four ample quartos. The labour of correcting I consider as nothing in comparison with that which I have had in the composition of this work; and hence, though much remains to be done, I find my mind lighter than it has been in the survey of the long and tedious road that lay before me. You, who have so often engaged in important literary undertakings, will know how to sympathize with a brother author in the enthusiasm of his pursuit, the apprehensions of disappointment, the lassitude of fatigue, and the cheering prospect of success, and will easily perceive that as the barometer rises or falls through these degrees, it constitutes the foul or fair weather of human life.

If you will take the trouble to desire any friend of

yours in London to call on Messrs. Cadell and Davies in your name, they will have directions to deliver him a copy of a poem called "The Press," written by my printer here, and published as a specimen of typography, which for the excellence of its mechanical execution, and the beauty of the engravings on wood, will certainly gratify the eye, whatever may be its effect on the understanding, about which I really believe the author himself is less anxious than about the sharpness of his letter and the blackness of his ink.

It would be heretical to conclude without a single word about botanical pursuits. Two of our new houses are completed and filled with plants, and the other five will now soon be finished. Shepherd says we may challenge all the kingdom in point both of elegance and convenience. He desired that when I wrote I would inform you that the Willows are safely arrived, for which, in the name of my brethren and myself, I beg to return you our best acknowledgements.

We join in every kind wish to Mrs. Smith and yourself.

Your very faithful Friend,

W. Roscoe.

I am highly delighted with your song: the two lines,

"Yon speck in night's retiring veil,
None but a lover's eyes could spy,"

are not excelled in poetry, feeling, and picturesque truth, by any passage I know. If this be your first attempt, you ought to exclaim, in the words of a wicked Italian, in allusion to a very different subject,—

> " Piango perchè di cio tardi un'accorsi
> A cui dovea piu di buon ora attendere;"

and I hope you will endeavour to make up for your lost time by producing other pieces of equal merit.*

* The editor is here tempted to give the entire song which was thus mentioned by a partial friend.

SONG.

The morning o'er the ocean breaks,
The orient waves are liquid fire,
While William from the topmast seeks
All that his fondest hopes desire.

Yon speck in night's retiring veil
None but a lover's eyes could spy—
" 'Tis land! 'tis England! messmates, hail!"
" 'Tis land!" the joyful crew reply.

" Oh Mary! is thy tender breast
Still to thy William fond and true,
As when we first our love confest?
As when with tears we bade adieu?"

Propitious gales and swelling waves
To William faint and tardy seem—
But now the lab'ring ocean heaves,
And thunders roll and lightnings gleam.

And now the shatter'd bark no more
Resists the sweeping whirlwind's sway;
It shivers on its native shore,
And death and horror close the day.

But love survives the wasting storm;
O'er William harmless thunders roll;
His Mary clasps his clay-cold form,
Her breath recalls his fleeting soul.

The following translation after Martial, from the same hand, and in a different style, may possibly not displease the reader.

Go mingle Arabia's gums
  With the spices all India yields ;
Go crop each young flower as it blooms;
  Go ransack the gardens and fields.

Let Pæstum, so fragrant and gay,
  Its roses profusely bestow;
Go catch the light breezes that play
  Where the wild thyme and marjoram grow.

Let every pale night-scented flower,
  Sad emblem of passion forlorn,
Resign its appropriate hour,
  To enhance the rich breath of the morn.

All that art or that nature can find
  Not half so delightful would prove,
Nor their sweets all together combined,
  Half so sweet as the breath of my love.

Sir J. E. Smith to Mr. Roscoe.

My dear Sir, Norwich, January 16, 1804.

I congratulate you sincerely on beginning to print your great work ; but I must not yet begin to think of reading it,—that will serve to make me anticipate *next* winter with pleasure. I hope very soon to bring out the third volume of my Flora ; it is small compared to your work, but the labour of the Mosses was very great. I can scarcely hope that you will find a moment to look at that part ; I must be content with the quiet approbation of half a score hard-working Germans, and they will I am sure see much to correct, even if they approve my *congenial* plodding. As to systematic niceties, I scarcely know any one that is competent, who will take the pains to follow me. As to *Columna*, the *Phytobasanos* is a mere curiosity, it being the first botanical book ever printed with copper-plates; but there is a modern edition of it nearly as useful. It is excessively rare. The *Ecphrasis* is inestimable for use, and daily in my hands. It has never been republished.

I am, my dear Sir, yours, &c.

J. E. SMITH.

From the same.

My dear Friend, Norwich, November 5, 1804.

What a long time it is since I wrote to you! Your letter is dated July 25th. Since that time we have spent a month at Lowestofft with my wife's relations, and returned to enjoy the company of our amiable friends and relations the Kindersleys. My mother and all of us are looking forward to next summer. May nothing happen to disappoint us! I am preparing an entirely new course of lectures for London and you. My Exotic Botany is to appear Dec. 1st. I shall have the pleasure of dedicating it to you, if you will allow me by that means to commemorate our friendship: though short, its growth has been vigorous; and if there be some vanity in the wish that my name may be associated with yours in some corner of fame's records, I hope it is a pardonable vanity. Those more private ties which have bound us together will at the last day atone for it. How does your Leo go on? Mr. Johnes writes to me that his second volume of Froissart is already in London, the third will be done by Christmas, and the fourth in June: it is very entertaining. Pray do you take the Annals of Botany, by Sims and Koenig? I think it a useful and amusing publication, and mean to send a paper to it, for the next number, (if it can be admitted,) on the decandrous papilionaceous plants of New Holland, which want clearing up, and I flatter myself I have found a clue to them.

Believe me, my dear Sir,

Ever yours most affectionately,

J. E. SMITH.

Mr. Roscoe to Sir J. E. Smith.

My dear Sir, Liverpool, November 22, 1804.

A letter from a friend after a long silence affects me with a feeling somewhat like returning home after a long absence. Yours came, too, at such a time that, if anything could have increased its value, certainly did so; as I received it from the hands of Mrs. Martin on Thursday last, just after enjoying a cheerful dinner with her and a family party of friends. To have buried it coolly in my pocket for two or three hours would have been a proof of more stoicism than I possess, and I therefore quitted the tea-party to converse with you for a few minutes *tête-à-tête*, which I concluded by resolving that I would instantly thank you for the additional instances of your remembrance and friendship which it contains.

Such, however, is the rapid flight of time with me at present, that some days have elapsed before I could find an opportunity of performing my rash vow. I shall now, however, hold him by the forelock till my engagement is performed.

I am sure I need not tell you that all our Liverpool friends who have seen Mrs. Martin are charmed with her. How, indeed, should it be otherwise? My wife is delighted beyond measure, and I really believe will contrive to get her good opinion in return. Surely no chemical affinity is stronger than that of similar hearts; distance itself forms no bar to it, as I feel at this moment.

The honour which you so kindly propose to confer on me on the publication of your Exotic Botany ought to give me unqualified pleasure;—to see our names united must always do so. Yet I cannot help feeling as if the world were turned upside-down, and am sure that my name, instead of occupying the top of the page, should be placed at the bottom. I cannot, however, find in my heart to say that if you can put a face upon it, I shall not be highly gratified, and shall endeavour to hold up my head as well as I can during the remainder of my days.

Mr. Shepherd sent Lady Hume a few plants about two months ago. I have since heard from her, with a specimen of the *Humea*, which I think well deserves the name of *elegans*, in reference both to the plant and its patroness. You will, I am sure, rejoice with me, when I tell you that Leo is now in the latter part of the fourth volume, and will yet, I hope, be nearly finished by the time I proposed,—the end of the year; such an exertion has seldom been made by a printer, as you will say when you see these ponderous tomes.

You will begin to think that when I once told you that I was a bad correspondent, I was acting the part of a coquette, who pretends to be shy of her favours only that they may be more highly valued; but however bountiful you may think I have been of late, I must beg you to present my love to Mrs. Smith, and tell her in return I transmit my wife's love to you. And now, my dear friend, let me seriously tell you that I love you both better than I once thought it would ever have been my lot to have loved again; and that the friendship which I have been so happy as to form with you (for I will not call it acquaintance,) will always be numbered among the most fortunate events in the life of

Your ever affectionate and faithful Servant,

W. Roscoe.

From the same.

My dear Sir, Allerton, May 26, 1805.

I have long intended to write to you, but have been prevented by a continual succession of unavoidable occupation and bodily indisposition, and sometimes by the junction of both.

Leo's reckoning is now made, and he must be sent to his account with all his imperfections. In the course of a few days after this comes to hand, you will receive a copy, which from its size would terrify a man of much less occupation than yourself, and which you will naturally lay aside till you can muster courage and find time to make so formidable an attack. Of the reception of this work, I am in many respects doubtful; but I do not suffer my apprehensions to render me miserable. I have taken all the pains in my power to make it deserving of the public notice, and have endeavoured to express the peculiar opinions which it may contain with decency, though with freedom. If all this will not do, I cannot help it; nor would I alter or suppress those opinions to obviate censure or obtain applause. In one place or another I have found an opportunity of expressing my sentiments on the great subjects of politics, morals, religion, and taste, as well as on a variety of inferior topics, which I hope are not impertinently introduced; and by these sentiments I am content to be judged of as long as my book may continue to be read.

Be assured, my dear friend, that I look forward to our meeting again at Allerton with a satisfaction that I should find it difficult to express. Let me hear from you soon, and believe me ever affectionately and truly yours,

W. Roscoe.

Sir J. E. Smith to Mr. Roscoe.

My dear Friend, Norwich, October 12, 1805.

I send you a budget about the ***Plantæ Scitamineæ***, which will, I hope, be of some use and amusement to you. May you feel as much pleasure in studying it for *my* sake, as I have had in preparing it for *yours!*

We had a good journey; and at Cambridge saw the famous *Ceres.* Ely and its fine cathedral detained us two or three hours. The tower is noble, and the west front as sharp and perfect as if just now finished. The chapels are beyond any thing for rich carving.

Inclosed are two pods of *Vanilla*, picked up in some shop when I was in Italy;—you will find their perfume still very powerful.

We saw Raphael's picture at Okeover; 'tis an undoubted original;—the countenances fine, colouring dark (perhaps changed), some parts incorrect in drawing. What I most wondered at is the exquisite effect of the bed-clothes and pillow of the cradle, equal to any Dutch picture, yet not laboured in the same way. It is one of those Holy Families of which Raphael at one time painted many, before he arrived at his last historical perfection.

My Exotic Botany has been much more fortunate in Baldwin's Literary Journal than your Leo; you know the fable of the Oak and the Bramble.

Pray remember us most kindly to Mrs. Roscoe and all your beloved children. I flatter myself I shall always have an interest in their hearts.

I know not how to finish my letter, for it seems like taking leave again. Let us lessen the distance between us as much as possible by correspondence.

Colonel Hardwicke comes to us on Tuesday for three or four days.

I am always, my dear Sir, with the truest esteem,

Yours, J. E. SMITH.

Dr. Bostock to Sir J. E. Smith.

Dear Sir, Liverpool, January 26, 1806.

I thank you for your letter containing the specimens of native Camphor. At present the nature of the substance appears to me quite incomprehensible. I have not yet performed all the experiments upon it that I propose to do. You will be happy to hear that our worthy friend Shepherd is well, and continues his exertions, with his usual success, in the garden. Even since you left us we have got some considerable additions to our collection. It would, I am confident, give you great concern to observe the unprincipled attack which is made upon Leo X. in the Critical Review,—not that it will affect the equanimity, or perhaps the reputation of the author; but it is painful to observe that such a spirit of illiberality can be sanctioned, by being admitted into the pages of one of our most popular periodical works. Surely the writer must be a stranger to our friend; it would be impossible for any one who knew him to treat him so ungenerously.

Believe me your faithful Friend,

J. Bostock.

Sir J. E. Smith to Mr. Roscoe.

My dear Sir, Norwich, April 2, 1806.

I hope you have not been uneasy for the fate of your Essay on Monandrian Plants. I now sit down to thank you for your letter of the 1st of March, and to tell you what I have done to your paper. I should not be worthy of your friendship if I did not execute my trust faithfully. First, then, you may be assured, as far as I have any judgement, (and I have been privy to many a consultation with Dryander about these plants,) that your paper, both in its aim and execution, will do you great honour as a botanist, and render a material piece of service to the science. No one before you has given any plausible account of these genera, as you well know. Your treatise is a plain, clear, unaffected discourse. Perhaps I was, at first reading it, disappointed that *you* had not made something more of it in the way of composition, with respect to ornament or episode, because I know that you are one of the few, who, with real science, could have done so. But this might have been "soaring above the path of true simplicity," and the more I read your paper, the less I feel any want of such adventitious merits;—I only mention all that has occurred to my mind. The leading qualities of these plants, their beauty, their fragrance, their affinity or resemblance to Palms or *Orchideæ*, whether real or supposed, the hot and moist climates which they prefer,—these things might have been alluded to. I should not, however, regret that scarcely any other pen than yours had passed them over.

You and I may congratulate one another on our good-humoured critics. I have composed a new introductory lecture for the Royal Institution. It will refute Salisbury, and yet not honour him with apparent notice. I think your critic an angel (or at least only a *tawny* devil,) compared with mine; yet I have been so forgiving as to make a motto for his *Paradisus.*

What malice lurks beneath this fair disguise!
The devil again in paradise tells lies!
But now, how plausible soe'er his tale is,
We always take his words *cum grano salis.*

So, my dear friend, as we are both got to the devil, we can be no worse, and may as well enjoy all that this world still affords us; which that we may long do is the prayer of your affectionate Friend,

J. E. Smith.

From the same.

My dear Friend, Blackheath, June 2, 1806.

Your paper was highly approved by everybody. Dryander (no complimenter,) was excited by it to take up the subject afresh; and he and I went carefully over your whole paper, turning to references, specimens, &c. *I* have *already* received the highest compliment possible from Aiton and Lambert, who, on hearing your paper, conceived that you could scarcely have had knowledge or materials to compose it, and that *I* must have written it!—See what it is to have a name! I certainly *could not* have written it. I have satisfied them with the real truth, and Dryander also. With this view I *purposely* let him read over your *original* MS., and see all my scratches and alterations, which being so few, and chiefly about technical matters, could do you no discredit. I wish you, therefore, carefully to preserve this original manuscript as it is; it had best be printed from this present copy, which must then be preserved in the archives of our Linnæan Society.

J. E. SMITH.

Mr. Roscoe to Sir J. E. Smith.

My dear Friend, Allerton, October 5, 1806.

By an omission of the bookseller's it was not till yesterday that I got a sight of the Exotic Botany for September, and had the pleasure of seeing the fine plant figured, to which you have done me the great and unmerited honour of annexing my name.

I have accurately examined your delineation and description, and certainly think it a good genus. It is most nearly allied to *Hedychium ;* and if, according to the system which I have attempted to lay down, I cannot admit that the irregularity of the exterior limb of the corolla affords a genuine distinction, yet the very singularly curved anthera, with its two appendages, is amply sufficient to establish it as a genus; which the irregularity of the corolla may be allowed to confirm. To *Kæmpferia* its affinity is merely in its exterior appearance; in its botanical characters it is wholly distinct. I therefore flatter myself that this nymph of the Asiatic mountains will, like a faithful spouse, retain the name which you have imposed upon her; and not, like too many of her sisterhood, elope to some more favoured admirer.

W. ROSCOE.

Sir J. E. Smith to Mr. Roscoe.

My dear Friend, Norwich, Oct. 16, 1806.

*   *   *   *   *   *   *   *

I am glad you assent to the character of *Roscoea;* but please to observe the irregularity of the corolla is no less *essential* than the form of the filament, &c. We must never confine our *essential characters* to one part of the fructification, though one part may give the leading mark. Corrêa has an idea that in every natural order there are one or more genera with an irregular flower, and one or more with a regular one. Since I heard this I have kept it in view, and have seen much to confirm it; yet I cannot think of a regular flower among the *Leguminosæ,* nor of an irregular one in the *Caryophylleæ;* but some may occur.

Saxifraga sarmentosa is the irregular flower in its natural order, and surely a distinct genus from *Saxifraga.* So *Celsia* differs from *Verbascum,* I think essentially. *Iberis* is the irregular flower in its order, where one would least expect to find any.

I have got a copy of Sir Joseph Banks's sketch, made in his voyage, of the *Hura Siamensium,* which, with the description in Retzius, are all we have to know that plant by. It is a *Globba,* with two lobes of the corolla hoisted up half way of the filament!

So parliament is dissolved! I hope we shall have no opposition for Norwich.

J. E. Smith.

Mr. Roscoe to Sir J. E. Smith.

My dear Friend, Nov. 8, 1806.

As it will cost you nothing, I write you a line to say that I have this day been returned as one of the members for Liverpool, by a majority over both the other candidates. A fortnight ago I should as soon have thought of being the Great Mogul.

Present my kind remembrances to Mrs. Smith, who will rejoice with me on this occasion; and believe me, my dear Friend, at all times most truly yours,

W. Roscoe.

Sir J. E. Smith to Mr. Roscoe.

My dear Friend, Norwich, Nov. 11, 1806.

You are beforehand with me; for it was my intention to have written by this post to congratulate you on the event of which the newspaper today gave us the first certain intelligence; but the hopes and fears belonging to it have occupied our minds ever since the first mention of your nomination. After a long and very pleasant country walk this fine day, I found your letter on my table. How kind and flattering was it of you to write to me the very day, when you must have been so busy, so agitated, and fatigued! Your letter, and especially your *first frank*, will be kept by me as relics. My brother and sister Martin have been very good in writing frequently concerning the state of the poll, and all the circumstances of this "almost painfully interesting event," as she justly termed it while it was in suspense; and this day brought my wife a most delightful letter from her*. Yet I assure you all this intelligence was by no means equal to our anxiety. What with telling and inquiring, and seeking every day and almost every hour for the most probable, authentic newspapers, where we might find the poll or any mention of you, my house has been turned topsy-turvy. Now we have enough to do to give and receive congratulations. No other event than this could have consoled us in any degree for the loss of Mr. Wm. Smith. He, good man! bears it with the equanimity and dignity worthy of himself.

But, my dear Friend, your election seems like a dream. I say frequently to myself, "Can this be?" and during the contest I have endeavoured to think

* Inclosing the following lines, written by Mrs. Martin,

On the Election of Mr. Roscoe for Liverpool.

Unstain'd is the glory of him, who, elected
By the voice of the people, with honour is crown'd;
The oak shall wave o'er him, the rose shall bedeck him,
And the loudest applauses be echoed around.

From Arno's sweet vale with the Muses retiring,
Where Painting and Poetry hail'd him their son;
Now greater his glory, midst senates haranguing,
To plead for humanity's cause as his own.

Give the shout, O his townsmen! o'er France let it echo;
Through the valleys of Tuscany loud let it roll;
But soft let it float o'er the far-spreading waters,
And whisper sweet peace to the *African's* soul.

I should be satisfied if you had but a respectable poll, and merely by the influence of what I scarcely thought this bad world could feel and value in any just degree:—this is beyond all hope! Yet all my joy, and even my affectionate feelings towards you personally, are absorbed in one consideration, too awful and transcendently important to be mixed with any other,—I mean the effect of this blessed event upon the fate of thousands, nay millions of happy Africans! Generations will succeed one another in peace and tranquillity, who will never know your name, indeed, but it will be treasured in

"The bosom of their Father and their God."

I feel that nothing could give so decisive a blow to the slave-trade,—such a public avowal that even Liverpool can do better without it, as the choice of you, the open and powerful enemy of that trade, to be the guardian of the interests of the town which is supposed to live by it! If ten righteous persons could have saved a town from destruction, surely 1157 will wipe away all reproach; and the rest of the nation will be ashamed to countenance this national sin any longer.

How deeply do we enter into the sensations of all your family on this occasion! I trust the sacrifices which must in some degree be made of domestic tranquillity, will be compensated by the unlimited increase of genuine honour and respect with which you will be surrounded, and the extended value of *even your* existence. I look far beyond the present moment, and I see one or more of your sons trained up to be your colleagues and successors. If Shenstone could see in a village school,

> " A little bench of heedless bishops here,—
> And there a chancellor in embryo,"

so can I see in my friend Mr. Henry, chaired (as doubtless he now is,) round the gardens at Allerton, the embryo of what his father is now become.

I know not whether your chairing is like ours at Norwich,—but when the poles of the chair are tossed, as they often are, out of the bearers' hands, and the candidate, with his limbs sprawling and coat flying, gives somewhat of the idea of a frog or a kitten thrown from one unlucky boy to another, I have been reminded of what Sir Thomas Brown says on another occasion; " There is nothing that will more deject a man's cooled imagination, when he shall consider what an odd and unworthy piece of folly he hath committed." But it seems your triumphant progress was majestic and graceful in the highest degree,—at least so I hear from an eye-witness.

Among all my selfish joys on this occasion, one, not the least, is the hope of meeting you, and perhaps Mrs. Roscoe, &c. in London in the spring. Forgive me for intruding so long on your time, but really I want to say much more. This free postage will cost you many a minute; but your friends must learn discretion by degrees.

All who know you here join in most heartfelt congratulations. Present them for us to Mrs. Roscoe and all your family; and believe me ever, my dear Sir,

Your affectionate and devoted Friend,

J. E. Smith.

Mr. Roscoe to Sir J. E. Smith.

My dear Friend, Allerton, June 25, 1807.

I have for some time past rejoiced in the thought that I am likely to see you in Lancashire in the course of the present summer. I already anticipate the happiness I shall have in your society at Allerton, where I must at least claim some portion of your time, and where I shall be delighted to stroll and saunter with you through the fields in an evening, instead of being locked up balloting for committees in St. Stephen's. In truth, my dear friend, it requires but little of the efforts of others to drive me from public life. The only wonder is, that I was ever brought into it; and I sink back with such a rapidity of gravitation into my natural inclination for quiet and retirement, that I totally despair of ever being roused again to a similar exertion. Add to this, that the one great object which was continually before my eyes is now attained, and I shall have the perpetual gratification of thinking that I gave my vote in the assembly of the nation for abolishing *the slave-trade to Africa*. Though not insensible to the state of the country, yet I see no question of equal magnitude; and am fully aware how little my efforts could avail in the political struggles of the times. Come then, my friend, and let us again open the book of Nature, and wander through the fields of Science. Your presence will increase my reviving relish for botanical pursuits; and when we are tired with those subjects, we will call in the aid of the poets and philosophers to vary our entertainment.

Pray have you seen Mr. Wordsworth's new poems? Whatever things are whimsical, whatever things are childish, whatever things are odd,—these, and many other things, are to be found in these volumes: but after all I like them, and listen to him with a pleasure something like that of an infant to the prattle of an old nurse. They are to be read in listlessness and leisure, like that in which they seem to have been written; and if you can bring your mind to the proper tone, depend upon it you will not find an hour or two misemployed in their perusal.

Let me have the pleasure of hearing from you as soon as possible, and believe me ever most truly and affectionately

Yours,

W. Roscoe.

Sir J. E. Smith to Mr. Roscoe.

My dear Friend, Norwich, July 4, 1807.

After all the agitation and anxiety of mind which I have felt for some weeks past on your account, how delightful is it to find, by your most welcome and interesting letter of the 25th of June, that you still possess yourself in undisturbed tranquillity; not like a reed that has bent before the storm, but like a palm-tree, around whose polished and upright stem the winds have whistled, without ruffling the lofty honours of its head! Such a plant can no more be nursed in St. Stephen's chapel, than the Norfolk Island pine, 250 feet high, in any of our stoves. You are now in your proper element, and very long may you continue so! The world is not worthy of you, "nor the world's law." The line of conduct you have pursued secures you from regret, and I trust you will soon look back on all that is past with no less satisfaction, on every account, than self-approbation. I wished it rather for your triumph than your happiness; and really triumphs of any kind are worth but little:—"One self-approving hour," &c.—you know the rest, and *that* a good man has, independent of triumphs founded on the accidental justice of the world.

Believe me ever, my dear Sir,

Most affectionately,

Your faithful Friend and Servant,

J. E. Smith.

Sir J. E. Smith to Mr. Roscoe.

My dear Friend, Norwich, Nov. 2, 1807.

What a long while I have been without acknowledging your kind and very interesting letter of invitation! I want to talk to you about many things. *Imprimis,* I want to know your *Cannæ;* for I wish to write a paper on the *species of Scitamineæ,* following you as your humble squire, "haud passibus æquis;" and it is highly desirable I should not be distanced out of sight of my leader at the very first steps of my course. If you cannot help me to specimens, figures, or synonyms of your species of *Cannæ* made out of *C. indica,* I must leave that section, and write only on *true Scitamineæ,* which I should not much regret, as I have some difficult new plants of the *Canna* section.

I this day sent a botanico-physiological paper to the Linnæan Society, on the germination of seeds, disproving the *vitellus* of Gærtner, which seems to me only a subterraneous *cotyledon,* and correcting many things respecting monocotyledonous plants, *so called,* as the quakers say. I like it better than many things I have done; but whether the half-dozen people who can judge of it will like it as well, may be doubtful. Do you agree with my rule about *shall* and *will* in the Athenæum for October*?

I hope very soon to send you my Introduction to Botany, on which I shall most earnestly request

* In reply to the inquiry in this letter concerning the use of *shall* and *will,* Mr. Roscoe tells Sir James that he has read with great pleasure his grammatical paper, and is much inclined to agree with him. "Of this," he says, "I am sure that it has thrown more light on the subject than anything I have before met with; and till a greater prophet arises, you may reckon on me as one of your disciples."

your friendly opinion, corrections, and advice. Many parts of it are but sketches, and many perhaps wrong; but I hope it will promote the study. Some parts are new and original. I am about the *Prodromus Floræ Græcæ*. You would hardly think that I can scarcely do more than ten or twelve species in that work in a whole morning from ten o'clock till three.

I heard a curious anecdote today from indubitable authority. At the conference at Tilsit, the King of Prussia, who, it seems, is a beau, had a pair of long pantaloons. Buonaparte said, "If your majesty puts on those yourself, it must be very troublesome. Pray do you begin buttoning them at the top or bottom?" He told the Queen, "he was too old to be influenced by her *beaux yeux*;" which, it seems, the Emperor of Russia does not find so innocent. Adieu, my dear Sir.

Believe me ever your affectionate

J. E. SMITH.

Sir J. E. Smith to Mr. Roscoe.

Hall Place, near Maidenhead, June 12, 1810.

My dear Friend,

Tired with the bustle of town,—suffocated and sleepless with the smoke and east wind,—I am come here for a few days to breathe and sleep, and to enjoy the society of some valuable old friends in their antique spacious mansion, amid vast avenues of limes, beech woods, abounding with rare *Orchideæ,* and a most beautiful surrounding country. How can I bestow some of my leisure better than in chatting with you, in reply to your kind letter of the 21st of April? I wish you could know Sir William * and Lady East, with whom I now am. I think we have mentioned them to you before; at least we have made *them* well acquainted with *you* by report. I am going with Lady East in search of *Monotropa Hypopitys,* which I never yet saw growing, but which I hear grows in the woods at Bisham Abbey hard by, where the unfortunate Plantagenets lie buried. Among them are the famous Earl of Warwick, the king-maker, and the last Earl, who died young in the Tower. There are no monuments of them. The Abbey is now a house, inhabited by Mr. Vansittart, whose charming daughter married Mr. Augustus East, a son of my friend's. They are now here, and we all botanize together in the fields or gardens.

I rejoice to hear of your intended botanical paper, which I presume relates to natural systems, and to Jussieu's especially. You threw out some excellent hints once, which I have often regretted I did not put down at the time. I am very anxious you should pursue the subject: you will no doubt treat Jussieu with the candour and respect he so highly deserves. I had a letter from him lately, and answered it. He is a very worthy, amiable character, and I would not hurt him, though he in conversa-

* A letter from this estimable friend is inserted page 108.

tion calls the Linnæan system *lèse nature.* I hope you agree with me as to the name of Linnæus,—see two or three last Monthly Magazines*. Jussieu in his French letter writes *Linnæus ;* so do all the French now.

Believe me, with best respects to Mrs. Roscoe and all your family, very affectionately yours,

J. E. Smith.

Mr. Roscoe to Sir J. E. Smith.

My dear Friend, Allerton, Jan. 2, 1811.

I have been so long without hearing from you, that I have at length brought down my obstinate hand to the paper, from an irresistible desire of knowing how you are, what you are doing, and when I may hope for the pleasure of seeing you once more in this part of England. For my own part, I have been kept in the house nearly three weeks by indisposition, from which I am now pretty well recovered. This twilight of convalescence is an excellent season for turning one's mind to favourite studies and amusements ; and when these happen to be intermixed with sentiments of affection and friendship, they afford a better *restorative* than most that the pharmacopœia can supply. Since I last wrote to you (I believe) we have made a purchase of the late Col. Velley's most beautiful collection of plants, including not only his marine specimens, but many others, all of which are pre-

* For the letter here mentioned see Appendix.

served in the most exquisite and perfect manner. I mention this, that it may be an additional inducement to you to hasten your visit, though I fear that if the living claims we have upon you be insufficient, we shall have little chance from any inanimate attractions.

Be so good as to tell me whether you have yet fired off my Congreve rocket against the French botanists. I hope not, as I think I could put some more combustibles into it. In particular, I think the distinction between a natural and an artificial system might be more fully explained. In other respects I have no objection to give them a broadside, and wish you could prevail upon both nations to confine their animosities within such harmless limits; but the business of cutting throats must go on, and seems even to be considered not as an accidental, but as a permanent state of society.

I presume you have seen the fine numbers of the *Flore Portugaise* of Count Hoffmansegg, in which I find honourable mention of you, and your works generally quoted. The Count's system affords another instance of the necessity of adhering to some established plan of arrangement, otherwise it will soon be impossible for those who cultivate the science to understand each other. His system is founded on that of Jussieu, but with such alterations as render it even worse for common use, or rather wholly impracticable, the primary distinctions being founded on the *tissu cellulaire* of the plant. Such of them as are formed upon the plan he approves of, he calls perfect plants; and such as have their cellular membranes *different or indistinct,* he calls imperfect plants; and his perfect plants are again subdivided by their cotyledons, but somewhat different from Jussieu, as he calls the dicotyledonous plants of Jussieu *plantes cotyledonées*; and the monocotyledonous, *plantes acotyledonées.* This is altogether so new, and so microscopic, and so difficult, that I doubt not but it will be very generally adopted, and that instead of quoting Linn. or Juss. we shall shortly cite Hoffm., unless some great reformer should start up and put him out of fashion.

I have not been idle of late, having been employed in devising a mode of putting an *effectual end* to the African slave trade, as at present

"We have scotch'd the snake, not kill'd it."

I am also looking into the state of *the arts* during the middle ages, for a memoir of which I have good materials.

W. Roscoe.

Sir J. E. Smith to Mr. Roscoe.

My dear Friend, Norwich, Sept. 7, 1812.

Be not alarmed at seeing a letter from me so soon after my last. We have been spending ten days at Holkham, and I write now at the earnest desire of Mr. Coke, to try to persuade you to come and see him and us. He says you have given him some hopes, but have as yet only disappointed him. Now I can conceive nothing more delightful than spending a fortnight with you under his roof, and have promised him to do so whenever you come. To contemplate his pictures and statues, to rummage among his books, drawings, manuscripts, and prints, (where we every day find treasures unknown before,) is extremely agreeable, and he kindly entrusts all his keys to me in full confidence. I found a case of the *earliest* printed books, which no one had examined since the time of his great-uncle Lord Leicester. Such MSS. of Dante, drawings of the old Italian masters, treasures of European history, —you have no idea! The house is one of the finest in Europe, and its riches are inexhaustible. But of all things its owner is the best worth your seeing and knowing. He is so amiable, with all the *first gloss* of human affection and feeling upon his heart; so devoid of all selfishness that, with the early and constant prosperity he has experienced, his character is next to a miracle; and he has such an agreeable liveliness and playfulness of manners, that nobody is more entertaining. You would exactly suit in all your ideas of men and things. Do give me some hopes that you will come over this autumn with Mrs. Roscoe or some of your family. We will meet at Holkham; and if you can descend (without breaking your neck) to our "low estate," we will strive to rival even Holkham in the heartiness of our welcome. I shall show you the Linnæan reliques, and we shall consult you about a new botanic garden now projecting. Do, my dear friend, think of all this;—but do not let it be in frost and snow. I was laid up there last December by a cold caught in going.

Ever most affectionately yours,

J. E. SMITH.

Mr. Roscoe to Sir J. E. Smith.

My dear Friend, Allerton, Sept. 13, 1812.

I ordered a Liverpool Mercury to be sent you yesterday, by which you will see we have been giving a splendid dinner to Mr. Brougham, at which several of your friends took an active part. The lords present all spoke with great spirit, and the reading of the letters from the principal members of both houses went off extremely well. On the whole this meeting will do some good, and tend to draw closer that connexion between the commercial and manufacturing interests and the nobility and great proprietors of land, which is become essential not only to the prosperity, but the safety of the country.

And now, my dear friend, for your last letter.

" So cunning was the apparatus,
The powerful pothooks did so move him,
That, will he, nill he, to the *great house*
He went as if the devil drove him."

It would not, however, be so much for the sake of the great house, nor for all it contains, though nothing in its way could be more attractive, that I should wish to visit Holkham. It would be with the view of paying my respects to its excellent and distinguished owner, and of meeting you under his roof,—temptations which I feel I shall hardly be able to resist. At present, however, I cannot speak very decisively. Neither my wife nor myself are at present very well; and when I shall be able to venture abroad so far before winter I really dare not say. Mrs. R. and I ought to have gone to Hafod this year; but the same causes rendered it impossible. In the mean time be assured, my dear friend, that nothing upon earth could afford me a greater treat, and I shall consider it as a misfortune if I should not be able to meet you.

Your ever affectionate Friend,

W. Roscoe.

Sir J. E. Smith to Mr. Roscoe.

Norwich, Oct. 26, 1812.

Ah! my dear friend, the vision which I ventured to contemplate was too bright to be realized! Yet your letter almost tempts me to hope,

"Though hope were lost."

As to this disappointment, *you* may well bear it, because you know not what it is. 'Tis not the enjoyments of taste, literature, luxury, and novelty that I lament,—it is the losing the pleasure of seeing you beloved by so excellent a creature, so congenial to you. You would each find *a mine* of happiness of which you know scarcely anything. Mr. Coke is out of the sight and conception of the world in general; and I can give you no higher proof of my esteem and affection for yourself than by speaking of him to you without reserve.—How well did he express his own character the other day when he said he "dreaded nothing so much as self. Self is the worst tyrant in the world." Mr. Coke is, I think, one of the most *gracefully kind* and benevolent men, to all in their proper places, that I ever saw.

As to politics, mankind, farming, pictures, you and he would never fail of conversation; but these are smaller matters.

My friend Mr. Fountaine of Narford, (another fine house,) has numerous Italian manuscripts, copied at Florence for his ancestor Sir Andrew Fountaine, by permission of Cosmo III. These he offers to your entire "use and behoof." We are also to look over his Raphael ware, the finest except at Loretto. See my Tour,—*Loretto.*

You see *hope* will revive in my breast;—do not blast it again. Our Bishop you must come and admire: I fear he goes to town before the spring: he is here *now*, alas! You will smile at this letter, and some would think me stark mad;—if so, God grant I may never be cured! Let me hear from you. I have fifty things more to say; but for the present, Adieu.

Yours ever and ever,

J. E. Smith.

Mr. Roscoe to Sir J. E. Smith.

Allerton, February 27, 1813.

My ever dear Friend,

* * * * * * *

I have lately done something towards a catalogue of my pictures, drawings, prints, &c., which are become much more numerous since you saw them, and which I promise myself great pleasure in submitting at no distant period to your examination and criticism. Amongst the pictures is the portrait of Leo X. with the Cardinals de Rossi and Giulio di Medici, which I have been assured by many persons is the celebrated copy of Andrea del Sarto which was sent as the original of Raffaelle to the Duke of Mantua. From Mantua it went to Parma, and was thence transferred with the rest of the collection to Capo di Monte near Naples *. You may probably have seen it there in the year 1787, although you do not mention it in your Travels; and when you see it again, will perhaps be able to clear up my conjectures. That the picture was carried off from Capo di Monte several years since is certain; and as I can hear no account of it from any quarter, and have been told by several persons that mine appears to be in every respect a perfect resemblance of the picture of Raffaelle now in the Louvre, I flatter myself with the probability of its being the copy of Andrea, at the sight of which Giulio Romano was

* This picture was purchased by Mr. Coke, and is now at Holkham.

deceived; and on being assured it was a copy, declared that he did not value it less than the work of Raffaelle himself.

Throughout the whole of my troublesome complaint I have had many sleepless hours by night, in some of which I strung verses together, which I wrote down in the morning,—of these I send a specimen, which I beg you to present to Mrs. Smith, with my kind remembrances:—should she approve of them, she will perhaps do me the favour of sending a copy to Miss Coke. I long to hear from you again, and am at all times, my dear Friend, most affectionately yours,

W. ROSCOE.

SONG

On the Ball given by the Friends of Mr. Brougham and Mr. Creevey. Liverpool, November 1812.

The fair face of morning when sudden clouds cover,
And tempests and darkness envelope the day,
Shall the gloom of the moment deter the true lover
Who hastes to the home of his mistress away?

When heaved from its base proudly swells the vext ocean,
And danger rides high on the crest of the wave,
Undaunted the mariner views the commotion,
And bares his bold bosom the sea-storm to brave.

Then say, shall the Patriot e'er prove a recoiler?
Shall the champion of freedom e'er stoop to despair?
Shall he basely resign to the hands of the spoiler
The prize that high Heaven has consign'd to his care?

No! still to his task with fresh vigour returning,
He shall wage the bold war with corruption again,
As the lion, that, roused by the beam of the morning,
Shakes off the slight dew-drops that hang on his mane.

If he falls—like the warrior he falls on his duty,
Whilst his country shall hail him and angels approve;
If he conquers—he wins from the bright hand of beauty
The wreath wove by Liberty, Friendship, and Love.

Sir J. E. Smith to Mr. Roscoe.

My dear Friend, Norwich, March 21, 1813.

I heartily wish I could see your pictures or any thing that belongs to you; but this year *you* must be Mahomet, and *I* the mountain,—another year we will try to reverse the matter. I have no recollection of the picture you mention at Capo di Monte: that collection was in such dirt and confusion (all on the floor) that we saw it to disadvantage. Pray is the white speck (indicating light) on the pupil of the eyes? Andrea del Sarto I think never put it;—but he might, perhaps, *in copying*. Lady Rockingham had a fine Cornelius Jansen without that speck. Is it usually omitted by him?

My wife answers for Miss Coke, and sends you in return what, I think, is not unworthy even of *your* verses; because, like them, it is an effusion of a good heart. I subjoin an epigram on Holkham, which was sure of pleasing my fair young friend, who is worthy of her father,—I need say no more for her.

Leicester, high priest of fortune and of taste,
Raised fairy scenes amid the desert waste:—
Holkham's chief grace owns not his magic rod;
"An honest man's the noblest work of God."

J. E. Smith.

Sir J. E. Smith to Mr. Roscoe.

My dear Friend, Norwich, October 3, 1814.

I had intended writing to you a few days since, but many things have prevented me. I hope I am still time enough for the main purpose of my letter, which is to make another attempt, at the desire of Mr. Coke, to induce you to visit Holkham. We have spent a delightful fortnight there lately, and two hours almost every day were devoted to an examination of the manuscripts. I am going there on Monday with our good Bishop, for a few days, for the express purpose of looking further into these treasures, and if you could join us, you would complete the joy of the whole party.

I must tell you a part of our discoveries:—besides beautifully illuminated MSS. on vellum, of many of the Latin classics; a most exquisite Boccaccio; a very old and fine Dante; a Chronique d'Hénault, in two immense folios, richly illuminated; and other things of that kind;—there is a very valuable collection of historical Italian MSS., fairly copied at Florence, Venice, &c., for Lord Leicester; they are partial or local chronicles, memoirs, &c., *very curious*. Among others is a complete copy of Burchard's Diary. This *delectable* treasure will surely tempt you of itself. I think you knew nothing of it but what Gordon has printed:—am I right in this? There is one *original* monastic chronicle, itself of the date of 1300 or 1400. There are also many things which we want you to tell us the value of.

The printed books are inestimable in value and number.

How many things have I to talk over with you, and not a few to show you, before we go to Holkham! You must contrive not to limit your time; you have found it difficult to get to that charming place, but you have no idea how difficult it is to get away. Yours, &c.,

J. E. Smith.

Soon after this letter was written, Mr. Roscoe paid his first visit to Norfolk, as will appear by the succeeding letters from Mr. Coke and the Bishop of Norwich, to whom Sir James had the satisfaction of introducing his friend.

Mr. Coke to Sir J. E. Smith.

My dear Sir, Holkham, December 15, 1814.

I cannot forward the inclosed, which reached me by yesterday's post, without acknowledging all your friendly assistance, and expressing the great pleasure afforded me by your own and Mr. Roscoe's visit.

The more I saw of him, the more I was delighted with the benevolence of his mind, the rectitude and liberality of his principles, as well as with his superior acquirements.

Believe me, with our united kind regards to Lady Smith, ever, my dear Sir, very faithfully

Yours,

Thos. Wm. Coke.

The Bishop of Norwich to Sir J. E. Smith.

My dear Sir James,

I feel, if possible, more proud of being indebted to your friendly partiality for the favourable opinion which Mr. Roscoe is so good as to entertain of me, than I do even of his approbation; and yet the esteem of such a man is a source of higher gratification than any which it is in the power of kings or ministers to bestow. Many thanks for your kind invitation. To wait upon you, and to meet Mr. Roscoe, are certainly very great temptations; for men like him are rare beings—

"Numero vix sunt totidem quot
Thebarum portæ vel divitis ostia Nili."

Old as I am, I cannot therefore but feel anxious to say, before I die, "Virgilium vidi." Adieu!

Yours sincerely and affectionately,

H. Norwich.

Mr. Coke to Sir J. E. Smith.

My dear Sir James, Holkham, Feb. 9, 1815.

Leo X., most magnificently bound, made his appearance yesterday, and will be more highly prized than any manuscript in my possession. To you, I may fairly say, I am more particularly indebted for this most inestimable gift; I should probably never have known Mr. Roscoe, if it had not been for your kindness in bringing us together; it has established a mutual regard between us, which I am satisfied will be pleasing to us both during the remainder of our respective lives. To say the truth, he is a most extraordinary personage ;—such a head, such a heart, such suavity of disposition, such courage in the pursuit of what is right, such pure philanthropy, are seldom combined in one individual ;—imagine then, my dear Sir, the store I shall set by the present of his book. How preferable such a testimony of esteem from such a man, to the baubles which may be derived to a cringeing sycophant from a profligate court! If I live and have my health, I will do myself the pleasure of passing a few days with him at Allerton in September or October next. Could you not accompany me? I will not keep you from home more than five or six weeks.

You will be pleased to hear that I have consigned to Roscoe's care four dozen of my MSS. to be bound.

Eliza and the ladies unite with me in kind regards to yourself and Lady Smith; and believe me, my dear Sir, with great esteem,

Yours most sincerely,

THOS. WM. COKE.

From the same.

My dear Sir, Holkham, August 8, 1815.

You must be here one whole day at least, before our departure, to look over the manuscripts our friend Roscoe has had bound, and his remarks, which are very interesting. I need not say with what delight I look forward, in having so cheerful a companion and inestimable a friend to accompany me in my travels, for they will be nothing short of a moderate tour upon the Continent.

Remember me in the kindest manner to Lady Smith; and, with great regard and esteem,

I remain yours most faithfully,

THOS. WM. COKE.

From the same.

Dear Sir, Holkham, Jan. 17, 1816.

I have within these few days received a most delightful and gratifying letter from our worthy friend, who writes as under, after having mentioned his having sent a large parcel of MSS. "In particular I must beg your attention to a most beautiful manuscript of Cæsar's Commentaries, and a small thick volume lettered *Præses fid.*,—two of the *finest I ever saw*. The *Livy*, in the chest just mentioned, is beyond all *estimation*, being, as I can clearly demonstrate, the *individual book* sent by *Cosmo de' Medici* to *Alfonso* king of Naples, as a peace offering on the termination of hostilities between them," and which is mentioned in his Life of Lorenzo de' Medici. "That I should have had" (he says,) "the good fortune of seeing and turning over at my leisure such a book, is almost incredible."

He mentions Mr. Jones having got into a regular system,—three binders besides himself constantly employed; so that you see, my good friend, your hint was not thrown away; and the satisfaction which Mr. Roscoe over and over again assures me he takes in looking over and arranging these various MSS., and in noting the collection, removes from my mind the great trouble I was once fearful I had imposed upon him.—He has sat for his portrait to Shee, and I am told it is very like *.

I have been blessed with all my grandchildren and three daughters for some time, who are all well, and unite in kind regards with, dear Sir,

Yours,

THOS. WM. COKE.

From the same.

My dear Sir, Holkham, Oct. 5, 1816.

I know you will feel interested to hear of my purchases at Mr. Roscoe's sale of pictures, which I rejoice to tell you went off at much higher prices than had been expected. I have the Leo, the Head of Christ, the *chiar'oscuro* by Michael Angelo; the portrait of a Venetian Lady and her Son by Giorgione, and a Holy Family by Vinci; which were all favourites of our inestimable friend's, and he expresses much satisfaction in the thoughts of seeing them here.

I shall have the pleasure of seeing you and Lady Smith at the sessions. With our united kindest regards to both, I remain, my dear Sir,

Ever most faithfully,

THOS. WM. COKE.

* This portrait of Mr. Roscoe hangs in the manuscript library at Holkham.

Mr. Roscoe to Sir J. E. Smith.

My dear Friend, June 10, 1817.

I have now to speak to you on a subject which I flatter myself may eventually lead to my enjoying somewhat more of your society than I had promised myself. We are establishing an institution here for education and lectures on a large scale, embracing the whole circle of literature, science, and the arts. Our proposed capital is 30,000*l.*, towards which upwards of 22,000*l.* is already subscribed. We have prepared a capital building, with lecture-rooms, school-rooms, exhibition-rooms, &c., upon a commodious and extensive scale; the whole of which are now ready for use. I have promised to give an introductory lecture, which I believe will be followed by a regular course by Dr. Traill, our very worthy and scientific townsman, and Dr. Vose, an excellent physiologist. But our committee are desirous, in publishing their first report, to be able to state to the proprietors the names of such celebrated scientific characters in other parts of the kingdom, as they think may be induced to favour them with their assistance; and it is with great pleasure I have undertaken to solicit you on the occasion.

Our buildings have cost 10,000*l.*

I trust you will allow us to say in our report that we have hopes of your assistance.

You will receive another letter from this post, so

that you will have a double assurance that I am, my dear Friend, ever truly yours,

W. Roscoe.

Sir J. E. Smith to Mr. Roscoe.

My dear Friend, Norwich, Sept. 28, 1818.

I have not much to say, except to express the pleasure I have had so lately in your society, and my happiness at seeing you so well. On Sunday the treacherous weather just permitted me to climb from Matlock to the high rocks on Cromford Moor, often celebrated in English Botany: I had not been up to them since 1792. I was quite enchanted with the wide extended view; and the balmy air among heath, bilberries, ferns, mosses, &c. seemed

"Redolent of joy and youth,
To breathe a second spring."

I met with all my old friends among the Lichens, &c.; but the long dry summer seems to have kept back their fructification.

I have written to Mr. Coke about your intended visit.

The worthy Bishop of Winchester writes that he is "exceedingly delighted" with my Cambridge pamphlet. 'Tis curious that four Oxford bishops should decidedly approve of my pretensions.

I have just got a work of Sprengel's on *Umbelliferæ,*—which I think he has reformed almost as successfully as you have the *Scitamineæ.* In it I am called—"*μεγα κυδος βριταννων*"—*φευ αιδοος!*

J. E. Smith.

Sir J. E. Smith to Mr. Roscoe.

Norwich, March 27, 1820.

My dear and very kind Friend,

I blame myself for not having sooner replied to your welcome and consolatory letter, received above a month since.

I am truly thankful on my dear mother's account that she "fell asleep" so happily, as really never to have known what death was,—nor did she ever know the fear of it: her religion was of the most cheerful kind,—no gloom, no uncharitableness had any share in it. She was quite prepared, and had talked to me about every thing connected with her departure, long ago. I had been in the habit of almost daily calls, to chat a minute or two with her, and I miss her with a degree of sadness I did not expect. I vainly thought I had fortified myself beforehand: my only resource is the reflexion that I have nothing to regret for her sake, and I am thankful she did not survive me, which her unimpaired health and the probability of her living as long as anybody ever does, made me often dread for her. *She* also sometimes dreaded it,—not from any alarms about my health, but because of her own probable long life.

March 28.—I have just learnt by a letter from my wife, who is at Lowestofft, that her excellent mother is no more;—such are the ravages of such a severe winter!

I would add a word of literature. I send today

a great budget of Linnæan correspondence to "Master John Nichols" to print,—very curious and interesting. Have you seen the letters he has printed of Dr. Richardson, Sloane, Lord Petre, and the great Sherard? The originals are now in my hands, with many others;—a mine of botanical anecdote.

Your ever affectionate
J. E. SMITH.

Mr. Roscoe to Sir J. E. Smith.

My ever dear Friend, April 15, 1820.

I need not say how truly I sympathize with Lady Smith and yourself in the grief and anxiety you have lately had to sustain, from the loss of your very near and dear relatives.

The crown of a happy life is a peaceful death; and if we could take our departure with the composure of your late excellent mother, it would seem almost desirable for us to escape from this "sea of troubles." But though we may have sufficient causes to induce us to wish to lay down our burthen, yet something still occurs to provide us either with a reason or a pretext for wishing to remain a little longer; and I fear, if put to the test, we should be like our poor friend Fraser, who, after having been seventeen times across the Atlantic, and brought more plants into this kingdom than any other person, complained to my son, when he called on him in his last illness, that "Providence had cut him off in the midst of his labours." Yet as such things will occur, I see with pleasure you still continue to add to the great mass of knowledge and information on scientific, and particularly on botanical subjects, for which the world is already so highly indebted to you; and *that* not merely by your own valuable labours, but by giving to the public the productions of your eminent predecessors, and rendering them immortal in the offspring of their own mind.

Almost the only relaxation which I have for some time past been enabled to take, has been in extending and examining my collection of *Scitamineæ,* which are now so numerous as to form an important class, capable of as correct an arrangement as any in the system. In the course of this, I have diligently studied and copied your excellent remarks in Rees's new Cyclopædia.

By the liberality of Dr. Carey of Serampore, and Dr. Wallich of Calcutta, we have for some time past been furnished with all the living plants of this tribe that India can afford; many of them from the remotest parts of Napal, and a splendid collection of seeds chiefly from Napal and Subhatoo.

My present arrangement of *Curcuma* is the finest and most numerous of the whole order. Of *Hedychium* I have 15 or 16 species; 5 of *Roscoea,* &c., which wait for your sanction and name.

I beg my most affectionate respects and condolence to Lady Smith, and am, dear Sir James,

Yours,

W. Roscoe.

From the same.

My dear Friend, Liverpool, July 14, 1821.

You will probably have seen by the public papers that I have undertaken to give my assistance to Mr. Valpy, in publishing a collection of the Italian poets in 48 volumes; in which it is intended that the works of each author should be introduced by biographical and critical dissertations, extracted from the best literary historians and critics of Italy,—a mode which I proposed. You will perhaps be more surprised to hear that I have also acceded to a proposal made to me to write a new life of Pope, and publish a new edition of his works,—an undertaking of much more labour than the other, and at the present time, in which a sharp contest is carrying on both as to his moral and poetical character, attended with some peculiar difficulties. From a pretty close examination of the later editions of his works, I am not, however, greatly discouraged; nor can I help thinking that, however deficient I may be found in some respects, I shall be able to give an edition more just to the character of the author, and more accommodated to the use of the general reader, than any of those hat have been published since the time of Warburton.

When to these I add my additional volume to the Life of Lorenzo, you will have great reason to accuse me of presumption, and perhaps of folly: but I know not how it is,—I never can accomplish any thing unless I have a great many other things that call for my attention at the same time, when I am as diligent in what I am about as obstinacy and perseverance can make me. Whether this arises from an attachment one acquires for a particular subject, or to a perversity of disposition that delights to be employed in any thing but what it ought to be, I shall not venture to determine.

With respect to our botanical concerns, the *Hedychium excelsum* has flowered with us this year in grand style; and several others are going into flower, some of which we expect to be new. We have also been greatly surprised by the appearance, a few days since, of a flower of a new and beautiful species of *Roscoea*, from a plant sent us as a species of *Orchis* from Sillet? This plant is entirely different as well from the *purpurea* as from the four others of which I have dried specimens. I shall send you a slight figure of it, in order that you may add it to your list of species, and give it a specific appellation. What do you think of either *speciosa*, or *lucida*? The first it deserves, as being, I conceive, the largest and finest flower known of the genus; but the second would perhaps be more appropriate, as alluding to the extreme delicacy and transparency of the petals, which no drawing can express. The whole flower is of a pale purple, changing in some parts to a clear watery white. The flowers arise in succession as in *Kæmpferia*, to which and to *Hedychium* it appears to have the nearest affinity.

Since my return home I have been enabled to

make some important additions to the genus *Canna*, and have drawn up a synoptical table of twenty species, of which I printed a few copies for the correction of my friends, but which I am unwilling to part with till I know your opinion upon it. The tables of *Hedychium* and *Curcuma*, which I have also nearly finished, will each of them amount to about the same number, and most of the plants are now growing with us.

Accept, my dear friend, my best thanks for the additional memorial of your friendship in the present of your two valuable volumes of the "Correspondence of Linnæus and other Naturalists," a work highly worthy of you, and which will always be grateful to every true lover of science, and to every pious and candid mind. Your anecdote of the identity of the peach and nectarine, reminds me of a circumstance equally extraordinary which has occurred to us here respecting *Hedychium.*—Happening to meet a friend in the botanic garden, he informed me he had a plant in flower in his hot-house, about five miles from Liverpool, which he did not know, but which from his description I concluded must be a scitaminean. With his permission, Henry Shepherd and I immediately went and brought it to Liverpool; when it appeared to us all to be quite a new species, wholly different from any of those growing with us, and from my dried specimens from India. The plant was not more than three feet high,—the flower a dull yellow. As it increased freely by the roots, the plant was afterwards divided, and several specimens were distributed under the name of *Hedychium flavum*, when to our great astonishment in the following year it grew to twice its former height, and turned out to be no other than our old acquaintance *H. coronarium* with a white flower at least three times the size it had before produced,—a result which I assure you will render me very diffident in future in deciding on the species of this beautiful genus.

And now, my dear friend, I think it is high time I should inquire after your health and present avocations; and particularly how Lady Smith and yourself intend to dispose of your time during the ensuing summer (for in Lancashire it is not yet begun).

Adieu, my dear friend, and believe me, with the most affectionate respect and kindest remembrances to Lady Smith and yourself,

Your ever faithful

W. Roscoe.

Mr. Coke to Sir J. E. Smith.

Holkham, August 10, 1825.

Shame would justly, my dear Sir, attend me, if I delayed answering your kind letter the first moment I could snatch from the whirl in which I have been writing.

Lady Anne accompanied me to Norwich assizes on Monday. We took up our residence at the Palace, where we were most hospitably and kindly invited; and I am sure it will give you and Lady Smith pleasure to hear that I never saw the Bishop in better health or spirits, though I had been told he was not. In our way back we stopped two nights with the Mr. Ansons, at Lyng.

I have now all hands at work with my harvest: within a fortnight or thereabouts I hope to be forward enough to enable me to leave home with a contented mind for Cannon Hall, having promised my dear Eliza that I would spend a fortnight with her this autumn. It was our intention to have gone into Scotland, had we not been prevented by Lord Hastings, who is coming to us with all his family the very beginning of September.

Happy should I be if you and Lady Smith would give him the meeting, as Lady Loudon is very fond of botany, and Lady Anne having now a gardener to her liking. I think you would be pleased with the improvements that have taken place since we had the pleasure of seeing you. I will frank your letter to Roscoe, and accompany it with a line which will give both him and you much gratification, to hear that within this fortnight Blaikie has found in his office some highly valuable MSS. of Lord Chief Justice Coke's. When Chantrey has fixed his time of coming to Holkham, I will not fail to let you know.

Yours most faithfully,

Thos. Wm. Coke.

Sir J. E. Smith to Mr. Roscoe.

Henbury Hill, near Bristol, August 6, 1825.

Yes, my good friend, Henbury Hill, near Bristol! The latter place I know you have heard of, for I find your name inscribed on the roll of a Society, to which mine has also been added; but of the little village of Henbury, however worthy of celebration by your classical pen, you perchance may not have heard. Perhaps, nevertheless, you may reply,

> "The place itself is neither new nor rare;
> I wonder how the devil you came there:"—

so I will proceed with my narrative. I left my own home April 30th, and passed a fortnight with my worthy friend T. Forster, at Walthamstow, and as much with his brother Edward at Hale End. I then passed two weeks in Chapel Place, near Cavendish Square. Meanwhile I gave a course of ten botanical lectures at the London Institution, attended to much business of the Linnæan Society, visited several friends, and enjoyed myself sufficiently, "being in sound health of mind and body." I passed a morning at Paddington with Mr. Coke, Lady Anne, and their two fine boys,—Dr. Davy of Caius accompanying me. There could not be a more delightful sight. I dined one day with our friend Lady Anson,—Mr. Coke, Mr. W. Coke, and Lord Suffolk being of the party. I never was better in my life than during this visit to London: but on coming to Bristol by a day coach, June 12th, in great heat, and after giving my first lecture next day, my old inflammatory complaint, with an affection of the lungs, attacked me; and after giving three lectures I was obliged to resign myself to bleeding, James's powders, and starvation; all which being vigorously used, under the inspection of my most excellent and skilful friend Mr. Estlin, I was enabled to finish my course. We remained nearly a month at the Hotwells, enjoying the fine air and water, and amused by the numerous passing vessels, steam-boats, &c.; the noble views of wood and river, accompanied by music on board the boats, and often in the woods,—all close to our windows. Here I rapidly regained my health, sleep, and strength, insomuch that I took walks of three or four miles in the day.

Last Wednesday we removed to a very smart cottage, completely furnished, even with a library, in which, among other first-rate books of the present age, is the quarto edition of your Lorenzo.

This cottage was recommended to me by Mr. and Mrs. Brooke, who inhabit a very pleasant house near it and Blaize Castle, and from whose garden and grounds, open to us, we enjoy delightful views of the Welsh Hills, the Severn, &c., and on all sides a most beautiful country. Our cottage, with a field and garden, looks over a rich and beautifully wooded valley, extending to the Avon, on whose opposite bank Mr. Bright has a noble house and grounds. In October we mean to prosecute our journey leisurely homeward, visiting some friends in our way.

I am sorry to hear of your having suffered much by the rheumatism; but hope the warm weather has removed it. To the excessive heat I perhaps owe my illness.

Your ever affectionate Friend,

J. E. Smith.

Mr. Roscoe to Sir J. E. Smith.

My dear Friend, Toxteth Park, Sept. 3, 1825.

Your most kind and welcome letter from Henbury Hill should have been sooner acknowledged, had not continual interruptions, combined with a state of unaccountable indolence and debility, prevented me from turning my thoughts to any subject but such as had irresistible claims upon me, and from which I extricated myself the first moment it was in my power. If, in return for the narrative you have so kindly given me of your peregrinations and transactions since you left home, I should furnish you with mine for the same period, they would appear like the track of a snail compared to the flight of an eagle, or the journal of a pedlar to the history of some mighty traveller. You pass from county to county, visit your friends, and take up your abode where you please; whilst I remain on the same spot, without emigrating even from the blue bed to the brown, and whenever I am disturbed only exclaim,

"Let me, let me rest."

The only object that excited my exertion was the publication of my Monandrian Plants.

I need not say there are many things in these on

which I long for your opinion ; but I will not intrude on you till you have seen and examined them, further than to say, that I am in hopes of clearing up the difficulties that have existed in ascertaining the precise limits of the genera *Maranta*, *Phrynium*, and *Thalia*, and of settling some other important points that have occurred in the course of the work, especially as to *Costus*, of which four fine species have flowered with us this summer, all of which will shortly appear in my work.

When shall we meet again? For my own part I can only repeat my own words, that

> "....Hope strives in vain through futurity's gloom
> To descry one bright moment in seasons to come."

Yet I will not despair. In the uncertainty that attends this earthly state, in which we cannot foresee the consequence of our placing one leg before another, I would gladly flatter myself that something may occur to draw us together again, and enable us to enjoy each other's society—if not with all the life and vivacity, with all the warmth and affection we ever experienced; which good wish I hope Lady Smith will not refuse to share: at all events present to her my kindest remembrances, and believe me always, my dear Friend, most faithfully yours,

W. Roscoe.

Mr. Roscoe to Sir J. E. Smith.

My dear Friend, Toxteth Park, Dec. 14, 1826.

It was with great pleasure I received your letter giving me so satisfactory an account of your health and of the progress you are making in your great work, which will fill up a *desideratum* in the botany of this country that no other hand could have supplied. I rejoice also in the disposition you feel for the continuation of your labours, it being a strong impression on my mind that nothing is more conducive to life and health than some employment that calls for our continued attention, and prevents a moment from being irksome on our hands. For my own part, I feel as if my existence were twined round my employments, and when those have finished I shall have finished too.

* * * * * *

I thank you also, my dear friend, for your kind information respecting *Amomum*; but since I last wrote to you I have had the good fortune to succeed beyond my utmost expectations, in having had specimens of one of the largest and finest of the tribe sent me from Demerara,—the flowers preserved in spirits, and the fruit in an air-tight bottle.

This I conceive to be the true Malaguetta Pepper, or Grains of Paradise, respecting which I expect to know more when I can obtain a sight of your article on *Malaguetta* in the Cyclopædia.

The capsule somewhat resembles that of Gærtner, but is upwards of six inches in length, of a deep orange colour, covering a pulpy rind, inclosing the receptacle of the seed, of the thickness of a man's finger, on which the seeds are beautifully arranged, and imbedded in a tomentose substance.

Frequently do I lament my distance from you; and severely do I feel, on numerous occasions, the loss of your able and friendly advice.

I follow your example, "and comfort myself with the bright spots in my horizon." Above all, I delight to preserve and cultivate those feelings of friendship and affection which have been the charm and happiness of my life, and few of which have returned me so ample a harvest as those on which I am at present employed. Adieu, my dear Friend!

I am most faithfully,

W. ROSCOE.

Sir J. E. Smith to Mr. Roscoe.

My dear Friend, Norwich, Jan. 8, 1827.

I have received lately your 9th and 10th numbers, and need not tell you with what interest I have looked them over. What a treasure is your plate and account of the original *Thalia!*

Your intended *Matonia* is, I presume, distinct from *Elettaria*, which last name may as well remain, as the French school will doubtless retain it. I should be highly gratified by the perusal of your manuscript materials.

Surely the true Malaguetta Pepper is from Africa, about Sierra Leone, and is the Grains of Paradise.

See *Amomum* in Rees's Suppl.; also *Mellegetta* in the body of the work.

The plant I mean must be altogether different from what you have lately got from Demerara. Is not the "tomentose substance," in which the seeds of your Demerara plant are imbedded, dried pulp?

Very truly yours,

J. E. SMITH.

Sir J. E. Smith to Mr. Roscoe.

My dear Friend, Norwich, Dec. 29, 1827.

Having a small corner in a frank allowed me, I cannot refrain from writing a word or two, though I have but little to say, but that I have done my fourth volume of English Flora, except printing the index, and am now getting on with *Flora Græca*, of which I hope to get towards the conclusion in the course of this winter; so that, whatever happens to me, I shall have done my part of the work.

My eyes are much recovered, so that I find no impediment from them in bright weather: all pain and inflammation are gone. I suffer from rheumatism and great consequent weakness in my legs: my stomach, too, is dyspeptic; but I totter on. Our excellent friend at Holkham is counting upon our visiting him in the spring. Whenever you can go, we will, if possible, meet you.

Yours,

J. E. SMITH.

Mr. Coke to Sir J. E. Smith.

My dear Sir James, Holkham, Oct. 2, 1827.

I cannot think of sending you the inclosed letter without accompanying it with one line to say that our friend, in his letter to me, writes in excellent spirits, and holds out hopes of my seeing him in the spring. Pray God his health may be sufficiently restored to enable him to do so, is my sincere wish.

With our tour into Scotland Lady Anne was delighted: but not being able to leave home so early by a fortnight as we had intended, owing to Lady de Clifford's kind and unexpected visit, we did not go beyond Blair Athol. We visited all the principal manufactories in our way through England,—Manchester, Bolton, &c. &c. In our way to Glasgow I saw the establishment at Lanark, and returned by Lord Rosebery's, which was by far the most picturesque and beautiful place we visited in Scotland; and upon our return home had the happiness to find our children in perfect health.

The sporting season having now commenced, I have taken the liberty of sending you a little game. Lady Anne unites with me in all kind remembrances to yourself and Lady Smith.

Ever, my dear Sir James,

Faithfully yours,

Thos. Wm. Coke.

CHAPTER XIV.

Of the religious, social, and scientific Character of Sir J. E. Smith.

WHEN the religious opinions and principles of a man have been such as to support him through the trials of life and in the hour of death, they form a part of his history deserving our notice: this memoir therefore would be imperfect without giving such a statement of Sir James's as may be relied upon for containing the most essential points.

His principles were these,—"That a man can be no Christian, *as to faith,* who does not judge for himself; nor *as to practice,* who does not allow others to do so without presuming to censure or to hinder them."

His opinions were formed from the same source whence many, with equal sincerity, derive very different ones. His creed was the New Testament, and he read it as a celebrated divine* recommends; that is, "as a man would read a letter from a friend, in the which he doth only seek after what was his friend's mind and meaning, not what he can put upon his words."

He was a firm believer in the divine mission of Jesus Christ; and in maintaining the doctrine of the strict unity of God, as one of the truths our great Master was commissioned to teach, he considered his opinion truly apostolical.

* Whichcot.

" I look up," he says in a letter to a friend, " to one God, and delight in referring all my hopes and wishes to him; I consider the doctrine and example of Christ as the greatest blessing God has given us, and that his character is the most perfect and lovely we ever knew, except that of God himself. This is my religion; I hope it is not unsound." *

He considered opinions and principles very distinct, though often confounded. The latter he looked upon as very important; the former no otherwise so, than as conducing to good principles,—and he esteemed one virtuous act, one honest determination, to be more worthy in the sight of God than any notion or discovery concerning the essence of that Great Being who is raised far above all human comprehension.

The writer has often heard him observe, that whatever in the sacred writings is not clear to the capacity of the humblest, most unlettered rustic, cannot be essential to salvation; and he was of the same opinion as our immortal Wickliffe, the morning-star of the Reformation, who contended " that wise men should leave that as unimportant which is not *plainly expressed* in Scripture."

With regard to opinions, he also agreed with the amiable Lindsey, " that Christians have yet to learn the *innocency of error*, from which none can plead exemption, and to *bear with each other* in their differing apprehensions concerning the nature of the First Great Cause and Father of all, and the person of Christ, and the manner and date of his deriving his

* Letter to Davall, April 25, 1790.

being and high perfections from God; and surely it must also be owned to have been left in some obscurity by God himself in the writings of the Apostles, (otherwise so many men, wise and good, would not have differed, and still continue to differ concerning it,) and so left, it should seem, on purpose to whet human industry and the spirit of inquiry in the things of God, and to give scope for the *exercise of men's charity* and mutual forbearance of one another, and to be one great means of cultivating the moral dispositions, which is plainly the design of the holy spirit of God in the Christian Revelation, and not any high perfection in knowledge, which so few can attain." *

Let it not be supposed that Sir James was indifferent to opinions, and considered all systems equally good; on the contrary, he preserved his own through good report and evil report, and no temptation of interest ever made him swerve one moment from the maintenance and vindication of those he had adopted: but among these, the first was *charity;* exclusiveness he considered as the very characteristic of Antichrist and pride. There was no sect of Christians, among the good and sin-

* "In matters of eternal concern," says the biographer of Sir William Jones, "the authority of the highest human opinions has no claim to be admitted as a ground of belief; but it may with the strictest propriety be opposed to that of men of inferior learning and penetration: and whilst the pious derive satisfaction from the perusal of sentiments according with their own, those who doubt or disbelieve should be induced to weigh, with candour and impartiality, arguments which have produced conviction in the minds of the best, the wisest, and most learned of mankind."

cere, with whom he could not worship the Great Spirit to whom all look up, enter into their views, excuse what he might consider as their prejudices, and respect their piety; and whether it were in the pope's chapel, or the parish church, he felt the social glow,

> "To gang together to the kirk,
> And all together pray;
> Where each to his great Father bends,
> Old men, and babes, and loving friends,
> And youths and maidens gay."

The affection he thus felt for others, he in general had the happiness of finding reciprocal, "for love must owe its origin to love." No one had less of a sectarian spirit; nor did he ever attempt to make converts, except to christian charity.

Where speaking, in his Tour, of some customs in the catholic church, "The stocks and stones," he observes, "which the people are taught to worship, are dressed out to their imagination with attributes of rectitude and benignity, borrowed from the pure idea of an intellectual Deity; for so congenial are virtue and benevolence to the human mind, that no system of worship could support itself without their semblance; and even those most corrupt in principle could have little success in practice, without a constant appeal to the eternal law written in our hearts:—as to forms, the mind will associate its conceptions with visible objects. The devotion of some persons is best excited in a choir, of others in a conventicle, and of others in the holy house of Loretto; but 'one is their Father, even God'."

He was adverse to such a view of the Supreme Being as is injurious to the *perfect goodness* of his character, which, *because* his *power is unbounded,* has supposed it might please him to exercise that power to the subversion of his no less immutable attributes, *justice* and *mercy.* Such ideas of our Creator appeared to him dishonourable to that parental character which makes our adoration spring from the heart, and delight in obeying his commands: such a view of God is to invest him in the evil passions, the imperfection and weakness of humanity. He believed that "in no being is the sense of right so strong, so omnipotent, as in God; and that his almighty power is entirely submitted to his perception of rectitude. He ascribed to him not only the name, but the disposition and principles of a father; that he has a father's concern for his creatures, a father's desire for their improvement, a father's equity in proportioning his commands to their powers, a father's joy in their progress, a father's readiness to receive the penitent, and a father's justice for the incorrigible;—that God's justice has for its end the highest virtue of the creation, and *punishes* for *this end alone;* and thus it coincides with *benevolence,* for virtue and happiness, though not the *same,* are inseparably conjoined. He looked upon this world as a place of education, in which God is training men, by mercies and sufferings, by aids and temptations, by means and opportunities of various virtues, by trials of principle, by the conflicts of reason and passion, by a discipline suited to free and moral beings, for union with himself, and for a sublime and ever-growing virtue in heaven."

"It is not," continues Dr. Channing, "because he is our Creator merely, but because he created us for good and holy purposes; it is not because his will is irresistible, but because his will is the perfection of virtue, that we pay him allegiance. We venerate not the loftiness of God's throne, but the *equity of goodness* in which it is established."

A sentiment resembling the last may be found in Mason's Memoirs of Gray, in a paper which contains some very pertinent strictures on the writings of Lord Bolingbroke. In speaking of the Deity, Gray remarks:

"His eternity, infinity, omnipresence, and almighty power, are not what connect him, if I may so speak, with us his creatures. We adore him, not because he always did, in every place, and always will, exist, but because he gave and still preserves to *us* our existence by an extension of his goodness. We adore him, not because he *knows* and *can do* all things, but because he made *us capable* of *knowing* and of *doing* what may conduct us to *happiness*. It is therefore his *benevolence* which we adore, and not his *greatness* or *power;* and if we are only to bear our part in a system, without any regard to our own particular happiness, we can no longer worship him as our all-bounteous parent,—there is no meaning in the term."

"This unlimited power," says Mr. Locke, "cannot be an excellency without it be regulated by wisdom and goodness; and therefore, looking on God as a being infinite in goodness as well as power, we cannot imagine he hath made anything with a design it should be miserable. His justice is nothing but a branch of his goodness."

"Je ne vais pas si loin que St. Augustin, qui se fût consolé d'être damné si telle eût été la volonté de Dieu. Ma resignation vient d'une source moins désintéressée, il est vrai, mais *non moins pure*, et *plus digne* à mon gré de l'Etre parfait que j'adore. Dieu est juste: il veut que je souffre, et il sait que je suis innocent. Voilà la motif de ma confiance, mon cœur et ma raison me crient qu'elle ne me trompera pas. Laissons donc faire les hommes et la destinée, apprenons à souffrir sans murmure, tout doit à la fin rentrer dans l'ordre, et mon tour viendra tôt ou tard!"

It is delightful thus to find enlightened and disinterested men, who have no object but *truth*, concurring in their hopes and views from the spontaneous deductions of their own individual minds, and, in that perfect love which casteth out fear, reposing on their "Father and their God."

The subject of the present memoir cherished a perfect faith in the goodness of God. The goodness of God "was the reason of the hope that was in him." Believing that he framed the human soul for eternal duration and for happiness, he never troubled himself about the *time* or *manner* of his future existence, or what was to constitute it; considering himself incapable of forming any judgement, he relied implicitly on the benevolence of that parental Being who had "vouchsafed to call him hither to this great assembly and entertainment, and had permitted him to contemplate his works, to admire and adore his providence, and to comprehend the wisdom of his conduct."

The apparent evil, the partiality, the injustice, in our present life, were to him assurances, combined with revelation, of a more perfect state hereafter.

He believed virtue not to be communicated without effort; that the great object of our probation here is to *acquire* it, and that to possess ourselves of it we must cultivate and cherish it by our own unremitting endeavours, by the rules prescribed by him who was himself "the way, the truth, and the life."

The following paragraph was found after Sir James's death, among other papers, on his library table. It seems to be a memorandum of opinions which, had his life been prolonged, he had a design to use and enlarge upon.

"The attributes of God are,

Power,
Wisdom,
Goodness. } All infinite and eternal.

"Power and wisdom are communicable to his creatures in any degree he pleases. Goodness is in the very nature of things incommunicable, and must be *acquired*, else it is a nonentity.

"Man has the power of acquiring goodness. He must therefore be fully capable of comprehending or of knowing what it is. But man is manifestly incapable of comprehending the wisdom or power of God; nor is it of any use to him to attempt it.

"Hence all our reasonings respecting a future state, founded on the wisdom or power of God, or on natural principles, have always proved vain."

"Those only are conclusive which are deduced from his goodness; of which, as we are formed and commanded to imitate it, we are necessarily able to judge.—J. E. SMITH."

His devotional feelings were ardent, retired, confiding: like the pious Fenelon, he felt that "God is the true friend of the heart, and that there is no comforter like him." To say he was accustomed to the duty of prayer,. seems needless after this,—none was less restricted to a set of words, none expressed "the soul's sincere desire" more frequently or fervently. Devotion and benevolence were the marked characters of his gifted mind. His love of nature was the love of God.

The subject of the few foregoing pages naturally recalls to mind the events of the spring of 1818.

The public has already been in possession of the motives which induced Sir James to offer himself as a candidate for the botanical chair at Cambridge*, being neither a member of the university, nor of the church of England.

Among a great variety of letters from the vene-

* Published in two pamphlets entitled, "Considerations respecting Cambridge," (1818); and "A Defence of the Church and Universities of England," (1819).

rable Professor Martyn, beginning so early as 1813, and all tending to the same point, one only shall be selected, to remind the reader of the ground upon which he concluded he might safely stand.

One motive, which at least had as much influence with Sir James as any prospect of pecuniary advantage, was the anticipation of associating with a learned and polite body of gentlemen and scholars, among whom he had several personal friends and warm supporters; and with these he continued the same friendly intercourse as before, though with fewer opportunities, and in a different manner, from what would have been the case in a nearer connexion with the university. On this account, chiefly, the disappointment of his views hurt his affectionate and social feelings more than it wounded his pride.

My dear Sir, Pertenhall, March 14, 1818.

The season approaches when I feel an annual regret, that in consequence of my age and infirmities I am unable to fulfill my duty as Dr. Walker's reader, in giving a course of botanical lectures. If you could, consistently with your other engagements, undertake to read a course next term, I should esteem it a great favour done to me personally; and I have no doubt of its being well received by the university.

You are aware that you must have the sanction of the vice-chancellor, who, I am persuaded, will be ready to give the university an opportunity of profiting by your instructions, as he doubtless knows that you take the lead in the science of botany in this country, and that your reputation is too well established to need any recommendation from me. As far as my power extends, I am happy in giving you full authority to take such specimens of plants and flowers as you think requisite for your lectures, together with the use of the lecture-room, at any time or times that may be convenient, —always under the controul of the vice-chancellor, and with a complete reliance on your discretion in the use of the garden.

Sincerely wishing it may suit your convenience to comply with this my request,

I remain, dear Sir,

Your ever faithful Friend,

THOS. MARTYN.

In a subsequent letter the Professor makes the following observations: 12th April.

Bradley was never of any university, and had no education: my father admitted only with a view to the professorship, and never took any degree. Vigani, the first professor of chemistry, was an Italian; Dillenius, as you know, was a German. What was Erasmus and many others who were invited to gi e lectures merely from their learning, independent of any consideration of their opinions in religion, or of their having been educated on the spot. T. MARTYN.

For Professor Martyn Sir James entertained the warmest regard, and their friendship was uninterrupted through life: mutual good will and good offices subsisted between two men, who equally experienced that "*il y a dans la botanique un charme qu'on ne sent que dans le plein calme des passions, mais qui suffit seul alors pour rendre la vie heureux et douce.*"

Sir James's sentiments towards the clergy of the Establishment cannot be better shown than in a letter he addressed, some years ago, to a dignitary of the church, upon his tendering his resignation as Fellow of the Linnæan Society:—

Dear Sir,

I found your letter on the table of the Linnæan Society at our anniversary meeting, and read your resignation with much regret, while I felt obliged by your having so largely stated your reasons. Had I been previously informed of your intention, I should have been anxious to have altered it, and might therefore have presumed to controvert some of those reasons. I have always been particularly happy to see my favourite study cultivated by persons eminent in character or station, and especially by the clergy, who, whether of dignified rank, or in the humble, not less venerable, walk of country pastors, may, in following this pursuit, be eminently useful to those around them; while they relax their own minds from severer studies, and derive health of body, with tranquillity of mind, from one of the purest of all sources. Indeed they are but following one precept of *Him* who, as I have always thought, conferred the highest honour on our study that it ever received, and exalted a pleasure into almost a duty, when he said, "Consider the lilies of the field how they grow."

Forgive me for citing this authority, so familiar to you; for I hold it as high as you can possibly do, though we may differ on some points connected therewith. I have always been no less pleased to see the study of nature lead good people to lay aside noxious prejudices and antipathies. I love to see the controversialist, like a cultivated plant, dismiss his thorns or his acrimony, and make an Eden of the garden of science. But I encroach too much on that province whose duties you so highly adorn. I am not ignorant of your high clerical character, nor of the very extensive calls you must have upon you; nor do I mean to interfere: I only contend, like Ray, that the study of the book of nature is inferior to none in dignity or utility.

I remain, with every sentiment of respect,
Dear Sir,
Your very faithful and obedient Servant,
J. E. Smith.

A philosophical consideration of a future state of existence beyond death and the grave, was a frequent topic of discourse with Sir James; not for the purpose of expelling doubts, for he entertained none, but for the enjoyment of recurring to the first, the last, and most unconquerable desire implanted in the human heart.

"I grieve," he says in a letter to Davall, "to see how much imbecility clouded the latter days of Haller; yet even *that* is an argument for the immortality of the soul, had we no other light; for what kind of deity must govern the world, if beings capable of what Haller was were only, as it were, a flash of existence, and all their acquirements turned to no account, and their hopes and powers were excited only to be disappointed?"

"Design," he observes elsewhere, "is evident throughout nature. Some who unhappily doubt every thing else, allow this; but it is sufficient to build every thing upon. We see wisdom employed for beneficent ends. If we are indulged with powers to catch even a glimpse of the Divine Wisdom, is it not enough to prove we are something more than the clod of the valley? But is there no *design* in this permission?—is it intended to call forth our powers and hopes, only to destroy them? Where would be the wisdom or the beneficence of this? If natural religion goes thus far, is there no *design* in the further sources of information with which our Maker has favoured us? Is it not as evident in these as in the other? Nature plans the happiness, the beauty, the perfection of material beings;—the revealed will of God considers the interests of immortal creatures; starving amid the richest treasures of nature, if they have no hope beyond!"

Such were the ideas it gave him pleasure to cherish, and to impart to others,—to build up their hopes by every aid, and persuade to the practice of virtue. His introductory lectures are imbued with such sentiments, and express them in terms which bring their author before the mental eye of those who recollect his conversation, as a portrait reminds us of those we knew. His solemn thoughts were such as gave him freer spirits, and while they elevated the tone of expression enlivened it also.

"One advantage" which Sir James mentions as having found in the study of nature is, "that it is inexhaustible; but" (he continues) "it boasts a still greater,—that, as far as I have been able to observe, it never loses its relish at the decline of life. Several botanists have continued the pursuit with undiminished fondness after the loss of sight, a misfortune one would think the most fatal to all their enjoyments. Many more have derived from this soothing study the best alleviation they could find for the bitterest domestic losses and calamities. With what delight did Linnæus in his last illness turn over and over the acquisitions of his pupil Thunberg in Africa!—and how have I seen the countenance of Scopoli, suffering under the immediate pressure of an unmerited attack upon his honour and all his means of support, resume all its wonted animation and pleasure in talking on the subject of botany! As a taste for the beauties of nature, or in other words, an admiration of the works of God, raises the mind and character above the troubles and cares of this world, may we not hope that such a temper of mind may be far more highly gratified and exalted in a future state? Such a hope is in harmony with all our best feelings, and may surely be humbly indulged without mischief or blame; so far at least as it does not interfere with that absolute reference of every thing connected with futurity to the wisdom and goodness of our Creator, which ought to be not only our duty, but our consolation and delight. In this I conceive the very essence of faith and piety to consist. But under the controul of this disposition every enjoyment and every hope is enhanced; and He who has presented us with a garden as the seat of primæval innocence and delight, cannot be offended by our associating the admiration of his works with any ideas or hopes concerning the happiness in store for us hereafter."

"Another great recommendation of natural history is the habit it necessarily gives of arranging our ideas and exercising our powers of discrimination. In this it vies with the study of grammar and the mathematics. It is the pursuit of truth,—a love of which is as inherent in every sound mind as the love of life. It is a science of facts; and the only way by which it can of itself be advanced, or contribute to the improvement of our understandings and powers, is by practical observation and inquiry. It teaches us to see and discriminate, and then to reason. The worm that crawls on the ground can *perceive,* the bird that flies in the air and builds its curious nest can *contrive;* but man only is allowed to contemplate, compare, and weigh the designs of Infinite Wisdom.

"The exercise of this high privilege soon brings its own reward. We cannot long 'walk with God in the garden of creation,' without admiring the beauty and partaking of the felicity which the Creator delights to bestow and to display.

"A superficial observation of either, however,

will afford but imperfect enjoyment or instruction. This privilege, so consoling to our nature, so encouraging to our hopes, and so improving to our intellects, may of itself be said to whisper immortality. Nor is it restrained to the eminently wise or profoundly learned. Like all manifestations of the divine goodness and mercy, it is plain and open to all. Who does not at once perceive that the Author of nature must be wise and good? Has he not then evidently intended that we should perceive it,—that we should attend to and cultivate this knowledge, and derive wisdom and happiness from so rich a source? The further we proceed with a right mind to prosecute our inquiries, the more are all these suggestions confirmed. This I trust is so obvious, that I shall consider it as a tacit sentiment between us, more easily perceived than expressed; for all

> '.... our highest notes the theme debase,
> And silence is our least injurious praise.'

"As we discover the admirable fitness of every part for its destined purpose, we become far more sensible of its beauty; and as we learn how partial evil mercifully operates for the prevention of more extensive ills, and even for the production of general good, we bow in adoration to the wisdom we cannot fathom, but in which we soon see we may safely and cheerfully confide. As soon as we enter upon the deliberate contemplation of the works of nature, we feel the necessity of classing and arranging them for our convenience.

"Natural history conducts us to a knowledge of

the manners and œconomy of different animals, and their places and dependencies on each other in the great scheme of creation. Here the most insensible mind cannot fail to be struck with the infinite variety of means by which similar ends are accomplished; while amid the richest profusion of variety and beauty nothing is superfluous, nor any end attained but by the most advantageous and compendious means.

"We are told that 'God formed man of the dust of the ground,'—and there is much more contained in that text than we usually perceive. In the creation of the animal kingdom, and of man at its head, the production of intellect was undoubtedly the primary object of our Maker.

"An ingenious writer has said of the sublimest statue in the world, the Apollo Belvidere, that 'the artist has made so much use of matter only as was necessary to execute and give a body to his thought.' This is literally true of the creation of animated beings. Their corporeal part is manifestly but the vehicle, the instrument, the subordinate agent of that sentient existence in which their individual being consists, and which, in the *lowest* link of this living chain, acts and enjoys, and in the *highest*, reasons, adores, and is even permitted to love and imitate, its divine Creator. 'What can be a greater example of infinite power,' says the celebrated Linnæus, whose piety was equal to his knowledge, 'than a little portion of inactive earth rendered capable of contemplating itself as the work of infinite wisdom; and of considering the innumerable effects of that wisdom displayed in the surrounding creation?'

"The exertions of the intellectual part of all living beings are spontaneously and invariably directed to the pursuit of happiness;—it may mistake the means, but never loses sight of the end. It is in its turn made subservient to the preservation of its material organs, by a dread of pain, an inherent love of life, and fear of destruction. It is obliged to attend to the calls of hunger and thirst, and is most bountifully rewarded for all such attention, not only by the gratification of the senses, but by the endearments of social affection, the ties of parental love, and of individual attachment; in all which the brutes are allowed to share, and thus to partake of some of the best enjoyments of *human* nature; as we, by the cultivation of our higher powers, and especially of our benevolent affections, raise ourselves towards the *divine*.

"Even the exertions necessary in the animal world for self-preservation and self-defence, which at first sight seem an evil in the creation, will be found greatly overbalanced by good in the pleasures derived from exertion, activity, and contrivance, and the happiness which results from success. The apparent cruelty by which some animals tyrannize over their own species, or prey upon others, keeps up the perfection of nature by removing the weak and the sickly, and prevents the brute creation in general from ever knowing the miseries of protracted sickness or old age. When they can no longer defend themselves, they are quickly removed.

"It is only we who have higher aims that can turn such evils to advantage; and were it not so, we also should most assuredly have been spared all similar sufferings.

"If we experience the severest of evils from which the brutes are exempt, must it not be for purposes which do not concern them? Else we should be 'of all creatures the most miserable.' Thus Nature confirms truths which she is not of herself competent to teach. Even we ourselves are made subservient to the good of the animals below us. Whole tribes of them, in their own nature gentle and defenceless, are brought up in ease and security to serve us for food or for clothing, and never know want, or age, or infirmity. It is our duty that they should never know pain, and I wish we fulfilled it. Much is required from the legislature, even in this country, to guard and favour these helpless beings, to whom we are so much indebted. It is no part of the plan of Providence that they should be uselessly tortured. He who endowed the horse and the dog with obedience and affection, such as may often put their masters to shame, designed they should be happy, not miserable, in our service. 'A merciful man regardeth the life of his beast,' and not only its life, but its ease and comfort. I wish the thoughtless to consider this, and that the cruel might be made to feel it."

Speaking, in another place, of the characteristic properties of animals, vegetables, and minerals, Sir James observes, "If it be asked what is this vital principle so essential to animals and vegetables, and of which fossils are destitute, we must own our complete ignorance. We know it, as we know its omnipotent Author, by its effects. The effects of this vital energy are continually going on in every organized body, from our own elaborate frame to the humblest moss or fungus. Those different fluids, so fine and transparent, separated from each other by membranes as fine, which compose the eye,—all retain their proper situations (though each individually perpetually removed and renewed,) for sixty, eighty, or a hundred years, or more, while *life* remains. So do the infinitely small vessels of an almost invisible insect, the fine and pellucid tubes of a plant,—all hold their destined fluids, conveying or changing them according to fixed laws, but never letting them run into confusion so long as the vital principle animates their various forms. But no sooner does *death* happen, than, without any alteration of structure, any apparent change in their *material* configuration, all is reversed. The eye loses its form and brightness, its membranes let go their contents, which mix in confusion, and yield to the laws of chemistry alone. Just so it happens, sooner or later, to the other parts of the animal as well as vegetable frame :—chemical changes, putrefaction, and destruction immediately follow the total loss of life ; the importance of which becomes instantly evident when it is no more. I humbly conceive therefore, that if the human understanding can in any case flatter itself with obtaining (in the natural world) anything like a glimpse of the *immediate agency* of the Deity, it is in the contemplation of this *vital principle,* which seems independent of material organization, and an impulse of his own divine energy."

In another introductory lecture, recurring again to his own more particular study, Sir James mentions that "A gentleman * of the first rank in talents, as well as in society, lately suggested to me a lively illustration of the advantages of scientific botany as a mere amusement. 'Suppose,' said he, 'a person were possessed of a secret by which he could in the busy throng of this great town, on the sight of every one he meets, immediately recognise their name, their residence, their occupation, their connexions, without trouble or uncertainty, it would be thought an enviable talent; but if to this were added a power of learning, at the same time, their characters, dispositions, and powers, there are few who would despise the acquisition of such an advantage.' Just such is systematic botany with regard to plants. But we may take a far higher ground for the recommendation of all natural science: a taste for the beauties of nature, or in other words an admiration of the works of God, waiving the consideration of all worldly profit and advantage, cooperates with the highest means of divine instruction in elevating and improving the mind.

"The christian philosopher, the more deeply he studies nature, the more certainly finds every thing in harmony with his best hopes and dependencies. He well perceives that the same power which raises a plant from its seed, can abundantly perform its

* Sir Thomas Frankland, Bart.

promises to him. The caterpillar that resigns itself to its temporary tomb, should be a model of his own glad submission to the natural and moral laws of his Maker; and if in his inquiries and pursuits he is sometimes baffled, sometimes shocked or distressed by apparent unintelligible evil, amid a profusion of good, let him conceive as addressed to him those words which his Saviour addressed to his disciples on an extraordinary manifestation of his person, 'It is *I*, be not afraid!'"

The concluding passage of the Preface to his English Flora breathes the same sublime spirit, and may be referred to as the latest expression of those cheering hopes he was permitted to utter.

"He who feeds the sparrows, and clothes the golden lily of the fields in a splendour beyond that of Solomon himself, invites us his rational creatures to confide in his promises of eternal life. The simplest blade of grass, and the grain of corn to which he gives 'its own body,' are sufficient to convince us that our trust cannot be vain. Let those who hope to inherit these promises, and those who love science for its own sake, cherish the same benevolent dispositions. Envy and rivalship in one case are no less censurable, than bigotry and uncharitableness in the other. The former are as incompatible with the love of nature, as the latter with the love of God; and they altogether unfit us for the enjoyment of happiness here or hereafter."

Sir James composed at different times several Hymns; and these, in like manner with what has gone before, express his feelings and ideas. On such occasions they were warmed and elevated by a devotional fervour which those alone can know, who feel that at the approach of death,

> " They sink into a Father's arms,
> Nor dread the coming day."

From nine that are extant, the three which follow are inserted as specimens of his talent in this species of composition.

HYMN.

How shall my mortal powers aspire
To soar above this barren clod?
How join on earth the heavenly quire,
And hymn my Saviour and my God?

Can his transcendent grandeur bow
To hear a feeble creature's praise?
Can I propitiate, with my vow,
"The Ancient of Eternal Days?"

Yet what, but his almighty power,
Could first from dust and ashes bring
My humblest longings to adore
The heaven and earth's all-glorious King?

Would his supreme perfections shine,
Though veil'd, yet radiant, to my sight,
Were nought but sin and sorrow mine,
And my last refuge endless night?

Stretch then, my soul, th' advent'rous wing;
And dare to hope, and love, and praise:
The God who prompts thy voice to sing
Confirms thy hopes, and claims thy lays.

Thy love, a spark of heavenly fire,
His grace will raise and still refine,
Till certainty absorbs desire,
And heaven's eternal year is thine.

HYMN.

It is I, be not afraid.—Matthew xiv. 27.

When power divine in mortal form
Hush'd with a word the raging storm;
In soothing accents Jesus said,
"Lo, it is I! be not afraid."

So, when in silence nature sleeps,
And his lone watch the mourner keeps,
One thought shall every pang remove;—
Trust, feeble man, thy Maker's love.

Blest be the voice that breathes from heaven
To every heart in sunder riven,
When love and joy and hope are fled;
"Lo, it is I! be not afraid."

When men with fiend-like passions rage,
And foes yet fiercer foes engage;
Blest be the voice, though still and small,
That whispers—"God is over all."

God calms the tumult and the storm;
He rules the seraph and the worm;
No creature is by him forgot,
Of those who know, or know him not.

And when the last dread hour shall come;
While shuddering nature waits her doom;
This voicé shall call the pious dead;
"Lo, it is I! be not afraid."

The next, written after the loss of a mother he tenderly loved, and, in the course of a few weeks after, that of another relative, no less dear to the present writer, cannot be inappropriate in this place. Filial affection was one of the strongest which Sir James possessed, and it extended to them both in no common degree.

HYMN.

Thou shalt sleep with thy fathers.—2 Samuel vii. 12.

As o'er the closing urn we bend
Of each beloved and honour'd friend,
  What tears of anguish roll!
In vain in death's unconscious face
The living smile we seek to trace
  That spoke from soul to soul.

But shall not memory still supply
The kindly glance, the beaming eye,
  That oft our converse blest;
That brighten'd many a prospect drear,
Revived our virtue, soothed our care,
  And lull'd each pain to rest?

And when these frail remains are gone,
Our hearts the impression still shall own,
  Our mortal path to cheer:
O God! to point the way to heaven
These angel-guides by thee were given;
  How blest to meet them there!

Had the strength of Sir James Smith's bodily frame been commensurate with the powers of his mind, his friends might have anticipated length of days as his portion: but his health, which had never been unbroken, visibly declined for the last five or six years, and on the 17th of March 1828, after the illness of a single day, it pleased God to remove him from the world.

The quality which rendered the social character of the revered subject of these pages so engaging, arose not less from the benevolence of his heart than from the stores of his understanding. He shed a joy around him, springing from the pure sympathy he felt for those he knew to be his friends.

Absolutely free from the vanity of display, he sought rather to relax his overwrought mind in the company of such as cast aside their wisdom for a while, or in the presence of the young and innocent to study nature in another form.

As a naturalist he had a benevolent delight in infancy itself, and interpreted the mute language of smiles and struggles, the rudiments of expression and intelligence which met his own, and gave him the happy sensation of looking upon beauty and innocence in creatures destined to act a part in a new existence, claiming compassion and direction, and as the fairest semblance of the kingdom of heaven.

This temperament has been obvious to notice in the foregoing letters. However different he profession, age, character, or situation, of those who corresponded with him, we perceive that each addresses him as a friend upon whose fidelity he can rely;—upon various occasions we have observed these men of science appealing to the heart of their friend for sympathy in distress, or participation of prosperity, as if he were the intimate and sole depository of their affections and confidence.

Towards the young and friendless especially these feelings were directed; and in every instance it may with truth be asserted that Sir James was no repecter of persons, but quite unbiassed by the station, whether high or low, of the individual towards whom his kindly feelings were attracted. As might be expected, this strength of attachment in his disposition was productive of a corresponding one in the objects of his esteem. In later life more particularly, his love for the young was productive of much benefit to their moral character. He did not present virtue before their eyes, a stern forbidding form; but never omitted an opportunity which offered, to display the force of some moral precept, some religious truth, which his own experience had confirmed; and his own indulgent temper rooted such precepts more deeply in the heart.

Of selfish gratifications he had *none;* "his mind was formed for friendship, and could not exist without it."

But there is no point wherein he appears more amiable, than in the pleasure it gave him to promote or hear of the happiness of others; envy and jealousy were passions unknown to him, and he always considered them effects of conscious unworthiness. If he was in any respect wanting in charity, it was towards the malignant dispositions of the world.

Of the poor and humble, it gave him heartfelt delight to observe and enter into their scanty pleasures, their little vanity, or even weakness; but the knowledge of the sacrifices they make to humanity and duty, their kindness to each other, their fortitude in distress, melted his heart, and willingly would he have have wiped all tears from their eyes. He truly felt that "God hath made of one blood all the families of the earth," and his benevolent sympathies extended to the whole human race.

"A man of a kinder heart," observes his friend Dr. Davy, Master of Caius College, "more amiable dispositions, and purer moral habits, cannot exist; and it was his felicity to have his mental abilities proportioned to his moral qualities. In his own science he had from his youth stood at the head of the botanists of this kingdom, and I believe equal to those of any other; (but on this subject I must leave to those of his friends who are more skilful in that department, the office of making a more appropriate panegyric;) but of him it may be truly said, that none could enjoy the world possessing or deserving more friends, and none could quit it with better hopes."

"This great and irreparable loss," says Mr. Roscoe, "I too much remember as one of the weightiest misfortunes of my life; for though I was sensible that the health of my dear friend was precarious, yet I had flattered myself that, being younger by so many years than myself, I should have left him my survivor. I cannot however but rejoice in his calm and happy departure, his great worth fully understood, his fame established, and his most valuable work just finished. When I consider these circumstances, together with his pure and pious mind, I cannot repine at the result; and if it were not presumptuous, I would express an earnest wish that my latter day might be like his."

"To express," says Dr. Maton, "the sincere and deep concern which the melancholy intelligence created in my mind, and in that of every one who knew our departed friend, seems to be almost superfluous; for I am sure that you cannot imagine how it should be otherwise than that I should to my heart lament such a loss. The lovers of science will join in lamenting it throughout the civilized world, and our Society must feel it to be irreparable. *They* have lost their corporate *parent*, as well as one of their greatest ornaments. My friendship with the late excellent Sir James Smith was of almost *forty* years' standing,—it was cordial, it was *constant*. I shall never cease to reflect on it with peculiar pride and satisfaction."

The nobleman who has succeeded to the president's chair in the Linnæan Society, expresses a regard which could only be the consequence of the same friendly feelings and high esteem which Sir James entertained for his lordship.

"I cannot," writes Lord Stanley, "allow a single moment to pass without expressing my sincere thanks for the very flattering attention shown me in communicating the loss we have sustained by the removal of our valued friend Sir James Smith, —an attention the more flattering as I cannot but consider it a sort of testamentary addition to the many acts of kindness I have received from him in life, and as a proof that his surviving connexions do me the justice to consider me as one of those friends who truly regarded, and now as sincerely regret him.

"His loss, indeed, in the friendly and social circles in which he so much delighted, must be long and deservedly felt; nor will the place he filled in the estimation of the public and literary world, with so much advantage to their objects and such high credit to himself, be soon or easily supplied.

"I cannot but fear the effect which this shock may produce upon the health, already too much shaken, of another valuable friend,—I mean Mr. Roscoe, who, I believe, is at this moment at Holkham. It will indeed be singular if the lives of both these men, amiable and worthy, as well as able supporters of the great principle of religious freedom, should be fated to close just at the moment when that great point, as far as regarded at least all Protestant dissent, appears to be achieving its last final victory."

This misfortune did not follow immediately,—Mr. Roscoe survived his friend more than three years. He survived to add another testimony to those already adduced of the value he attached to the scientific attainments of his friend. In the preface to his splendid work upon the Monandrian plants, while speaking of the recent loss, Mr. Roscoe adds the consolatory reflection, "that it did not happen till Sir James had been enabled to terminate his great work, the English Flora, a work which perfected the system of English botany as far as present discoveries admit; and has, together with his other learned writings, conferred upon its author a name and station which will remain pre-eminent as long as the science itself exists."

"If," as we are told by the Rev. E. B. Ramsay, "much of the secret of human happiness consists, as Paley observes, in the formation of habits of observation, a knowledge of botany largely contributes to that happiness: for in a solitary walk, in a journey, or in the absence of those with whom we can converse, objects are constantly occurring to interest or amuse. Thus botany has sources of enjoyment similar to those so well described by Cicero when speaking of the happiness arising from the study of letters. 'Hæc studia adolescentiam agunt, senectutem oblectant, secundas res ornant, adversis perfugium ac solatium præbent, delectant domi, *peregrinantur, rusticantur.*'

"It is a noble and delightful office of the man of science to spread around him the happiness of knowledge, and to put into the power of others the gratifications which science can so liberally afford. But this is a power not granted to all who are in possession of knowledge; for the communication and the possession of wisdom are by no means always united; and surely the value of any man's knowledge is to be estimated very much according to the happiness he diffuses around him. The character of the late illustrious President of the Linnæan Society will thus live in connexion with science and its pleasures, and his name be repeated with gratitude by thousands, who will consider him as a benefactor, for having given them habits of observation and attention to the natural objects around them, by which their sources of enjoyment were multiplied and pleasures made to spring up at every step."

"To his extensive botanical acquirements," observes Professor Hooker, "he added the high attainments of an elegant scholar, and a talent of composition which has rendered his writings universally popular, and has been the means of throwing a charm over his botanical writings scarcely known to the science before."

A writer in the Philosophical Magazine * thus concludes his Obituary: "In summing up the scientific character of Sir James Edward Smith, it may be comprised in a few words. As a naturalist he contributed greatly to the advancement of the science, and stood pre-eminent for judgement, accuracy, candour, and industry. He was disposed to pay due respect to the great authorities that had preceded him, but without suffering his deference for them to impede the exercise of his own judgement. He was equally open to real improvement, and opposed to the affectation of needless innovation. He found the science of botany, when he approached it, locked up in a dead language;—he set it free by transfusing into it his own. He found it a severe study, fitted only for the recluse;—he left it of easy acquisition to all. In the hands of his predecessors, with the exception of his immortal master, it was dry, technical and scholastic;—in his, it was adorned with grace and elegance, and might attract the poet as well as the philosopher."

* May 1828. Vol. iii. p. 397.

APPENDIX.

INTRODUCTORY LECTURE,

Read at the London Institution, May 2nd, 1825.

THE history of botany as a science has often been given in various forms and languages. It makes a principal part of an Introductory Discourse which I had the honour of delivering at the opening of the Linnæan Society in 1788, and which is published in the Linnæan Transactions and elsewhere. The subject, under a rather different point of view, is continued in the article "*Botany*" of the Supplement to the Edinburgh Cyclopædia. It would ill become me to take up your time with what is detailed in those essays, though I have heard much of the former introduced into the lectures of other teachers; and if it tended to entertain or inform their hearers, my principal ends in the original composition were answered. I conceive that a writer on scientific subjects is most honoured, when his observations and discoveries are used, without mistrust or particular acknowledgement, as current coin, whose value is undisputed, and whose stamp and date are easily ascertained by those who are curious in such particulars. But an author himself ought to be the last person in the world to look for such authority or to repose upon it. The more he has done, the more he will find to do, and, if the world should be disposed to give him credit for any thing, the more he will be aware of the duty and the difficulty of not misleading its confidence or disappointing its expectations. When science has been long and extensively cultivated, and is become an object of great popular attention, it assumes a very different character from the abstruse pursuits of the cloister or the schools, or the speculations of a few recluse and abstracted proficients. No study has undergone a more remarkable change in this respect than botany. From its earliest dawn as a science, in the writings of the Greeks and Arabians, almost to our own time, it has been considered in no other light than as a branch of medical study; and even the highest praise bestowed upon this, his favourite pursuit, by the great Haller, is, that "it equals every branch of medical science in utility, and surpasses every one in agreeableness." The chief object of the earliest and most learned botanists, after the revival of literature, was to ascertain the plants used in medicine by the ancients; and however imperfectly described in their works, the vegetable kingdom was to these botanical physicians a great storehouse of remedies, whose recorded qualities were scarcely to be contested or examined, provided the individual plants could be settled beyond dispute. This indeed was the great difficulty,—to the more or less complete removal of which we are indebted for the existence of botany as a learned pursuit; which having engaged, and often baffled, the powers of many a first-rate genius, has taken its rank among the more distinguished studies of philosophers. The attention of many following ages was devoted to the mere discrimination of one kind of plant from another, before any ideas of the necessity or the principles of arrangement, much less of the constitution and œconomy of vegetable bodies, or of sound principles of inquiry into their natures and properties, had ever entered into the mind of man. Such objects are comparatively of recent date. Some of them have arisen almost under our own observation. They are now become the most important and instructive part of botanical science; and when united to an elegant and disinterested love of the beauty of nature, as the work of God and an emblem of his own boundless perfection, they constitute, in all their branches, one of the most refined, improving, and unexceptionable pursuits, that can claim the notice or employ the leisure of persons of either sex, or of any age or condition. Can it be necessary for me, before I go any further, to expatiate on the recommendations and advantages of my favourite study, to which I have devoted my life? I take it for granted that none of you would take the trouble to come here without some prepossession in its favour; and if your partiality be not increased by the views, however limited, which I shall have time and opportunity to lay before you, I might now raise your expectations only to disappoint them.

We cannot proceed far in even the most general and comprehensive views of the physiology of plants, without a perception that such inquiries are eminently useful in teaching us *to think*, to consider the most common objects in a perfectly new light, and with the help, as it were, of new senses. How many of the more intelligent and improved of rational beings have enjoyed, and daily do enjoy, the beauty and perfume of flowers, the verdure and grateful shade of trees, in all their luxuriancy of foliage,—without considering by what means or for what purpose those lovely forms and colours are so infinitely varied! How many taste the delicious variety of fruits, and even see them grow gradually to perfection, without bestowing a thought on the possible means by which these various exquisite scents and flavours are extracted from the same common soil; or how the air, the warmth, or the light of heaven, lend their respective assistance to the production of such wonders! To these contemplations the young mind may most beneficially be directed; but there is no mind above being improved by reflection, and the requisite exertion soon brings its own reward. But if the inquiries to which I have just alluded, rarely enter into the contemplation of a common observer, how much more seldom do the most reflecting minds spontaneously notice, with due attention, those almost infinitely varied circumstances of form, situation, or colour, by which individuals of the vegetable as well as of the animal creation, and their several parts, are distinguished; observing, at the same time, those delicate and intricate combinations and coincidences by which the whole of nature is harmonized, and her manifold irregularities and luxuriances reduced to the most perfect order! Here indeed our limited faculties fail us, as we attempt to measure them against the operations of Omniscience; but we may derive abundant consolation from the reflection, that we are allowed and enabled to consider those operations at all, and to converse, at the most humble distance, with *Him*, who in perfect wisdom has made them all, and who, in condescending to render his laws in any measure intelligible to us, has manifestly designed that we should at least *try* to understand them. If he has not made the study of these natural laws, like those of his moral government, our indispensable duty, he has reserved no ordinary reward for such of his children as raise their thoughts to him even among these his lowest works. What he has made beautiful to us, he doubtless intended we should admire; and in whatever form we are allowed to trace the footsteps of his wisdom and beneficence, it must be, in every instance, for the great ends of our own moral and intellectual improvement.

Such considerations as these cannot but suggest themselves to every mind directed to the observation of nature. They stimulate us to exertion, and they reward our perseverance. They furnish us also with a ready answer to all who doubt the importance or the utility of our pursuit, and who think it a degradation of their own sublime talents or characters, to submit either to the guidance or the admonition of infinite wisdom, however manifested to us.

How far the attentive study, or even the slightest observation of nature, is delightful and salutary to the mind, let those say who have, by a natural taste or an acquired habit, given themselves to such contemplations. Is it better to walk abroad with the eyes open or shut? Is the interchange of ideas in human society delightful and instructive? Are the imitations of nature in the finest works of art admirable? And shall it not be thought a privilege to hold converse with the source of all thought and wisdom and perfection? Do not the changes of seasons, and the endless variations in the aspect of nature and her productions, excite perpetual attention, and entertain us with never-ceasing variety? If our deepest inquiries lead us so far as to understand why a bud unfolds itself in the spring, and a leaf falls in autumn, we shall have learned enough to convince us of the existence of design in nature, and of the application of the most wise and compendious laws to the most decided and satisfactory ends. One fact, thus established and understood, may serve the philosopher as a basis, not by his machinery to overturn the world, but to raise a structure that shall truly reach from this world to another. The study of nature is undoubtedly, above all others, a science of practical observation; but to withhold the exercise of our reasoning faculties as we pursue it, would be a strange example of intellectual blindness. The old physicians indeed, to whom I have alluded, were often content to adopt the opinions and follow the practice of men who had lived a few centuries before them; and some have declared they would rather be *in the wrong* with an ancient than in the right with a modern. But such an abject prostration of the best faculties and duties of man has never long existed, except where the most sordid interest has acquired an exclusive dominion. In medicine and natural science such a tyranny must soon work its own cure;

for the physician or the philosopher who will not think for himself, is not likely to be long trusted or respected by others.

Systematic botany, as a branch of natural history, is conversant with the external forms of things; but even this is advanced by the study of their internal nature and œconomy. So the qualities of plants are better understood, the more attentively we scrutinize the laws of their exterior configuration. These various inquiries, therefore, may go hand in hand with great advantage; though they have seldom done so,—as will appear to any one who pays a little attention to the history and progress of each.

The discrimination of the species of plants, which so long occupied the earlier botanists, whose labours are recorded for our use in the numerous descriptions and often excellent figures of the sixteenth and seventeenth centuries, could scarcely be pursued without leading to some ideas of arrangement or classification. And yet it is a curious fact, that such principles as these early botanists put forth were singularly deficient in accuracy and ingenuity. To class plants according to their being eatable or poisonous, bulbous or fibrous in their roots, arborescent or herbaceous in their constitution, is, in every point, almost equally futile and uninstructive. Nor is it less curious, that philosophers of great eminence, who long after undertook the investigation of the anatomy and philosophy of the vegetable kingdom, were (to their own great disadvantage,) as little attentive to the natural affinities or diversities of plants, in the prosecution of their various experiments and the theories founded upon them. In like manner, the class of professed systematic writers, in their earliest attempts at arrangement, seem to have been led more by rules of technical discrimination than of natural or philosophical combination. None of these three descriptions of persons thought of deriving any aid from each other. Hence the reproach not altogether unjustly cast upon systematic botany, or at least upon some systematic botanists,—that their study teaches nothing of the uses of plants. On the other hand, physiologists, in their investigations of the vegetable as well as animal kingdoms, have fallen into errors and laboured under many disadvantages, for want of accurately knowing one thing from another; just as mere practical or empirical observers or collectors of plants, so far from noticing any circumstance that might lead by reason or analogy to assist their inquiries, very often, from the neglect of the clearest principles, commit the most absurd mistakes. Thus I have known Chervil gathered for Hemlock, the purple *Lythrum* for the *Digitalis*, and Creeping Crowfoot for the Fiorin Grass. The great physiologist Spallanzani made observations upon young aquatic snails, which had not yet acquired their shells, mistaking them for the vegetable *Tremella;* and proved, by learned deduction, that the said *Tremella*, or rather the snails, were of an animal nature; nor was he at all pleased at being set right in this matter of fact.

We are indebted to the Italian philosophers of the seventeenth century, especially to Malpighi, for the first considerable information respecting the anatomy or internal structure of the vegetable body; though our ingenious countryman, Dr. Grew, nearly about the same time was intent on the same subject. In the next century their facts and observations, assisted by new ones, were made good use of by physiologists, properly so called,—such as Hales, Blair, Bradley, and our immortal Newton himself, in England; and more recently by Du Hamel and Bonnet, with a number of less original writers, on the continent. Labourers in this field of science have, for some time past, not been wanting, though much yet remains to be done. Every hand is not alike successful in advancing the progress of real knowledge. Those who have had some particular object in view, as the uses of the stamens and pistils of plants, or the theory of their reproduction, have paid little regard to the physiology of other parts,—as the calyx, corolla, or foliage. So, with respect to classification, some have entered very deeply and successfully into the accurate principles of generic and specific distinction, as Dillenius, who hardly gave a thought to general arrangement; while many systems have been planned by men who had no correct ideas of the natural principles of discrimination or affinity.

It seems as if the science of botany were no sooner destined to emerge from obscurity and confusion, by a just perception on these various subjects, than it becomes in danger of plunging again into darkness by a neglect of them. The natural principles of arrangement for a scientific knowledge of plants, and a permanent discrimination of their families and species, have been no sooner distinguished (at the suggestion of Linnæus) from an artificial scheme, for their convenient investigation, than these different objects are confounded. Local Floras, for beginners, are disposed according to natural orders; as if plants could be made out by them in an analytical manner,—which the most learned could scarcely attempt. This would perhaps be of small consequence, did not the depreciation of the artificial system of Linnæus induce a neglect of those great and important principles, that didactic precision independent of all systems, to which his own fame and the advancement of every branch of natural science are owing. If any of his principles should prove to be ill-founded, or his laws objectionable, either with regard to the characters or the names of plants, let those who are competent by all means amend them. Let Linnæus himself be brought back to his own wise laws, which he too often transgresses.

But let not inadvertence or ignorance take place of his sagacity and experience, and publish their crudities without reading his works; founding genera without knowing any sound rules for their discrimination, and giving names according to futile principles which have been scouted over and over again.

The French nation, always aiming at a paramount authority in matters of science, grudging England her Newton, and Sweden her Linnæus, has attempted to make a schism in botany, as in other departments of knowledge. But these are the contemptible aims of second-rate characters.

The illustrious Frenchman Bernard de Jussieu, the early and confidential friend of Linnæus, conceived a philosophical idea of natural orders of plants, and conferred on the subject with the only man whom he found capable of appreciating or understanding them. He, in his letters to Linnæus, even gives him the honour of having first formed a scheme of natural orders of plants. However this may be, they certainly laboured for some time in conjunction. Afterwards Linnæus pursued his own course alone; and the first and most important result was the assertion of a principle to which I have already alluded,—that natural and artificial classification are in themselves essentially distinct, and calculated for purposes altogether incompatible with each other. This may be easily illustrated. That the distinction of one species of plant from another is strictly natural, no one will deny. Such a distinction is not only obviously founded in nature, but is confirmed, as clearly as possible, by experience, in the vegetable as well as the animal kingdom. Varieties among plants do, indeed, often arise from seed, originating in various causes; which causes are, as yet, very imperfectly known. But the experience of every day teaches, that such varieties are but of limited duration. They are aberrations from the regular plan of Nature, which she is ever checking and defeating. As far as any observation or record extends, no permanent change has taken place in any known species of plant. On the contrary, the history of those variations which have most attracted notice, as in eatable fruits and ornamental flowers, shows very distinctly that each has but a limited and uncertain duration. This fact being more clearly demonstrated by Linnæus than it had previously been, he advanced a step further, and asserted, that certain assemblages of species, of plants as well as animals, agreeing in their appropriate characters and qualities, are as much founded in nature as the species themselves. These are called *genera,* or families of plants. Each is marked by an assemblage of natural characters, in which our idea of such genus is founded; while it is distinguished from all others by some essential character. A few familiar examples will explain this. In the animal kingdom the great tribe of Monkeys forms a natural genus or assemblage of species, which no one will dispute. The Lion, Tiger, Leopard, Cat, and many more, constitute another, whose rough tongue and retractile claws exhibit an essential character, that confirms their other peculiarities and common habits. Parrots, also, happily termed "the monkeys of birds," are no less similar to each other than different from the rest of their tribe. In plants, the genus *Rosa,* under all the elegant differences of its species and their several varieties, is no less well marked. You will find its indisputable characters in every botanical book. The valuable genus *Quercus,* the Oak, known by the acorn and its scaly cup; the very curious and extensive genus *Ficus,* the Fig, so peculiarly distinguished by the concealment of its flowers in a cell or bag, which afterwards becomes the fruit; are both no less obviously defined by the hand of nature than a dog and a horse.

But there are other natural assemblages not less evident than natural genera.

These are termed natural orders.

The Rose, the Bramble, the Strawberry, and the Cinquefoil, have numerous characters, as well as properties, in common, which stamp them all together as a natural order; while each is marked by a peculiarity in its fruit, not unattended by other differences, by which its individual generic distinction is just as clearly defined. So beasts of prey and birds of prey, properly so called, are known to everybody as natural tribes, consisting of several no less natural genera or assemblages of species. Nor are Grasses, as one clear natural order, and Umbelliferous plants as another, less strikingly evident among vegetables. Bernard de Jussieu and Linnæus concurred in tracing out several of these natural orders. But Linnæus, who long survived his learned friend, at length became persuaded, that neither our knowledge of the species and genera of plants, nor the discriminative powers of the human mind, were as yet adequate to the forming a general system of such orders; nor even to define, perhaps any one of them, beyond a possibility of doubt or exception. Genera may, in a vast number of instances, be clearly and technically defined with almost mathematical precision; but scarcely any natural order, however distinctly perceptible, has yet been reduced to irrefragable definition in words. This being the case, it is manifest that no assemblage of such orders can serve for the analytical investigation of unknown plants.

That object is attainable by simple and indisputable principles only, which keep clear of the ambiguities of natural orders, and which require no previous knowledge of any such combinations. The purpose of an artificial system, or any system for the use of a learner, is, to make out the characters, name, and history of some unknown plant.

The chief object of the study of natural orders is to enable us to understand and to teach others the limits of natural genera.

It is of great consequence that these two very distinct objects should not be confounded, for they cannot be assimilated. A simple and ready artificial arrangement, on the plan of an alphabet, is necessary for those who, having found a plant, want to discover its name and history, by investigation of its technical characters, according to rules which are easily learned, and require no previous experience.

Such is the sole aim of the popular system of Linnæus. Being founded on parts which every plant in perfection *must have,* it must be universally applicable. These parts are the stamens and pistils of flowers. Their number, situation, and proportion, afford the principles by which the twenty-four classes of Linnæus, and their respective orders, are discriminated. This alphabet is the key to our botanical dictionaries, whether they be universal systems, or *Species Plantarum;* or whether they be partial or local catalogues, termed *Floras,* of particular countries, or catalogues of limited collections or gardens.

Such works as these last have multiplied prodigiously since the promulgation of the Linnæan system, and have tended very essentially to promote a practical knowledge of plants. To this end, they should be accompanied with as much useful information as their concise plan admits; or, at least, with references to more ample treatises.

Much scorn has been bestowed, by some affected philosophers, upon these "*mere catalogues,*" as they call them, which, as they allege, teach nothing but names. But these philosophers have never attempted to instruct us how we are to perpetuate or communicate a knowledge of things, without the very commodious invention of distinguishing them by names. The subjects of the vegetable kingdom are indeed so immensely numerous (sixty thousand Phænogamous), that this study of nomenclature is become of itself no less arduous than necessary; but be this as it may, it cannot be dispensed with.

Voyages and accounts of distant countries are now no longer written or read for mere amusement, like novels or fairy tales. We are not now to be satisfied with reading that particular nations or people use some "certain herb or tree" for food or medicine. We expect to know the scientific name, or the botanical affinities, of every curious or important plant, whence alone we can derive any real instruction.

I hope it will in time be thought no less essential, that the distinguishing characters of every such object, in the animal as well as vegetable kingdom, should be indicated by travellers of competent information.

When we have stored our minds with the names, and a sufficient practical knowledge of the technical distinctions of a number of plants, we may try to study their natural affinities and properties.

We shall find where we are most likely to meet with an eatable fruit, a medicine of a particular quality, or a texture or substance that may serve any one œconomical purpose. Thus a sound scientific knowledge is acquired step by step; and, accordingly, Linnæus read lectures on the natural orders of plants to his most assiduous and accomplished pupils only. To begin to teach botany by these orders, would be like putting Harris's *Hermes* into the hand of an infant, instead of his horn-book.

Linnæus, to the last, professed himself but a learner in this abstruse science; and was so well aware of his limited knowledge, that he never would attempt to construct any general system of natural classification. He did indeed, as his friend Bernard de Jussieu had done, attempt ranging his orders in a simple series, there being a manifest affinity between some of them rather than others. He went a step further, and formed a sort of map, indicating in a geographical manner the proximity and the bearings of the several orders with regard to each other. This was an original idea of Linnœus, and it greatly facilitates a comprehensive notion of the subject. In this state he left it; and we are indebted to two or three of his pupils, who took notes of his lectures, and with his leave published them, for almost all our knowledge of his principles or opinions. These are often so imperfectly or incorrectly detailed, that we cannot but lament the work was not undertaken, or at least revised and corrected, by himself.

Meanwhile, however, the ancient rivalship between the schools of Upsal and of Paris took the commendable and beneficial form of a just and worthy emulation. To the French botanists we are indebted for the great attempt of moulding into a regular shape the philosophical speculations of Linnæus and of Bernard de Jussieu. The latter was succeeded by a nephew, capable of advancing his undertakings, and worthy to inherit his fame. The present Professor de Jussieu has constructed a regular system, or methodical arrangement, upon the natural orders of his uncle: a system which is necessarily imperfect, because there are numerous genera whose orders remain undetermined; and necessarily artificial also, because human knowledge, as well as human intellect, being limited, no man can fully comprehend, much less preconceive, the entire plan of the all-wise Author of Nature, without exception or ambiguity. One celebrated character only, the famous Sir John Hill, has laid claim to this power.

When Dr. Garden discovered in Carolina what was thought to make a new order of animals, Sir John Hill wrote to congratulate him, as having verified a sublime speculation of his own.

"I have disposed," says he, "all the works of creation in a series or chain, and have distinguished each by its appropriate number. In doing this I have perceived that there must be some intermediate beings, not yet discovered, to fill up breaks in my chain. Your new animal comes precisely between two of my numbers where there is one number unoccupied." Had I not seen the original letter of Sir John Hill in Dr. Garden's hands, I could scarcely have believed the existence of so much presumption. The opinion, and its pretended anticipation, prove equally unfounded; for the animal supposed to constitute, not only a new genus but a new order, proves really a fish of the well-known genus *Muræna.* In acceding to this opinion of the learned Professor Camper I must ingenuously confess, that, in this instance at least, the sagacity of an able physiologist has rendered a material service to systematic naturalists; though I cannot submit to the contempt with which he in general regarded men as able and learned as himself, though in a line of science for which he had little taste or ability. The natural orders of plants, therefore, not being as yet all determined, and the relationship of many of them to each other either very disputable or not at all perceptible; there being, moreover, a great number of genera which cannot be referred to any of them; the chain of connexion by which they have, nowever ingeniously, been linked together, constitutes, in my opinion, but an artificial system, infinitely less easy and commodious than the professedly artificial system of Linnæus, and making combinations, in several instances not more natural. This has been well illustrated by my able friend Mr. Roscoe, in the eleventh volume of the Linnæan Society's Transactions. No one can deny that the present system of Jussieu offers the whole vegetable kingdom to our consideration under a most instructive view, as bringing together every acknowledged point of affinity, and indicating even the most remote. I only contend that it is but an artificial and imperfect *assemblage* of natural orders, and that its scheme of arrangement is totally unfit for the use of beginners, the most learned, who have no occasion to attend to it, being scarcely competent to unravel its numerous contradictions and exceptions. What seems to me most paradoxical is, that botanists, of the French school, who are strenuous for the use of natural orders to the exclusion of an artificial method, object altogether to the principle of Linnæus, that all genera are, or ought to be, natural; they assert, that all genera are mere artificial combinations, existing in the mind of man only; and yet they will have their own orders, which are assemblages of such genera, to be indisputably founded in nature. In the one point or the other they surely must err. Both appear natural to me, though I am very far from supposing that we have brought the knowledge of either to perfection. We can only advance towards that point by long experience and close observation.

The best practical botanist, is he who knows the greatest number of species, and combines them into the most natural genera. The best theoretical one, is he who conceives the most clear and comprehensive idea of natural orders, separately or combined; who perceives best what constitutes an order or a genus, and, above all, from what particular principles, in each different order, generic distinctions are best to be derived. But most of all I should wish to attempt the demonstration of some of his or Linnæus's natural orders, for they are very often the same, that you may be enabled to see in what their distinctions consist, and to pursue the ideas of these great men, for your own pleasure and instruction. The subject is boundless, and perfection is unattainable. But we may derive great assistance from considering nature under various points of view, and from applying to practical use what others have, with great pains and difficulty, been able to establish.

All systematic botanists, since the subject was first attentively considered, have concurred in one sentiment, that the parts of fructification, or, in other words, the organs which compose the flower and fruit, are the only ones which can, with any certainty, be resorted to for the essential principles of arrangement. Linnæus extends this maxim to the generic characters of plants, in which botanists of the French school do not concur.

I cannot but esteem this one of their greatest errors, which, as far as I can perceive, may always be avoided by a competent exertion of the mental powers; the parts of fructification, well considered, being abundantly sufficient in every case to characterize genera that are really and naturally distinct. For those that are not so, of which examples in books are but too frequent, I am ready to allow that no good distinctions are to be found. The first great difference in plants universally considered as dividing the vegetable kingdom into two most natural and essentially distinct sections, is founded on the structure of the seed, and its mode of germinating or sprouting. If its form be simple, and the bud or embryo undivided, the plants so circumstanced are said to be *Monocotyle-*

dones, or having but one cotyledon. If the embryo divides, in sprouting, into two * primary lobes, which originally made up the bulk of the seed, as in a Pea or Bean, and between which the young plant springs, such are termed *Dicotyledones*. These are the most numerous.

The obscure tribes of *Fungi, Mosses, Ferns*, and others, whose parts of fructification are but partially known, and which Linnæus referred to his class *Cryptogamia*, are supposed, with a few doubtful additions, to make up a third division, called *Acotyledones*, as having no cotyledon at all. But this is an extremely doubtful point, scarcely capable of demonstration, or even of investigation; and which the further we look we find less reason to adopt. It would be more consonant with fact to say, that what are denominated *Monocotyledones*, have really no cotyledons. As, however, the *Cryptogamous* tribes just mentioned, are too imperfectly known in their germination, to allow us to say any thing decisive about them, it is best to leave them apart, and to adopt for the bulk of the vegetable kingdom the popular terms and ideas of *Monocotyledones* and *Dicotyledones*. This distinction is confirmed by others, well worthy of remark. The *Monocotyledones* are most simple in their whole structure, external as well as internal. Their vascular system is on a larger scale, and more easily examined. Their bark and wood are scarcely, if at all, deposited in concentric or annual layers. Their stems are less branching, their foliage generally simple, undivided, and without indentation. The parts of fructification are disposed according to the number 3 and its compounds, which I believe the learned Mr. Brown either first observed, or, at least, has much confirmed and illustrated.

The *Dicotyledones* are in every respect more complex and various. Their seed-lobes either remain under ground, or more commonly are elevated in the form of a pair of leaves, till the real foliage comes forth. The wood of

* Rarely many-cleft.

their stems and roots is composed of annual circles, in the youngest or outermost of which the whole process of their circulation, nourishment, and increase, for the present year, is transacted. The number 5 and its compounds generally prevail in their parts of fructification.

These great primary distinctions being, in the main, established, we proceed to principles of subdivision. In these, what is technically called *insertion,* or the mode in which one part is situated with respect to another, takes the lead. Chiefly is to be considered whether the *Germen* (the rudiment of the fruit or seed,) is below the other parts of the flower or above them, *inferior* or *superior.*

Next we are to notice how the *Petals* (coloured leaves of a flower) if present, and the *Stamens* (organs of impregnation) are connected with each other, or with the *Calyx* (external covering). Lastly, whether the Stamens and Pistils are situated in the same flower or in two distinct ones of the same species.

These several principles are the key to Jussieu's system. The comparative value, or certainty, of each, we shall investigate as we proceed. We must always keep in mind that not one of them is, in every case, strictly absolute, or without exception, in its interference with natural orders, perhaps not even the great division of *Monocotyledones* and *Dicotyledones.*

Hence, an ingenious French author, M. Richard, rejecting this, has divided plants, according to their mode of germination, into *Endorhizæ* and *Exorhizæ.*

The first, as the name imports, produce their root from an internal tubercle, evolved in germination. In the others, the lower point of the embryo itself becomes the root.

Endorhizæ are generally analogous to *Monocotyledones,* and *Exorhizæ* to *Dicotyledones.* But besides the great obscurity of these distinctions, nothing is gained by them. They are liable to as many exceptions as the cotyledons.

I have only now to add, that nothing like mathematical certainty is to be found in the classification of plants.

Methodical arrangement of natural bodies is but a choice of difficulties. Specific characters, indeed, very often possess this absolute certainty, independent of any casual variation. Generic characters also are, not unfrequently, absolute, and capable of precise limitation. Sometimes they are marked by one most elegant and peculiar character. But classes and orders, however apparently obvious, are to be defined but in general terms; they are rather to be perceived than described. If, therefore, I shall be obliged, in explaining the method of Jussieu, to point out so many anomalies and inconsistencies, as may seem to render it too difficult and uncertain for the most learned to make any use of, I wish my hearers to attribute those imperfections rather to the intricacies of nature, than to the inability of the writer. The difficulties in question do not affect an experienced botanist; who, knowing the orders themselves, neither wants a clue to make them out, nor a very precise definition of their limits, concerning which he must judge for himself. But for these very reasons, it is clear as the day, that no student can enter upon the first knowledge of plants, and the investigation of their genera and species, by so intricate and precarious a path.

ON THE SURNAME OF LINNÆUS.—See page 343.

Sir J. E. Smith to the Editor of the Monthly Magazine.

Sir, Norwich, March 10, 1810.

In reply to your correspondent in the Monthly Mag. for March last, p. 123, I beg leave to give my reasons for continuing to write the name of Linnæus, in its original form, rather than Linné. The Swedes did not adopt the use of regular surnames till the early part of the last century. When each family took a name, literary people in general chose one derived from Greek or Latin; hence arose the family names of Mennander, Melander, Dryander, Aurivillius, Celsius, &c. Some gave a Latin termination to names of barbarous origin, as Bergius, Retzius, Afzelius, Browallius; and these became Swedish names even with that termination entire.

The name of Linnæus was in this latter predicament. Its termination, therefore, is by no means " boorish," or " plebeian," or " vile," but of classical origin; and these names have the peculiar felicity of being transferable into any language without inconvenience, and especially of entering spontaneously into Latin composition. If your correspondent be in the habit of writing or reading many scientific books in Latin, he will duly appreciate this last consideration. With respect to English writing, as we mention Titus, and Marcus Aurelius, in their original orthography, without following the French, who call them Tite and Marc Aurêle, no one has found any difficulty in making an English word of Linnæus.

When this great man became ennobled, I am well aware that, in conformity to the court ceremonies of the day, which were all French, a termination borrowed from the language of that people was, in his case, as in others, adopted, with the strange jumble of a Gothic prefix, and he became in Swedish von Linné, as in French de Linné, and in barbarous Latin à Linné. No one that I know of has adopted any of these in English, though some have called him Linné, but hitherto with little success. I presume no one would wish to Anglicize his name into Linny; and yet that, however ridiculous, would be the only correct and consistent measure, except we retain the *von*, the *de*, or the *à*.

I have therefore always used his original name, without any design, or surely any suspicion, of slighting the honours which his sovereign conferred upon him, and which, I will venture to say, reflected glory on his royal patron in return. By such a disposal of honours their lustre is preserved, as in the cases of a Marlborough, a Newton, and a Nelson, from that deterioration to which, from human imperfection and error, they are in their very nature otherwise prone, but from which it is the interest of every good citizen to guard them. I do not conceive, however, that any one needs to be reminded of the various dignities, whether courtly or academical, conferred on the illustrious Swede. His simple name Linnæus, recalls them all. We have no occasion to say *the Emperor* Julius Cæsar, *King* Henry IV. of France, *Mr. Secretary* Milton, or *the Right Honourable Mr.* Addison. Neither is it necessary to say Sir Charles Linnæus, or the Chevalier de Linné, to remind us that he was Knight of the Polar Star, and the first person who ever received that honour, equal to the Garter with us, for literary merit. I must, therefore, protest against any interpretation of intended slight in this case, for my meaning is the very reverse.

I believe the practice followed in England has decided the conduct of other nations. In Latin he is now always called Linnæus, even by the Swedes, and, what is still more striking, the French now write Linnæus, even in their own language. I presume your correspondent had never a design of recommending, for *Latin* composition, any thing but Linnæus; and I hope he will not hereafter think me pertinacious, nor in any respect blameable, if, for the above reasons, I continue the same practice in English, leaving every one to follow me, or not, at his discretion, and trusting to time and experience for a final decision. I must express my regret that the title of the Linnæan Society, as I would always write it, has in its charter been spelt Linnean. The latter had in view the name of Linné, and was so far proper; but I have always conceived the diphthong to be more classical, and if we preserve the word Linnæus in English, undoubtedly more correct. In this point, most certainly every writer may judge for himself, and in speaking there luckily is no ambiguity.

I remain, Sir,

Yours, &c.

Jas. Edw. Smith.

The grammatical article printed below is mentioned at p. 340, in a letter to Mr. Roscoe.

OBSERVATIONS ON *SHALL* AND *WILL*.

To the Editor of the Athenæum.

Sir, Norwich, Aug. 5, 1807.

The various writers on English grammar have scarcely explained with sufficient precision the proper use of the verbs *shall* and *will*, with the preterites *should* and *would*. Many persons consider the just application of these verbs as a sort of *shibboleth*, a test of a true-born Englishman, for which no rule can be given, and which no foreigners, not even our fellow-subjects of Scotland and Ireland, can ever learn. The latter is undeniable; and it is equally true that an Englishman, however uneducated, will never, except corrupted by an intercourse with strangers, commit an error in this respect. Nevertheless, before we indulge in triumph over other people, it becomes us to give, if possible, a reason for our own conduct; nor ought any reproach to fall on those who, in learning a language, err on a point concerning which no rule can be given.

Allow me therefore to suggest that *shall* and *will* are two distinct verbs, *which interchange their meanings in the different persons*. *Shall* and *should* express, in the first person, simple futurity, but in the second and third persons they imply a command or decision of the speaker. On the contrary, *will* and *would* in the first person imply the same command or decision, while in the second and third they express only simple futurity. There is, however, a third verb, *will*, in every one of the three persons, which expresses the decision of the person spoken of.

This is best understood by a table of each verb and its tenses.

No. 1. expresses simple futurity:	No. 2. the decision of the speaker:	No. 3. the decision of the party spoken of.
I shall	I will	I will
Thou wilt	Thou shalt	Thou wilt
He will	He shall	He will
We shall	We will	We will
You will	You shall	You will
They will.	They shall.	They will.
	Preterite.	
I should	I would	I would
Thou wouldst	Thou shouldst	Thou wouldst
He would	He should	He would
We should	We would	We would
You would	You should	You would
They would.	They should.	They would.

The first persons singular and plural in both tenses of the verbs No. 2. and 3. are necessarily the same, because the speaker and the party spoken of are one; and therefore No. 3. may be said to have no first person. The only apparent ambiguity lies between the second persons of No. 1. and 3; but the verbs are so essentially distinct in sense, that this is never perceived even in writing; and in conversation it is always avoided by the emphasis laid on No. 3. An Englishman says to his friend, "I *shall* be very glad to see you:" not, like the Scotch and Irish, "I *will* be very glad." The simple futurity only is here proper to be expressed, for an effort of implied volition destroys the civility. It is no compliment to any one to say, "I *will try* to be *glad* to see you;" and the expression itself is a solecism. Writers on experimental philosophy in Scotland often say, "If we make such and such an experiment, we *will* find the result, &c." though they mean simple futurity, that the result *will* be so. It is as ridiculous to say, "we *will* (that is, we are determined to) find the result so," as to say, "the result *shall* be so;" or else to understand the verb No. 3. instead of No. 1. "the result *will* (or is in its own mind determined to) be so." A true English writer is so much on his guard against the above error, that he sometimes falls into a contrary fault; for instance, in a note of civility written, as usual, in the third person:—"Mr. A. hoped to have sent Mr. B. the promised books this morning, but is disappointed. He trusts, however, that next week he *shall* have no difficulty in procuring them." The true meaning of which is, "he confidently thinks to himself, *I shall have no difficulty in procuring them.*" The writer could not so completely transform himself in idea into a third person, as to forget his own identity. If he had wanted to say, "he trusts that his bookseller *will* have no difficulty in procuring them," the third person being evident, no doubt could have arisen in his mind.

Perhaps the use of the three verbs in question may be best illustrated by a familiar conversation between A., B., and C., in which each verb is numbered wherever it occurs.

A. I *shall*[1] fall down the precipice if I go further in this slippery path. If C. goes on, he *will*[1] inevitably tumble, and so *wilt*[1] thou. We *shall*[1] all break our necks. I *will*[2] turn back.

B. We *will*[2] go on, and thou *shalt*[2] go with us, whether thou *wilt*[3] or not.

A. You *will*[1] both repent when too late, since you *will*[3] be so obstinate. What perverseness! They *will*[1] suffer, and I *shall*[1] be blamed; but their parents *shall*[2] know the truth. I *will*[2] bear this no longer: you *shall*[2] both turn back. I *should*[1] like to see you fall if I were sure you *would*[1] not hurt yourselves, and then you *should*[2] obey me whether you *would*[3] or not. Your parents *should*[2] have sent some one with you that you *would*[3] have attended to; but they *would*[3] not, even though I *would*[2] have had them. They *would*[1] be very sorry to see what I see, and they *would*[3] be obeyed.

C. We *would*[2] obey thee if thou *wouldst*[3] but be less loquacious; and whatever may happen, I insist that thou *shouldst*[2] not blame us. Thou *wouldst*[1] be most to blame for alarming us.

We must not confound with the foregoing verbs a fourth, *should,* as a sign of the potential mood, or, in other words, as *expressing the possibility of simple futurity.* This is unchanged in all the persons, whether singular or plural.—Thus C. might proceed in the above conversation:

C. If I *should*[4] fall, and thou *shouldst*[4] see me, and if B. *should*[4] be unable to help me, we *should*[1] all lament our mishap. If we *should*[4] be so unfortunate, and if you both *should*[4] go home without me, and if my parents *should*[4] be there, &c. &c.

An apparent exception to the above rule occurs in scriptural language: "The Lord shall judge the world in righteousness." "The Lord shall reign for ever and ever," &c. In such cases, though the fact ought, in ordinary language, to be announced as a matter of simple futurity, yet by a kind of poetical licence, the narrator assumes more than ordinary authority, which can only be expressed by the tone of *decision* or *determination* derived from inspiration itself. We overlook the speaker or writer in the consideration of this all-commanding decision, proceeding from the fountain of unchangeable truth. To reduce these sentences to grammatical exactness by substituting *will* for *shall,* as some injudicious and tasteless reformers of spiritual poetry have done, enfeebles their sentiment, without adding any thing to their precision, for they are used only on an occasion in which there is no possibility of mistake.

I cannot but esteem the irregularity of the verbs in question rather a beauty than a defect in our language. No Englishman is puzzled in their use, because the rule above given is absolute. Endless confusion arises from those who, not perceiving that rule, attempt to correct the language; and the matter is only further obscured by writers who, considering the whole but as one verb, explain it on false principles.

J. E. S.

For the following anecdotes concerning the great Haller, Sir J. E. Smith was indebted to Mr. Davall. Some of them are new, and any incident in the life of such a man cannot be a subject of indifference to the reader.

ANECDOTES RELATING TO HALLER,

Communicated by Mr. Wyttenbach to Mr. Davall: some extracted from Zimmerman's Work,—these are marked Z.; *others, from Wyttenbach's private knowledge, with* W.

Dans son enfance il étoit foible et maladif; ce qui a fait qu'il prit de très-bonne heure du goût pour la retraite, et une inclination insatiable pour la lecture. Aussi à l'âge de neuf ans il commença à compiler un dictionnaire Hebreu et Grec de tous les mots du Vieux et du Nouveau Testament, une grammaire Chaldaïque, l'histoire des vies de plus de 2000 hommes savants et célèbres selon les modèles de Bayle et de Moréri. Tout ce qu'il entreprit déjà dans ce tems-là, visoit au grand, et son génie vouloit tout embrasser, tout approfondir; mais sa vie retirée, la simplicité extrême et naïve de ses mœurs, le peu de goût qu'il montroit pour les amusements ordinaires de la jeunesse, firent qu'on le traita presque d'imbécile, et qu'on le méprisa même parmi ses parents.—Z.

Dès l'age de dix ans son amour pour la poésie s'éveilla et le domina si fort pendant quelques années, qu'il ne fit presque que composer des poésies qui quelques années après ont été brulées par lui-même.—Z.

Un trait assez drôle de sa simple naïveté est, que rencontrant une fois une pauvre personne, il lui donna une pièce de cinq ou de dix batz*, tout ce qu'il possédoit. Le lendemain, en se réveillant et en s'habillant, il se mit à pleurer,—Demandé pourquoi? sa réponse fut, "J'ai donné tout mon argent à un pauvre, et comme l'Evangile dit que les aumônes seront récompensées plus que doublement, je cherchois à-présent dans ma poche mon argent doublé, et je n'y ai rien trouvé."—W.

Une bonne partie de ses sublimes poésies ont été composées dans ses courses botaniques, ou après avoir admiré les beautés merveilleuses de la nature dans les plantes, et fatigué de ses courses, il se jettoit sous l'ombre d'un bois, d'un arbre, d'un rocher, sur le gazon, méditant et composant dans son imagination les plus belles pensées, qu'il ne méttoit sur le papier qu'étant revenu à la maison.—Z.

Le bois de Bremgarten près de Berne, le Glassbrunn si souvent cité dans son *Hist. Stirpium*, et où croit la belle *Jungermannia ciliaris*, le Drakau, l'un et l'autre dans le même bois, ont très-souvent été ses retraites botaniques et poétiques. C'est sur le Gurten, montagne près de Berne, " supra semitam cavam, quæ ducit ex Wabern ad villam Spiegel" (souvent mentionné dans son *Historia Cryptog.*), qu'il a composé le commencement de son poème " Sur l'Origine du Mal."—W.

Revenant de ses voyages en 1729 à Berne, il essuia très-souvent les effets du motto des républiques, "Ne quis emineat." On disoit de lui qu'il étoit grand théoréticien, mais foible praticien en médecine. Il se présenta pour la place de médecin praticien de l'hôpital, mais on disoit que cela ne convenoit pas à un poëte. Il se présenta après pour la chaire de professeur en belles lettres;

* *Une pièce de* 10 *batz*, worth 1*s*. 3*d*. English.—E. D.

mais le refrein contre lui fut, que cela ne convenoit pas à un médecin.—Z.

Haller sur la fin de sa vie, dans son *Usong*, a fait un tableau d'après nature de tous ces contretems et obstacles qu'il a essuiés à Berne.—W.

Ne sachant pas ce que Mr. Smith a déjà dit de Haller dans son Voyage, je ne puis pas savoir non plus ce que je pourrai lui envoyer d'anecdotes sur ce grand homme; mais quand une fois la première édition de ses Voyages aura paru, il me sera plus facile de l'augmenter par des extraits de Zimmermann, ou par d'autres anecdotes.

Ce Zimmermann lorsqu'il a composé l'histoire de la vie de Haller, a été si prodigieusement enthousiasmé de son Mécène, qu'une fois le Docteur Langhanns de Berne, en présence de Mr. Sprungli le naturaliste, qui furent alors à Gottingue, osant dire que Mr. Haller n'étoit pas infaillible, et qu'il avoit aussi ses foiblesses, ce docteur a sur cela été provoqué en duel par Zimmermann. Mr. Sprungli m'a souvent raconté cette histoire, et jamais sans rire de tout son cœur.

Vous savez que j'ai vû, sur la fin de sa vie, Mr. Haller presque tous les jours pendant près de quatre mois. Il avoit un sommeil qui le persécutoit tous les jours d'abord après midi, et s'il s'y abandonnoit, il devenoit sombre vers le soir et mélancholique, et alors il ne vouloit entendre parler que de religion, et préféroit en général les écrits de St. Jean, dont il falloit toujours lui citer des passages. Si au contraire on l'avoit empêché de dormir, et il prioit même tous les assistants de le secouer, de le réveiller, alors il étoit calme, serein, tranquille, s'entretenoit avec plaisir de sciences, de nouvelles littéraires, et sa conversation étoit très-intéressante ; mais il revenoit cependant toujours à la religion, et me disoit souvent, " A-présent, mon ami, quittons les bagatelles, le tems qu'il me reste est trop court, il s'ecoule fort vite ; dites moi quelque chose de notre Sauveur, car il n'y a que lui qui puisse me sauver et me rendre heureux."

Il me disoit souvent, "Je suis pourtant un être bien malheureux, autrefois j'avois des idées si sublimes de l'éternité, je sentois si vivement le bonheur d'être vertueux, j'avois tant d'enthousiasme pour ma religion; mais à-présent je me sens si froid, si pesant, si indifférent, que cela m'attriste véritablement." Quand même alors je disois à ce grand physiologiste, " Mais vous savez que vous n'êtes plus dans l'âge où vous avez pleuré la mort de votre première femme: votre jeunesse, où dans le Bremgarten vous avez composé votre poème sur l'éternité, s'est écoulée, vos forces ont diminué, votre corps est usé." Cela ne le consoloit pas assez, et il ne vouloit jamais comprendre que son imagination et ses sentiments ne pouvoient plus être aussi vifs que dans sa jeunesse.

Ce grand poëte sur la fin de ses jours, pendant bien des mois, ne trouvoit de nourriture et de consolation que dans le simple E'vangile, dans les poésies les plus simples, et si je voulois l'entretenir à son gré, il falloit lui parler de religion, comme j'en parle aux plus simples paysans, aux enfants mêmes.

Quelques jours après que l'Empereur fut chez lui, un bon ecclésiastique qui le voyoit souvent, et dont d'ailleurs il étoit très-content, lui dit : "Mais, Monsieur ! je dois à juste titre vous féliciter de l'honneur que l'Empereur vous a fait. Cela doit vous soulager, vous faire grand plaisir." Haller n'y répondit rien pendant quelque tems, et après quelques moments il regarda le pasteur, et lui dit en souriant, " Heureux ceux dont les noms sont inscrits au ciel," et dès lors il ne fut plus question de l'Empereur.

Malgré cette façon de sentir, de penser, et d'agir, au milieu même de ses douleurs continuelles, et voyant la mort s'approcher de fort près, il lisoit également des romans, des brochures assez légères, &c. pour en faire des extraits et critiques pour la Gazette de Gottingue. Je lui reprochai cela une fois en lui disant, dans un accès qu'il avoit de mélancholie, que j'étois fort étonné de le trouver si souvent occupé de pareilles bagatelles ; mais il me répondit sur le champ, "C'est mon métier, mon ami ! je fais cela pour le Journal de Gottingue, non pas pour m'en amuser."

Ne pouvant être un moment sans s'occuper et travailler, et ne voulant pourtant pas abuser du Sabbath,—il s'occupoit ces jours à composer ses lettres sur la religion, et la réfutation des ouvrages de Voltaire.

Il est mort en observateur, en médecin. J'étois le dernier jour de sa vie à la campagne ; il me fit faire plusieurs messages, que je ne sçus malheureusement que le soir en revenant en ville. J'y courus fort vite, mais je le trouvai déjà mort depuis quelques moments. Un ecclésiastique des plus respectables de notre ville lui avoit tenu la tête, et tout d'un coup Haller dit avec une certaine force, " Je meurs ! " et c'en étoit fait.

Le seul honneur qu'on a fait à Haller après sa mort à Berne, fut celui de placer son portrait dans la galérie où sont nos avoyers, entre les deux avoyers regnants ; mais je crois qu'il doit cette distinction plutôt aux sentimens de Mr. Sinner, qui étoit alors bibliothécaire, qu'à un ordre souverain.

Vous verrez de tout ceci que je viens d'écrire fort à la hâte ce qu'il vous plaira ! Et disez à Mr. Smith que je me fais véritablement un plaisir de lui rendre quelque service ; ne pouvant pas lui fournir des découvertes intéressantes en botanique, je voudrois bien lui prouver par d'autres services que je l'éstime infiniment.

Je viens de lire dans le Journal littéraire de Jéna que quelqu'un s'est déjà chargé de la traduction des Voyages de Smith, ainsi que je puis espérer de les lire bientôt.

A REVIEW OF THE MODERN STATE OF BOTANY,

WITH A PARTICULAR REFERENCE TO THE NATURAL SYSTEMS OF LINNÆUS AND JUSSIEU*.

[" *A Review of the modern State of Botany,*" *mentioned* Vol. i. p. 494, *not having been printed in the Transactions of the Linnæan Society, it may prove an acceptable addition to the present work to introduce this Essay, which from its situation in the* Encyclopædia Britannica, *not being accessible to general readers, is given entire in this place.*]

THE Linnæan system of botany, the principles upon which it is founded, with its application to practice, have all been amply elucidated in the fourth volume of the *Encyclopædia Britannica.* The reader will there find a general view of this celebrated system, including the generic characters, as well as some of the specific differences, of most plants hitherto discovered, with their qualities and uses. The terminology of Linnæus is explained; his arguments for the existence of sexes in flowers are detailed; his ideas of a natural method of classification, and of its utility in leading to a knowledge of the virtues of plants, are subjoined to a compendious history of botanical science.

The writer of the present supplementary article proposes to take a different view of the subject. This study has, within twenty or thirty years past, become so popular, and has been cultivated and considered in so many different ways, that no dry systematic detail of classification or nomenclature is at all adequate to convey an idea of what botany, as a philosophical and practical pursuit, is now become. The different modes in which different nations, or schools, have cultivated this science; the cir-

* From the second volume of the Supplement to the *Encyclopædia Britannica.*

cumstances which have led some botanists to the investigation of certain subjects more than others; and the particular success of each; may prove an amusing and instructive object of contemplation. In this detail, the history of scientific botany will appear under a new aspect, as rather an account of what is doing, than what is accomplished. The more abstruse principles of classification will be canvassed; and the attention of the student may incidentally be recalled to such as have been neglected, or not sufficiently understood. The natural and artificial methods of classification having been, contrary to the wise intention of the great man who first distinguished them from each other, placed in opposition, and set at variance, it becomes necessary to investigate the pretensions of each. The natural method of Linnæus may thus be compared with his artificial one; and as the competitors of the latter have long ceased to be more than objects of mere curiosity, we shall have occasion to show how much the rivals of the former are indebted to both. In the progress of this inquiry, the writer, who has lived and studied among the chief of these botanical polemics, during a great part of their progress, may possibly find an occasional clue for his guidance, which their own works would not supply. No one can more esteem their talents, their zeal, and the personal merits of the greater part, than the author of these pages; but no one is more independent of theoretical opinions; or less dazzled by their splendour, even when they do not, as is too often the case, prove adverse to the discovery of truth. Nor is he less anxious to avoid personal partiality. *Incorruptam fidem professis, nec amore quisquam, et sine odio, dicendus est.*

About the end of the seventeenth century, and the beginning of the eighteenth, the necessity of some botanical system, of arrangement as well as nomenclature, by which the cultivators of this pleasing science might understand each other, became every day more apparent. Nor was there any deficiency of zeal among the leaders and professors of this science. Systems, and branches of systems, sprung up over the whole of this ample field, each aspiring to eminence and distinction above its neighbours. Many of these, like the tares, that fell by the way side, soon withered for want of root; others, like the *herba impia* of the old herbalists, strove to overtop and stifle their parents; and all armed themselves plentifully with thorns of offence, as well as defence, by which they hoped finally to prevail over their numerous competitors. This state of scientific warfare did not, in the mean while, much promote the actual knowledge of plants, though it prepared the way for a final distribution of the numerous acquisitions, which were daily making, by the more humble, though not less useful, tribe, of collectors and discoverers. The success of the Linnæan artificial system is not altogether, perhaps, to be attributed to its simplicity and facility; nor even to the peculiar attention it commanded, by its connexion with the striking phænomenon, brought into view at the same time, of the sexes of plants. The insufficiency, or at least the nearly equal merits, of the many other similar schemes that had been proposed, began to be most strongly felt, just at the time, when the great progress and success of practical botany, rendered the necessity of a popular system most imperious. While the cause of system was pending, some of the greatest cultivators of the science were obliged to have recourse to alphabetical arrangement. This was the case with Dillenius, the man who alone, at the time when Linnæus visited England, was found by him attentive to, or capable of understanding, the sound principles of generic distinction. These he probably understood too well to presume to judge about universal classification. It was the fashion of the time however for every tyro to begin with the latter; and the garden of knowledge was consequently too long encumbered with abortive weeds.

Linnæus had no sooner published and explained his

method of arranging plants, according to that which is generally termed he Sexual System, than it excited considerable attention. His elegant and instructive *Flora Lapponica* could not be perused by the philosopher or the physician, without leading its readers occasionally aside, from the immediate objects of their inquiry, into the paths of botanical speculation, and awakening in many a curiosity, hitherto dormant, on such subjects. But the scope of that limited Flora is by no means sufficient to show either the necessity or the advantages of any mode of arrangement. Linnæus may be said to have grasped the botanical sceptre, when, in the year 1753, he published the first edition of his *Species Plantarum;* and the commencement of his reign must be dated from that period. The application of his system to universally practice, in this compendious distribution of all the known vegetables of the globe; his didactic precision; his concise, clear, and certain style of discrimination; his vast erudition displayed in synonyms; and, perhaps as much as any thing else, the fortunate invention of trivial or specific names, by which his nomenclature became as evidently commodious, and indeed necessarily popular, as any part of his performance; all these causes co-operated to establish his authority. An immediate impulse was given to practical botany. The vegetable productions of various countries and districts were marshalled in due array, so as to be accessible and useful. A common language was established throughout the world of science; a common stock of knowledge and experience began to accumulate, which has ever since been increasing, and can now never be lost. Of these partial Floras to which we allude, those of Lapland and Sweden, the productions of Linnæus himself, were the models of most of the rest, and have never, on the whole, been excelled.

Hence arose the Linnæan school of botany, which though founded in Sweden, extended itself through Holland, Germany, and more or less perfectly in other parts of Europe, though not without impediments of which we are hereafter to speak. In Britain it was firmly established by the influence of some of the most able pupils of Linnæus, and strengthened at length by the acquisition of his literary remains. But these are adventitious supports. The strength of philosophical, like political, authority is in public opinion, and the cement of its power is public good.

As we proceed to trace the practical influence of the Linnæan system, or rather of the facility which it afforded, in botanical studies, it will be useful at the same time to observe the effects of adventitious circumstances, which render botany almost a different sort of study in different parts of the habitable globe.

In those northern ungenial climates, where the intellect of man indeed has flourished in its highest perfection, but where the productions of nature are comparatively sparingly bestowed, her laws have been most investigated and best understood. The appetite of her pupils was whetted by their danger of starvation, and the scantiness of her supplies trained them in habits of œconomy, and of the most acute observation. The more obvious natural productions of such climates are soon understood and exhausted. But this very cause led Linnæus to so minute a scrutiny of Swedish insects, as had never been undertaken before in any country; in consequence of which a new world, as it were, opened to his contemplation, and the great Reaumur declared that Sweden was richer in this department than all the rest of the globe. Such indeed was its appearance, because it had been more carefully examined. When the ardour and acuteness of the pupils of the Linnæan school first sought matter of employment for their talents, some few had the means of visiting distant, and scarcely explored, countries; but this could not be the lot of many. The greater part were confined to their native soil; and it is remarkable that those who were longest so confined, have displayed in the sequel the greatest abilities, and have rendered the greatest services to science, independent of the accidents which made the labours of others imperfect or abortive. Such men as Ehrhart and Swartz were not to be satisfied with the general productions of the fields or gardens to which they had access. They had no resource but in the recondite mysteries of cryptogamic botany, in the first instance. To these they directed their microscopic eyes, and more discriminating minds, with the happiest success. When they had derived from hence an ample harvest, Ehrhart, limited in circumstances and opportunites, hindered moreover perhaps, in some degree, by a singularity and independence of charäcter, not always favourable to worldly prosperity, opened to himself a new path. The native trees of the north, and especially the hardy shrubs and arborescent plants of the gardens, had not, as he judiciously discovered, received that correct attention, even from his master Linnæus, which was requisite to make them clearly understood. Difficulties attending the study of these plants, the various seasons in which they require to be repeatedly scrutinized, and the obscurity or minuteness of the parts on which their differences depend, were by no means calculated to deter this laborious and accurate inquirer. He submitted the supposed varieties of the shrubbery, the kitchen garden, and even of the parterre, to the same rigorous examination, and, for the most part, with the happiest success. His discoveries have not received the notice they deserve, for his communications were deformed with asperity and pedantry, and he did not always keep in mind the concise and sober principles of definition, which his preceptor had both taught and practised, and to which he owed so large a share of his well-merited fame. Ehrhart died prematurely; but his name ought to be cherished among those whose talents have advanced science, and who loved Nature, for her own sake, with the most perfect disinterestedness.

The fate of Swartz has been far more propitious to

himself and to the literary world. Having thrown more light upon the cryptogamic productions of Sweden and Lapland than they had previously received, and which has only been exceeded by the more recent discoveries of the unrivalled Wahlenberg, he undertook a botanical investigation of the West Indies. Carrying with him, to this promising field of inquiry, so great a store of zeal and practical experience, his harvest was such as might well be anticipated. Whole tribes of vegetables, which the half-learned or half-experienced botanist, or the superficial gatherer of simples or flowers, had totally overlooked, now first became known to mankind. Tropical climates were now found to be as rich as the chill forests and dells of the north, in the various beautiful tribes of mosses; and the blue mountains of Jamaica rivalled its most fertile groves and savannas in the beauty, variety and singularity of their vegetable stores.

Nor must we pass over unnoticed the discoveries of another illustrious disciple of Linnæus, the celebrated Thunberg, who has, now for many years, filled the professorial chair of his master, with credit to himself and advantage to every branch of natural science. The rare opportunity of examining the plants of Japan, and of studying at leisure the numerous and beautiful productions of the Cape of Good Hope, as well as of some parts of India, have thrown in the way of Professor Thunberg a greater number of genera, if not species of plants, than has fallen to the lot of most learned botanists; except only those who have gone round the world, or beheld the novel scenes of New Holland. These treasures he has contemplated and illustrated with great advantage, so far as he has confined himself to practical botany. We lament that he ever stepped aside to attempt any reformation of an artificial system. It is painful to complain of the well-meant, though mistaken endeavours of so amiable and candid a veteran in our favourite science; but what we conceive to be the interests of that science must form our apology. We cannot but be convinced, and the experience of others is on our side, that discarding those principles of the Linnæan system which are derived from the *situation* of the several organs of impregnation, and making *number* paramount, has the most pernicious and inconvenient effect in most respects, without being advantageous in any. This measure neither renders the system more easy, nor more natural, but for the most part the reverse of both. We have elsewhere observed, (*Introduction to Botany, ed.* 3. 358,) that the amentaceous plants are of all others most uncertain in the number of their stamens, of which Linnæus could not but be aware. "Even the species of the same genus, as well as individuals of each species, differ among themselves. How unwise and unscientific then is it, to take as a primary mark of discrimination, what nature has evidently made of less consequence here than in any other case!" When such plants are, in the first place, set apart and distinguished, by their monoecious or dioecious structure, which is liable to so little objection or difficulty, their uncertainty with respect to the secondary character is of little moment; their genera being few, and the orders of each class widely constructed as to number of stamens. Linnæus, doubtless, would have been glad to have preserved, if possible, the uniformity and simplicity of his plan; but if he found it impracticable, who shall correct him? Such an attempt is too like the entomological scheme of the otherwise ingenious and able Fabricius. The great preceptor having arranged the larger tribes of animals by the organs with which they take their various food, and which are therefore accommodated to their several wants, and indicative of even their mental, as well as constitutional, characters, Fabricius his pupil would necessarily extend this system to insects. But nothing can be more misapplied. Feeding is not the business of perfect insects. Many of them never eat at all, the business of their existence through the whole of their perfect state, being the propagation of their species. Hence the organs of their mouth lead to no natural distinctions, and the characters deduced therefrom prove, moreover, so difficult, that it is notorious they could not generally be applied to practice by Fabricius himself, he having, in the common course of his studies, been chiefly regulated by the external appearance of the insects he described. This external appearance depending on the form and texture of their wings, and the shape of their own peculiar organs, the *antennæ*, affords in fact the easiest, as well as the most natural, clue to their arrangement and discrimination.

As we presume to criticize the systematic errors of great practical observers, it cannot but occur to our recollection how very few persons have excelled in both these departments. Ray, Linnæus, and perhaps Tournefort, may be allowed this distinction. We can scarcely add a fourth name to this brief catalogue. The most excellent practical botanists of the Linnæan school have been such as hardly bestowed a thought on the framing of systems. Such was the distinguished Solander, who rivalled his preceptor in acuteness of discrimination, and even in precision and elegance of definition. Such is another eminent man, more extensively conversant with plants, more accurate in distinguishing, and more ready in recollecting them, than almost any other person with whom we have associated. Yet we have heard this great botanist declare, that however he might confide in his own judgement with regard to a species, or a genus of plants, he pretended to form no opinion of classes and orders. Men of so much experience know too much to be satisfied with their acquirements, or to draw extensive conclusions from what they think insufficient premises. Others, with a quarter of their knowledge, find no difficulty in building systems, and proceed with great alacrity, till they find themselves encumbered with their own rubbish; happy if their doubts and uncertainties will afford them a tolerable screen or shelter! But we here anticipate remarks, which will come with more propriety hereafter. We return from the consideration of the labours of particular botanists, to that of the diversities of nature and circumstance.

While it is remarked that in the cold regions of the north, the skill of the deep and learned botanist is chiefly exercised on the minute and intricate cryptogamic tribes, we are not to infer that Nature is not everywhere rich in beauty and variety. Mosses and Lichens afford inexhaustible amusement and admiration to the curious inquirer, nor are more gorgeous productions entirely wanting. Even Lapland boasts her *Pedicularis Sceptrum*, never seen alive out of her limits, and Siberia offers her own beautiful crimson *Cypripedium*, to console for a moment the miserable banished victims of imperial caprice. Kotzebue, though ignorant of botany, did not pass this lovely plant unnoticed, even in the height of his distress. The authoress of the pleasing little novel called "Elizabeth," has represented in a just light the botanic scenery of that otherwise inhospitable country; yet it must be allowed that its rarities are not numerous, except perhaps in those microscopic tribes already mentioned.

Let us in imagination traverse the globe, to a country where the very reverse is the case. From the representations, or accounts, that have been given of New Holland, it seems no very beautiful or picturesque country, such as is likely to form or to inspire a poet. Indeed the dregs of the community which we have poured out upon its shores, must probably subside, and purge themselves, before any thing like a poet, or a disinterested lover of nature, can arise from so foul a source. There seems however to be no transition of seasons, in the climate itself, to excite hope, or to expand the heart and fancy; like a Siberian or Alpine spring, bursting at once from the icy fetters of a sublime though awful winter. Yet in New Holland all is new and wonderful to the botanist. The most common plants there are unlike every thing known before; and those which, at first sight, look like old acquaintances, are found, on a near approach, to be strangers, speaking a different language from what he has been used to, and not to be trusted without a minute inquiry at every step.

The botany of the Cape of Good Hope, so well illustrated by Thunberg, and with whose treasures he scattered a charm around the couch of the dying Linnæus, most resembles that of New Holland. At least these countries agree in the hard, rigid, dwarfish character of their plants. But the Cape has the advantage in general beauty of flowers, as well as in a transition of seasons. After the dry time of the year, when every thing but the *Aloe* and *Mesembryanthemum* tribes is burnt up, and during which innumerable bulbs are scattered by the winds and driving sands over the face of the country, the succeeding showers raise up a new and most beautiful progeny from those bulbs. The families of *Ixia, Gladiolus, Iris, Antholyza, Oxalis,* and many others, then appear in all their splendour. Some of them, the least gaudy, scent the evening air with an unrivalled perfume, whilst others dazzle the beholder with the most vivid scarlet or crimson hues, as they welcome the morning sun.

The lovely Floras of the Alps and the Tropics contend, perhaps most powerfully, for the admiration of a botanist of taste, who is a genuine lover of nature, without which feeling, in some degree of perfection, even botany can but feebly charm. Of one of these the writer can speak from experience; of the other only by report; but he has had frequent opportunities of remarking, that the greatest enthusiasts in the science have been Alpine botanists. The expressions of Haller and Scopoli on this subject go to the heart. The air, the climate, the charms of animal existence in its highest perfection, are associated with our delight in the beauty and profusion of nature. In hot climates, the insupportable languor, the difficulty of bodily exertion, the usual ill health, and the effects of unwholesome instead of salutary fatigue, are described as sufficient to counterbalance even the pleasure which arises from the boundless variety, and infinite beauty, of the creation around. The flowery trees of a tropical forest raise themselves far above the human grasp. They must be felled before we can gather their blossoms. The insidious and mortal reptile twines among their boughs, and the venomous insect stings beneath their shade. We who enjoy the productions of these climates in peace and safety in our gardens, may well acknowledge our obligations to the labour and zeal of those who, by arduous journeys and painful researches, supply us with the riches of every country in succession. We do not, indeed, enjoy them in perfection; but we can study and investigate at leisure their various beauties and distinctions. We can compare them with our books, and profit by the acuteness of former observers. We can perpetuate, by the help of the pencil or the pen, whatever is novel or curious. We can preserve the plants and flowers themselves for subsequent examination, and return to them again and again in our closet, when winter has fixed his seal on all the instruction and pleasure afforded by the vegetable creation abroad. Yet let not the sedentary botanist exult in his riches, or rejoice too heedlessly in the abundance of his resources. A plant gathered in its native soil, and ascertained by methodical examination, is more impressed on the memory, as well as more dear to the imagination, than many that are acquired with ease, and named by tradition or report. The labours of its acquisition and determination enhance its value, and the accompaniments of delightful scenery, or pleasing society, are recollected, when difficulties and toils are forgotten.

The western continent is, with respect to botany, almost a world in itself. There exists, indeed, a general affinity between the plants of North America and those of Europe, and many species of the arctic regions are the same in both; but there are few common to the more temperate climates of each. A considerable number communicated by Kalm to Linnæus, which the latter considered as identified with certain well-known plants of our quarter of the world, prove, on more accurate examination, to be corresponding, but distinct species. Instances occur in the genera of *Carpinus, Corylus, Quercus,* as well as in the *Orchis* tribe, and others. These points of resemblance are found mostly among the vegetable productions of the eastern regions of North America. Mexico, and what little we know of the intermediate space, abound with different and peculiar productions. So, in South America; Peru, Guiana, Brasil, &c. have all their appropriate plants, of which we know as yet enough to excite our curiosity, rather than to satisfy it. Whatever has hitherto been given to the world respecting American botany has had one considerable advantage. Each *Flora* has been founded on the knowledge and experience of some one or more persons, long resident, and in a manner naturalized, in the countries illustrated. Those regions commonly comprehended under the name of North America, have afforded materials for the *Flora Boreali-Americana* of Michaux, and the more complete and correct *Flora Americæ Septentrionalis* of Pursh. Michaux, Wangenheim, and Marshall, have particularly illustrated the trees of those countries. But all these works have been enriched by the communications and assistance of men who had much more extensive and repeated opportunities of observation than their authors, except Mr. Marshall, could have. Such are the venerable John Bartram, the Reverend Dr. Muhlenberg, Messrs. Clayton, Walter, Lyon, &c. The Mexican Flora has received, for a long course of years, the attention of the able and learned Mutis, who long corresponded with Linnæus, and whose countrymen have prepared the sumptuous *Flora Peruviana*; each of the authors of which has repeatedly traversed, at various seasons, the rich and interesting regions, whose botanical treasures make so splendid and novel an appearance in those volumes. Of those treasures, we have still more to learn from the unrivalled Humboldt. The French botanist Aublet, after having gained considerable experience in the Mauritius, resided for many years in Cayenne and Guiana, for the purpose of studying the plants of those countries, of which his work, in four quarto volumes, gives so ample a history and representation.

All the writers just named have been practical botanists. They have generally excelled in specific discrimination, nor have they neglected the study of generic distinctions. Any thing further they have scarcely attempted. It is remarkable that they have all followed, not only the Linnæan principles of definition and nomenclature, but the Linnæan artificial system of classification. This same system was chosen by the veteran Jacquin, in his well-known work on West Indian plants, entitled *Stirpium Americanarum Historia*, as well as by Browne, in his *History of Jamaica*; not to mention Swartz, in his *Flora Indiæ Occidentalis*, who only wanders a little out of the way, to adopt some of Thunberg's alterations. We cannot but observe, that in the very department of botany in which he has most signalized himself, and with which he is most philosophically conversant, the *Orchideæ*, he totally rejects the ideas of Thunberg.

If we now turn our eyes to the oriental world, we shall find that the seeds of Linnæan botany, sown by Koenig, have sprung up and produced successive harvests among the pious missionaries at Tranquebar, who still continue to interweave a sprig of science, from time to time, among their amaranthine wreaths, which are not of this world. India too has long possessed a practical botanist of indefatigable exertion and ardour, who has thrown more light upon its vegetable riches, with the important subject of their qualities and uses, than any one since the days of Rheede and Rumphius. It is scarcely necessary to name Dr. Roxburgh, whose recent loss we deeply lament, and whose acquisitions and learned remarks are given to the world by the munificence of the East India Company, in a style which no prince has ever rivalled. That enthusiastic admirer of nature, Colonel Hardwicke, and the learned botanist Dr. Francis Buchanan, have also contributed greatly to increase our knowledge of Indian botany. The latter has enjoyed the advantage of investigating, for the first time, the remote and singular country of Nepaul; so prolific in beautiful and uncommon plants, that few parts of the world can exceed it, and yet meeting, in several points, not only the Floras of the lower regions and islands of India, but those of Japan, China, and even Siberia. The only systematic work on East Indian plants, is the *Flora Indica* of Burmann, which is classed according to the Linnæan artificial method. We cannot but wish it were more worthy of the system or the subject; yet, as a first attempt, it deserves our thanks. In speaking of Indian botany, shall we withhold our homage from that great and sublime genius Sir William Jones? who honoured this study with his cultivation, and, like every thing else that he touched, refined, elevated, and elucidated it, with a beam of more than mortal radiance. No man was ever more truly sensible of the charms of this innocent and elegant pursuit; and whenever he adverted to it, all the luminous illustrations of learning, and even the magic graces of poetry, flowed from his pen.

But we must extend our view beyond the utmost bounds of India, and of the then discovered world, to trace the steps of those adventurous circumnavigators who sought out, not only new plants, but new countries, for botanical examination. The names of Banks and Solander have, for nearly half a century, been in every body's mouth. Their taste, their knowledge, their liberality, have diffused a charm and a popularity over all their pursuits; and those who never heard of botany before, have learned to consider it with respect and admiration, as the object to which a man of rank, riches, and talents, devotes his life and his fortune; who while he adds, every season, something of novelty and beauty to our gardens, has given the Bread-fruit to the West Indies, and is ever on the watch to prompt, or to further, any scheme of public advantage. With the recollection of such men must also be associated the names of the learned Forsters, father and son, of Sparrmann, and of Menzies, who have all accomplished the same perilous course, and enriched their beloved science. The cryptogamic acquisitions of the latter in New Zealand, prove him to have attended to that branch of botany with extraordinary success, and at the same time evince the riches of that remote country. Indeed, it appears that any country proves rich, under the inspection of a sufficiently careful investigator. The labours of these botanists have all been conducted according to the principles and classification of Linnæus. Forster, under Sparrmann's auspices, has judiciously pointed out, and attempted to remedy, defects that their peculiar opportunities enabled them to discover, but with no invidious aim. They laboured, not to overthrow or undermine a system, which they found on the whole to answer the purpose of readily communicating their discoveries, but to correct and strengthen it for the advantage of those who might come after them. It is much to be lamented that, except the *Nova Genera Plantarum*, we have as yet so short and compendious an account of the acquisitions made in their voyage. To the technical history of these, however, the younger Forster has commendably added whatever he could supply of practical utility, and has thus given us all the information within the compass of hist means.

Long since the voyages of these celebrated naturalists, the same remote countries have been visited, in our own days, by two learned botanists more especially; these are M. La Billardière, and Mr. Brown, Librarian of the Linnæan Society. The former has published an account of the Plants of New Holland, in two volumes folio, with fine engravings; the latter has favoured the botanical world with one volume of a most acute and learned *Prodromus* of his discoveries. As his voyage was made at the public expense, we may trust that the government will consider itself bound to enable him to publish the whole of his acquisitions, in such a manner as to be generally useful. His own accuracy of observation, illustrated by the drawings of the inimitable Bauer, cannot fail to produce such a work as, we will venture to pronounce, has never been equalled. M. La Billardière has disposed his book according to the system of Linnæus, a rare example in France, where any thing not French usually comes but ill recommended. Mr. Brown, on the other hand, has written his *Prodromus*, at least, on the principles of classification established by the celebrated Jussieu, the great champion of a natural system of his own. On this subject we postpone our remarks for the present. Before we can enter on the subject of natural classification, it is necessary to consider the state and progress of botany, for some years past, in the schools, and among the writers, of Europe.

Sweden has continued to maintain her long established rank in the several departments of natural science, nor has Denmark been behind-hand with her neighbour and ancient rival. The son and successor of the great Linnæus endeavoured to follow his father's steps, and was ambitious of not being left very far in the rear; a commendable aim, which his short life, to say nothing of his talents or experience, disabled him from accomplishing. He completed, and gave to the world, the unfinished materials which his father had left, for a Supplement to his *Species Plantarum* and *Mantissæ*; and having enriched the book with many communications of Thunberg and others, as well as a number of original remarks, he felt a strong desire, not altogether unpardonable, of being thought the principal author of the work. All uncertainty on this subject, wherever other helps fail, is removed by the original manuscript of the *Supplementum Plantarum* in our possession. Ehrhart superintended the printing of this work, and made some alterations in the manuscript, traces of which are perceptible in the affected Greek names, given to some species of *Carex*, *Mespilus*, &c., as well as in their sesquipedalian specific characters. But he had introduced his own new genera of Mosses; which the younger Linnæus thought so alarming an innovation, that he ordered the sheet which contained these matters to be cancelled. We are possessed of a copy, which shows the genera in question to be almost all well founded, and what are now, under Hedwig's sanction, generally received, though by other names. The descriptions of Ehrhart are precise and correct, though his terminology is exceptionable, full of innovations, and crabbed expressions. Two years, almost immediately preceding the death of the younger Linnæus, were spent by the latter in visiting England, France and Holland, and were employed to very great advantage, in augmenting his collection of natural productions, as well as his scientific skill. During this tour, he attached himself strongly, through the medium of his old friend Solander, to Sir Joseph Banks; and while in France, he almost planted, or at least greatly advanced, a Linnæan school in that kingdom. He had scarcely resumed his professorial office at home, when he was unexpectedly taken off, by an acute disease, in his forty-second year. Of the talents and performances of his successor Thunberg, who still with honour fills the chair of the Rudbecks and the Linnæi, we have already spoken. Dr. Swartz is the Bergian professor of Botany at Stockholm. The *Transactions* of the Upsal Academy, founded by the younger Rudbeck, are continued occasionally; and those of the Stockholm one, whose foundations were laid by Linnæus, are published regularly. Both are from time to time enriched with botanical communications, worthy of the pupils of so illustrious a school. A veteran in botanical science, Professor Retzius, still presides at the University of Lund. The worthy and accurate Afzelius, well known in England, who accomplished a hazardous botanical expedition to Sierra Leone, is the coadjutor of Professor Thunberg; and the difficult subject of Lichens, under the hands of Dr. Acharius, is become so vast and so diversified, as to be almost a science of itself.

Denmark has always possessed some acute and learned botanists, and has, more than most other countries, been supplied with dried specimens of plants, as an article of commerce, from her West or East Indian establishments. Oeder, the original author of the *Flora Danica,* and Muller its continuator, have distinguished themselves; but their fame is inferior to that of the late Professor Vahl, who studied under the celebrated Linnæus, and who is the author of several excellent descriptive works. He undertook no less than a new *Species,* or, as he entitled it, *Enumeratio, Plantarum,* an admirable performance, cut short by his death at the end of the second volume, which finishes the class and order *Triandria Monogynia.* It is almost superfluous to mention, that Afzelius and Retzius, as well as Vahl, in all they have given to the world, have followed the system of their great master. The *Flora Danica,* chiefly a collection of plates, with few synonyms and no descriptions, has come forth, from time to time, for above fifty years past, in fasciculi, without any order, and is still incomplete. It was undertaken by royal command, and, in a great measure, at the sovereign's expense; though regularly sold, except some copies presented to certain distinguished men, as Linnæus.

After the example of Denmark, Sweden, &c. Russia has been desirous of promoting, throughout its vast dependencies, an attention to natural knowledge. Nor was any country ever more fortunate in the possession of an active and intelligent naturalist. The celebrated Pallas successfully devoted a long life to these pursuits, and to the communication of his discoveries and observations. He prompted the Empress Catherine to offer an unlimited sum for the museum, library, and manuscripts of Linnæus; but, fortunately for their present possessor, the offer was made too late. A *Flora Rossica*, on the most magnificent scale, was undertaken by Pallas, his Imperial mistress proposing to defray the cost of the whole undertaking, not merely for sale, but for gratuitous presentation, on the most princely scale, to all who had any taste or ability to make use of the book. This well-intended munificence was the cause of the ruin of the project. Half of a first volume was bestowed as the Empress intended. But the second part, instead of following the destination of the first, got into the hands of interested people, who defeated the liberal designs of their sovereign, misapplied her money, and by the disgust and disappointment which ensued, prevented the continuance of the work. Those who wished to complete their sets, or to obtain the book at all, were obliged to become clandestine purchasers, buying, as a favour, what they ought to have received as a free gift; and were moreover, like the writer of this, often obliged to put up with imperfect copies. In like manner, the intentions of the great Mr. Howard, respecting his book on Prisons, were rendered ineffectual, by the disgraceful avarice of certain London booksellers, who immediately bought up, and sold at a greatly advanced price, the whole edition, which its benevolent author had destined to be accessible to every body at an unusually cheap rate. These examples, amongst others, show that it is the most difficult thing in the world to employ patronage, as well as gratuitous charity of any kind, to real advantage, except under the guidance of the most rigorous discretion. "All that men of power can do for men of genius," says Gray, if we recollect aright, "is to leave them at liberty, or they become like birds in a cage," whose song is no longer that of nature and enjoyment. The great and the affluent may foster and encourage science and literature, by their countenance, their attention, and a free, not overwhelming, liberality. But when princes become publishers of books, or directors of academies, they generally do more harm than good. They descend from their station, and lose sight perhaps of their higher and peculiar duties, which consist in promoting the general prosperity, peace, and liberty of their subjects, under the benign influence of which, every art, science, or pursuit, that can be beneficial to mankind, is sure to flourish without much gratuitous assistance.

Several of the immediate scholars of the illustrious Swedish naturalist were planted in different parts of Germany. Murray, to whom he entrusted the publication of that compendious volume, entitled, *Systema Vegetabilium,* and who printed two successive editions of the work, was seated as Professor at Göttingen. Giseke was established at Hamburgh, and, after the death of Linnæus, gave to the world such an edition as he was able to compile, from his own notes and those of Fabricius, of the lectures of their late preceptor, on the Natural Orders of Plants. His ideas on this subject Linnæus himself always considered as too imperfect to be published, except in the form of a sketch or index, at the end of his *Genera Plantarum.* The venerable patriarch, Professor Jacquin, still survives at Vienna, where he, and his worthy son, have enriched botany with a number of splendid and useful works. They have given to the public several labours of the excellent practical botanist Wulfen, and others, which might, but for their encouragement, have been lost. The highly valuable publication of Host on Grasses, is conducted on the plan of Jacquin's works. His *Synopsis* of Austrian plants is an excellent *Flora,* disposed according to the Sexual System, as is the more ample *Tentamen Floræ Germanicæ* of the celebrated Dr. Roth, one of the best practical European botanists, and more deeply versed than most others in cryptogamic lore. The best Linnæan *Flora,* as far as it goes, that the world has yet seen,—we speak it without any exception,—is the *Flora Germanica* of Professor Schrader of Göttingen, the first volume of which, comprising the first three classes of the sexual system, was published in 1806. The correct distinctions, well-digested synonyms, and complete descriptions of this work, are altogether unrivalled. If the whole should be equally well executed, for which the longest life would be scarcely sufficient, it must ever be the standard book of European botany. Its descriptions of grasses are worthy to accompany the exquisite engravings of the same tribe from the hand of Leers, published at Herborn in 1775, which excel every other botanical representation that we have examined. They will bear, and indeed they require, the application of a magnifying-glass, like the plants themselves. The purchaser of this little volume must however beware of the second edition, whose plates are good for little or nothing. The name of Schrader has long been distinguished in Cryptogamic Botany. In this pursuit, the industrious and accurate botanists of Germany, shut out from extensive opportunities of studying exotic plants, have had full scope for their zeal and abilities. In this field the Leipsic school has distinguished itself. Here the great Schreber first began his career with some of the most perfect cryptogamic works, especially on the minute genus *Phascum.* Here the same author published his excellent *Flora Lipsiensis,* his laborious practical work on Grasses, and finally his improved edition of the *Genera Plantarum* of his friend Linnæus. But, above all, Leipsic is famous for being the residence of Hedwig, whose discoveries, relative to the fructification and generic characters of Mosses, form an æra in botanic science. Under the hands of such an observer, that elegant tribe displays itself with a degree of beauty, variety, and singularity, which vies with the most admired herbs and flowers, and confirms the Linnæan doctrine of impregnation, which the more obvious organs of the latter had originally taught. Nor must we, in speaking of cryptogamic plants, neglect here to record the names of Weis, Weber, Mohr, Schmidel, Esper, and especially Hoffmann; the plates of the latter, illustrating the Lichen tribe, are models of beauty and correctness. His *Flora Germanica* is a most convenient and compendious manual, after the Linnæan system. *Fungi* have been studied in Germany with peculiar care and minuteness. The leading systematic author in this obscure tribe, Persoon, was indeed born, of Dutch parents, at the Cape of Good Hope; but he studied and published at Göttingen. Two writers, of the name of Albertini and Schweiniz, have published the most minute and accurate exemplification of this natural order, in an octavo volume, at Leipsic, in the year 1805, comprising the Fungi of the district of Niski in Upper Lusatia. If their figures are less exquisitely finished than Persoon's, or less elaborately detailed than Schrader's, their descriptions make ample amends.

The German school of botany has, for a long period, been almost completely Linnæan. This however was not always the case, for, in the earlier part of his career, the learned Swede was attacked more repeatedly and severely from this quarter of the world than any other; his ridiculous critic Siegesbeck of Petersburgh excepted, who would not admit the doctrine of the sexes of plants, because the pollen of one flower may fly upon another, and his purity could not bear the idea of such adultery in Nature. Numerous methods of arrangement appeared in Germany, from the pens of Heister, Ludwig, Haller, and others, and even Schreber adopted a system like some of these in his *Flora* above mentioned. It would be to no purpose now to criticize these attempts. They cannot rank as natural systems, nor have they the convenience of artificial ones. Part of their principles are derived from Linnæus, others from Rivinus. Their authors were not extensively conversant with plants, nor trained in any sound principles of generic discrimination or combination.

They set off with alacrity, but were soon entangled in their own difficulties, and were left by Linnæus to answer themselves or each other. We here mention these learned systematics, for learned they were thought by themselves and their pupils, merely because they will scarcely require animadversion, when we come to canvass the great question of natural and artificial classification, they having had no distinct ideas of a difference between the two. Hedwig used frequently to lament, that his preceptor Ludwig had never perfected his system of arrangement; but from what he has given to the world, we see no great room to suppose he had any thing very excellent in reserve. Unexecuted projects are magnified in the mists of uncertainty. We have ventured elsewhere, in a biographical account of Hedwig, to remark, that even that ingenious man "did not imbibe under Ludwig, anything of the true philosophical principles of arrangement, the talents for which are granted to very few, and are scarcely ever of German growth. We mean no invidious reflections on any nation or people. Each has its appropriate merits, and all are useful together in science, like different characters on the theatre of human life."

Germany may well dispense with any laurels obtained by the very secondary merit of speculative schemes of classification, when she can claim the honour of having produced such a practical observer as Gærtner. This indefatigable botanist devoted himself to the investigation of the fruits and seeds of plants. Being eminently skilled in the use of the pencil, he has, like Hedwig, faithfully recorded, what he no less acutely detected. The path he struck out for himself, of delineating and describing in detail, with magnified dissections, every part of the seed and seed-vessel of each genus within his reach, had never been explored before in so regular and methodical a manner. Botanists of the Linnæan school are justly censurable for having paid too little attention to the structure of these important parts, in their generic characters. Indeed it may be said, that if they were able to establish good genera without them, and, after the example of their leader, merely preferred the more obvious and distinct organs, when sufficient for their purpose, their conduct was justifiable. If generic principles be natural and certain, it matters not on what parts of the fructification they are founded; nor is the inflorescence, or even the herb or root, rejected by sound philosophers, but because they are found to lead only to unnatural and uncertain characters. It is therefore extremely to the honour of Linnæus, Gærtner, and Jussieu, that their conceptions of genera are almost entirely the same. They meet in almost every point, however different the paths by which they pursue their inquiries. Their labours illustrate and confirm each other. Even Tournefort, who conceived so well, on the whole, the distinctions of genera, which he could but ill define, receives new strength from their knowledge, which does not overturn his imperfect performances, but improves them. The accurate student of natural genera cannot fail to perceive, that where Gærtner differs from Linnæus, which is but in a very few material instances, such as his numerous subdivision of the genus *Fumaria*, and his distribution of the compound flowers, it arises from his too intent and exclusive consideration of one part of the fructification, instead of an enlarged and comprehensive view of the whole. In other words, he neglects the Linnæan maxim, that "the genus should give the character, not the character the genus." Such at least appears to us the case in *Fumaria*. In the syngenesious family, being so very natural in itself, the discrimination of natural genera becomes in consequence so difficult, that Gærtner and Linnæus may well be excused if they do not entirely agree, and they perhaps may both be satisfied with the honour of having collected materials, and disposed them in different points of view, for the use of some future systematic, who may decide between them. However exact Gærtner may have been in discriminating the parts of seeds, we believe him mistaken in distinguishing the *vitellus* as a separate organ, distinct in functions from the *cotyledons*. His readers will also do well, while they profit by his generally excellent principles, not to admit any of his rules as absolute. They may serve as a clue to the intricacies of Nature, but they must not overrule her laws. Still less is our great carpologist to be implicitly followed in physiological doctrines or reasonings; witness his feeble and incorrect attack on Hedwig's opinions, or rather demonstrations, respecting the impregnation of Mosses. His criticisms of Linnæus are not always marked with that candour which becomes a disinterested lover of truth and nature, nor can we applaud in general his changes of nomenclature, or of terminology; especially when he unphilosophically calls the *germen* of Linnæus, the *ovarium*, a word long ago rejected, as erroneous when applied to plants. These however are slight blemishes, in a reputation which will last as long as scientific botany is cultivated at all. Botanists can now no longer neglect, but at their own peril, the parts which Gærtner has called into notice, and to the scrutiny of which, directed by his faithful guidance, the physiologist and the systematic must often, in future, recur.

We shall close this part of our subject with the mention of the Berlin school, where Gleditsch, who, in 1740, repelled the attacks of Siegesbeck on Linnæus, was Professor, and published a botanical system, founded on the situation, or insertion, of the stamens; the subordinate divisions being taken from the number of the same parts; so that it is, in the latter respect, a sort of inversion of the Linnæan method. In the former, or the outline of its plan, the system of Gleditsch is in some measure an anticipation of that of Jussieu. Berlin has of late been much distinguished for the study of natural history, and possesses a society of its own, devoted to that pursuit. Its greatest ornament was the late Professor Willdenow, who, if he fell under the lash of the more accurate Afzelius, is entitled to the gratitude of his fellow-labourers, not for theoretical speculations, but for the useful and arduous undertaking of a *Species Plantarum*, on the Linnæan plan, being indeed an edition of the same work of Linnæus, enriched with recent discoveries. This book, left unfinished at the end of the first order of the *Cryptogamia*, by the death of the editor, wants only a general index to render it sufficiently complete. The *Musci*, *Lichenes*, and *Fungi*, are systematically treated in the separate works of writers devoted to those particular, and now very extensive, subjects, from whom Willdenow could but have been a compiler. With the *Filices*, which he lived to publish, he was practically conversant. His insertion of the essential generic characters, throughout these volumes, is an useful addition, and now become necessary in every similar undertaking.

Little can be said of Holland in this review of the botanical state of Europe for a few years past. The Leyden garden has always been kept up, especially during the life of the late Professor David Van Royen, with due care and attention; we know little of its fate in the subsequent convulsed state of the country. Botany has long been on the decline at Amsterdam, though we are indebted to that garden for having first received, and afterwards communicated to other countries, such acquisitions of Thunberg in Japan as escaped the perils of importation.

The botany of Switzerland may, most commodiously, be considered in the next place. Here, in his native country, the great Haller, after a long residence at Göttingen, was finally established. Its rich and charming *Flora* has been illustrated by his classical pen, with peculiar success. Everybody is conversant with the second edition of his work, published in 1768, in 3 vols. folio, and entitled, *Historia Stirpium Indigenarum Helvetiæ*, with its inimitable engravings, of the *Orchis* tribe more particularly. But few persons, who have not laboured with some attention at the botany of Switzerland, are aware of the superior value, in point of accuracy, of the original edition of the same work, published in 1742, under the title of *Enumeratio Methodica Stirpium Helvetiæ Indigenarum.* This edition is indispensable to those who wish fully to understand the subject, or to appreciate Haller's transcendent knowledge and abilities. These works are classed after a system of his own, intended to be more consonant with nature than the Linnæan sexual method. We can scarcely say that it is so, on the whole; nor is it, on the other hand, constructed according to any uniformity of plan. The number of the stamens, compared with that of the segments of the corolla, or its petals, regulates the characters of several classes, and those are artificial. Others are assumed as natural, and are for the most part really so, but their characters are frequently taken from Linnæus, even from his artificial system, as the *Cruciatæ*, and the *Apetalæ*. Lord Bute has well said, that Haller was a Linnæan in disguise. His classification however was merely intended to answer his own purpose, with respect to the Swiss plants; for he was not a general botanist, nor had he a sufficiently comprehensive view of the subject to form a general system, or even to be aware of the difficulties of such an undertaking. He ought not therefore to be obnoxious to criticism in that view. His method has served for the use of his scholars, as the Linnæan one serves English botanists, by way of a dictionary. Some such is necessary; and those who should begin to decide on the merits of a system, before they know plants, would most assuredly be in danger of appearing more learned to themselves than to others. We cannot exculpate Haller from some degree of prejudice in rejecting real improvements of Linnæus, which are independent of classification; such as his trivial or specific names, by which every species is spoken of at once, in one word, mostly so contrived as to assist the memory, by an indication of the character, appearance, history, or use, of the plant. What did the great Swiss botanist substitute in the place of this contrivance? A series of numbers, burthensome to the memory, destitute of information, accommodated to his own book only, and necessarily liable to total change on the introduction of every new-discovered species! At the same time that he rejected the luminous nomenclature of his old friend and fellow-student, who had laboured in the most ingenuous terms to deprecate his jealousy, he paid a tacit homage to its merit, by contending that the honour of this invention was due to Rivinus. In this he was not less incorrect than uncandid, the short names of Rivinus being designed as specific characters, for which purpose Haller knew, as well as Linnæus, they were unfit. Useful specific characters he himself constructed on the plan of Linnæus, with some little variation, not always perhaps for the better, as to strictness of principle, but often strikingly expressive. Here, as in every thing connected with practical botany, he shines. The most rigid Linnæan, whose soul is not entirely shrivelled up with dry aphorisms and prejudice, must love Haller for his taste and enthusiasm, and the Flora of Switzerland as much for his sake as its own. No wonder that his pupils multiplied, and formed a band of enthusiasts, tenacious of even the imperfections of their master. The line of demarcation is now no longer distinctly drawn between them and the equally zealous scholars of the northern sage. The amiable and lamented Davall strove to profit by the labours of both. The Alpine botanists of France and Italy have served to amalgamate the Swedish and the Helvetian schools. The Flora of Dauphiny by Villars is nearly Linnæan in system, and the principles of the veteran Bellardi of Turin are entirely so; though he has been, in some of his publications, obliged to conform to the method of his late preceptor, the venerable Allioni, who, in spite of all remonstrance, had the ambition of forming a system of his own. His *Flora Pedemontana* is disposed according to this system, an unnatural and inconvenient jumble of the ideas of Rivinus, Tournefort, and others. This work is also faulty in the neglect of specific definitions, so that its plates and occasional descriptions are alone what render it useful; nor would it perhaps, but for the uncommon abundance of rare species, be consulted at all.

We may glance over the botany of Italy, to whose boundaries we have thus been insensibly led, as the eye of the traveller takes a bird's-eye view of its outstretched plains from the summits of the Alps. We may pass from Turin to Naples without meeting with any school of distinction. The northern states are not without their professors and patrons of botany, nor are their nobles destitute of taste in various branches of natural knowledge. The names of a Castiglione of Milan, a Durazzo and Dinegro of Genoa, and a Savi of Pisa, deserve to be mentioned with honour, for their knowledge and their zeal. The unfortunate Cyrillo, and his friend Pacifico, at Naples, were practical botanists. There is also a rising school, of great promise, at Palermo. But since the time of Scopoli, Italy has contributed little to our stock of information; nor are the latter publications of this eminent man, while he resided at Pavia, commensurate in importance or merit with those earlier ones, the *Flora*, and *Entomologia, Carniolica*, which have immortalized his name. Scopoli, who at first adopted a system of his own, had the sense and liberality, in his second edition, to resign it, in favour of what his maturer experience taught him to prefer, the sexual system of Linnæus.

Spain and Portugal claim our attention; the former for being the channel through which the gardens of Europe have been, for some years past, enriched with many new Mexican and Peruvian plants; and likewise as the theatre of the publication of some important books, relative to the botany of those countries. In speaking of American botany, we have mentioned the *Flora Peruviana*, whose authors, Ruiz and Pavon, rank deservedly high for their industry and knowledge. The late Cavanilles, resident at Madrid, has also communicated to the learned world much information, from the same source. Spain seems anxious to redeem her reputation, which suffered so much from the neglect, or rather persecution, of the truly excellent but unfortunate Dombey, who, like many other benefactors of mankind, was allowed to make all his exertions in vain, and finally perished unknown, in the diabolical hands of English slave-dealers at Montserrat. Portugal is most distinguished at home by the labours of a learned benedictine, Dr. Felix Avellar Brotero, author of a *Flora Lusitanica,* disposed after the Linnæan method, reduced entirely to principles of number; and abroad by the valuable work of Father Loureiro, entitled *Flora Cochinchinensis,* in which the plants of Cochinchina, and of the neighbourhood of Canton, are classed and defined in the Linnæan manner, with valuable descriptions and remarks. It is undoubtedly a disgrace to the possessors of such a country as Brasil, that they have not derived from thence more benefit to the world, or to themselves, from its natural productions. But they are satisfied with what the bowels of the earth afford, and they neglect its more accessible, though perhaps not less valuable, treasures. The jealousy and innumerable restrictions of their Government render what they possess as useless to all the world as to themselves. A genius of the first rank in natural science, as well as in every thing which his capacious mind embraced, has arisen in Portugal, and has been domesticated in the schools of Paris and London, the amiable and learned Corrêa de Serra, now a traveller in the United States of America. What little impulse has been given to literature in Portugal, and particularly the foundation of a Royal Academy of Sciences, is owing to him; and though his name has chiefly appeared in the ranks of botanical science in an incidental manner, no one possesses more enlarged and accurate views, or more profound knowledge, of the subject.

In the extensive, though incomplete, review which we have undertaken of the recent history of botanical science, the individual merits of particular writers have chiefly hitherto been detailed and compared. The most difficult part of our task perhaps still remains; to contrast and to appreciate the influence and the merits of two great and rival nations, in the general school of scientific botany; to consider the causes that have led to the particular line which each has taken, and to compare the success, as well as to calculate the probable future consequences, of their respective aims. England and France have, from the time of Ray and Tournefort, been competitors in botanical fame, because each was ambitious of supporting the credit of the great man she had produced. This contest, however, as far as it regarded theoretical speculations, has entirely subsided on the part of Ray's champions. In practical science, likewise, the admirers of Ray and of Tournefort have shaken hands, like those of every other school. On the subject of system, the question is greatly changed; for though a phœnix has arisen from the ashes of Tournefort, its "star-like eyes," darting far beyond all former competition, have been met, if not dazzled, by a new light, rising in full glory from the north; a polar star, which has been hailed by all the nations of the earth.

The Linnæan system of classification, with all its concomitant advantages of nomenclature, luminous technical definition, and richness of information, was planted, like a fresh and vigorous scion, in the favourable soil of England, already fertilized with accumulations of practical knowledge, about the middle of the last century. If we may pursue the metaphor, the ground was entirely cleared for its reception; for all previous systems had been of confined and local use; the alphabetical index having become the resource of even the most learned; and the pupils of Ray being held to his method of classification, rather by their gratitude for his practical instruction, than any other consideration. Accordingly we have, in our own early progress, before they were all, as at present, swept off the stage, found them rather contending for his nomenclature, imperfect as it was, because they were habituated to it, than for his system, of which, it was evident, they had made little use. Hence the first attempt in England to reduce our plants to Linnæan order, made by Hill, was chiefly a transposition of Ray's *Synopsis* into the Linnæan classes, the original nomenclature being retained, while the specific names of the *Species Plantarum* were rejected.

Hill's imperfect performance was superseded by the more classical *Flora Anglica* of Hudson, composed under the auspices and advice of the learned and ingenious Stillingfleet, in which the botany of England assumed a most scientific aspect, and with which all the knowledge of Ray was incorporated. At the same time, the principles of theoretical botany, and the philosophical writings of the learned Swede, were studied with no ordinary powers of discrimination and judgement, in a small circle of experienced observers at Norwich. A love of flowers, and a great degree of skill in their cultivation, had been long ago imported into that ancient commercial city, with its worsted manufacture, from Flanders; and out of this taste, something like the study of systematic botany had sprung. These pursuits were mostly confined to the humblest of the community, particularly among the then very numerous bodies of journeymen weavers, dyers, &c. Towards the middle of the eighteenth century, several of the opulent merchants seem to have acquired, by their intimate connexion with Holland, not only the abovementioned taste for horticulture, but likewise an ambition to be distinguished by their museums of natural curiosities. The former sometimes extended itself, from the flowery parterre, and the well-arranged rows of tulips, hyacinths, carnations and auriculas, into no less formal labyrinths, or perhaps a double pattern of angular or spiral walks, between clipped hedges, exactly alike on each side of a broad gravel walk. Such was the most sublime effort of the art within the compass of our recollection. "Grove" could by no means be said to "nod at grove," for the perpendicular and well-trimmed structure was incapable of nodding; but that "each alley should have a brother" was an indispensable part of the design. Greenhouses of exotic plants, except oranges and myrtles, were at this time scarcely known; and the writer well recollects having seen, with wonder and admiration, above forty years ago, one of the first African Geraniums that ever bloomed in Norwich. If, however, the progress of natural science was slow in this angle of the kingdom, the wealthy manufacturers, become their own merchants, found it necessary to acquire a knowledge of various foreign languages, in order to carry on their wide-extended commerce. In learning French, Italian, Spanish, Dutch, and German, they unavoidably acquired many new ideas. Their sons were sent to the continent, and it were hard indeed if many of them did not bring home much that was worth learning. The society of the place, aided by some concomitant circumstances, and the adventitious acquisition of two or three men of singular talents and accomplishments, became improved. A happy mixture of literature and taste for many years distinguished this city above its rivals in opulence and commercial prosperity. Such Norwich has been in our memory; and if its splendour be gone by, a taste for mental cultivation, originating in many of the before-mentioned causes, still remains, and is fostered by the novel pursuits of chemistry and natural history, on which some arts, of great importance in the manufactory of the place, depend for improvement. We trust the reader will pardon this digression from the subject more immediately before us, to which we shall now return.

Some of the more learned students of English plants, among the lovers of botany in Norwich, had long been conversant with the works of Ray, and even the *Historia Muscorum* of Dillenius. They were prepared therefore to admire, and to profit by, the philosophical writings of Linnæus. Hence originated the *Elements of Botany*, published in 1775, by Mr. Hugh Rose; who was aided in the undertaking by his equally learned friend, the Rev. Henry Bryant, of whose acuteness and botanical skill no better proof is wanting, than his having found and determined, nine years before, the minute *Tillæa muscosa*, for the first time in this island. Numerous pupils were eager to improve themselves by the assistance of such masters, and amongst others the writer of these pages imbibed, from their ardour and their friendly assistance, the first rudiments of a pursuit that has proved the happiness and the principal object of his life.

London became, of course, the focus of this science, as well as of every other. Of the English Universities, Cambridge most fulfilled its duty, in rendering its public establishments useful to the ends for which they were founded and paid. The names of Martyn, both father and son, have long maintained a distinguished rank in botany; and the latter, for many years, has inculcated the true principles of Linnæan science from the professor's chair. A botanic garden was established, by a private individual, Dr. Walker, about the period of which we are speaking. A Linnæan *Flora Cantabrigiensis*, by Mr. Relhan, has renewed the celebrity of that field, in which Ray had formerly laboured; and there has always existed a little community of Cambridge botanists, though fluctuating and varying, according to circumstances. At Oxford, botany, so vigorously established by Sherard and

Dillenius, slept for forty years under the auspices of the elder Professor Sibthorp, at least as to the utility of its public foundations. Yet even there the science had many individual cultivators, and if others were forgotten, the name of a Banks ought to render this school for ever celebrated. The younger Professor Sibthorp well atoned for the supineness of his father and predecessor. He published a *Flora Oxoniensis,* and extended his inquiries into the classical scenes of Greece, finally sacrificing his life to his labours, and sealing his love of this engaging study by a posthumous foundation, which provides for the publication of a sumptuous *Flora Græca,* and the subsequent establishment of a professorship of Rural Economy. Edinburgh, under the auspices of the late worthy Professor Hope, became distinguished for the cultivation of botany, as a branch of medical education. The physiology of plants was there taught, more assiduously than in almost any other university of Europe; and the Linnæan principles were ably enforced and illustrated, not with slavish devotion, but with enlightened discrimination. Nor must the dissenting Academy at Warrington be forgotten, where the distinguished circumnavigator Forster, of whom we have already spoken, was settled. Here many young naturalists were trained. The neighbouring family of the Blackburnes, possessed, even to this day, of one of the oldest and richest botanic gardens in England, have steadily fostered this and other branches of natural knowledge. The same taste has spread to Manchester, Liverpool, and the country around. Westmoreland, Northumberland, and Durham have their sequestered practical botanists, in every rank of life. Scenes celebrated by the correspondents of Ray are still the favourite haunts of these lovers of nature and science, who every day add something to our information, and to the celebrity of other parts of the same neighbourhood.

We must now concentrate our attention to the London school, which for about forty years past has maintained a rank superior to most other seats of botanical science; the more so perhaps for its being founded in total disinterestedness, both with respect to authority and emolument. Truth alone, not system, has been the leading object of this school; unbiassed and gratuitous patronage its support; and a genuine love of nature and of knowledge its bond of union, among persons not less distinguished from each other by character and opinion, than by their different pursuits, and various ranks of life. The illustrious Banks, from the time when, after his return from his celebrated and adventurous voyage, he devoted himself to the practical cultivation of natural science for the advantage of others, as he had long pursued it for his own pleasure and instruction, has been the head of this school. Here he fixed the amiable and learned Solander, for the remainder of his too short life. The house of this liberal Mecænas has ever since been, not only open, but, in a manner, at the entire command of the cultivators and admirers of this and other branches of philosophy; inasmuch as his library and museum have been devoted to their free use; and his own assistance, encouragement and information are as much at their service, as if his fortune and fame had all along depended on their favour. With such an establishment as this, aided by the perpetual resources of the numerous public and private gardens around, botany might well flourish. The liberal spirit of the leaders of this pursuit gave a tone to the whole. The owners of nurseries, though depending on pecuniary emolument for their support, rivalled each other in disinterested communication. The improvement of science was the leading object of all. One of this latter description took his rank among the literary teachers of botany. Lee's *Introduction* was much approved by Linnæus, whose system and principles it ably exemplifies, and who became the friend and correspondent of its author. Travelling botanists were dispatched, under the patronage of the affluent, to enrich our gardens from the Alps, the

Cape of Good Hope, and the various parts of America. Every new acquisition was scrutinized, and received its allotted name and distinction, from the hand of the correct and classical Solander, who one day was admiring with Collinson, Fothergill, or Pitcairn, the treasures of their respective gardens, and another labouring with the distinguished Ellis, at the more abstruse determination of the intricate family of marine productions, whether sea-weeds, corallines, or shells. His own acquisitions, and those of his friend and patron, in the fairy land of the South-Sea Islands, the hazardous shores of New Holland, or the nearly fatal groves and swamps of Java, were at the same time recorded by his pen, as they were gradually perpetuating by the slow labours of the engraver. To this band of zealous naturalists the younger Linnæus was, for a while, associated, as well as the excellent and zealous Broussonet, who, though not unversed in botany, devoted himself most particularly to the more uncommon pursuit of scientific ichthyology.

The Banksian school, altogether intent upon practical botany, had adopted the Linnæan system as the most commodious, while it pursued and cultivated the Linnæan principles, as the only ones which, by their transcendent excellence, could support the science of botany on a stable foundation. In these Dr. Solander was, of course, well trained ; and, having added so wide a range of experience to his theoretical education, few botanists could vie with him, who had, as it were, caught his preceptor's mantle, and imbibed, by a sort of inspiration, a peculiar talent for concise and clear definition. Abstract principles of classification, or even such outlines of natural arrangement as Linnæus had promulgated, seem never to have attracted Solander. In following the chain of his ideas, discernible in the materials he has left behind him, one cannot but remark his singular inattention to every thing like botanical affinity, to which the artificial sexual system was, with him, entirely paramount. The genera which, for extemporaneous use, he named with the termination *oides*, comparing each with some well-known genus, till a proper appellation could be selected, are seldom thus compared because of any natural affinity, nor scarcely any external resemblance, but because they agree with such in their place in the artificial system, or nearly perhaps in technical characters. A great botanist therefore, it is evident, may exist, without that vaunted erudition in a peculiar line, which some would have us consider as the only road to knowledge and to fame. We allow that this sort of erudition is now, since the attention it has received from Linnæus, Jussieu, and others, become indispensable to a good theoretical or philosophical botanist, as is the study of carpology, in consequence of the labours of Gærtner; we only contend that it is possible to know plants extremely well without either.

The learned Dryander, less skilled than his predecessor as the coadjutor of Sir Joseph Banks, in a practical acquaintance with plants, exceeded him in theoretical lore and ingenious speculation, and far excelled every other man in bibliographic information, as well as in the most precise fastidious exactness relative to every subject within the wide extent of his various knowledge. He furthered, upon principle, and with unwearied assiduity, every object of the noble establishment to which he was devoted; but he, like Solander, now sleeps with his fathers, and his place is supplied by a genius of British growth, who unites talents with experience, and theoretical skill, in the most eminent degree, with practical knowledge.

Although it is almost superfluous to name the most eminent disciples of the London school of botany, it might seem negligent to pass them over without some particular mention. The ardent and ingenious Curtis has left a permanent monument behind him, in the *Flora Londinensis*, to say nothing of the popular *Botanical Magazine*, continued by his friend Dr. Sims. The *Flora Scotica* of Lightfoot first offered, in a pleasing and familiar garb, the botanical riches of that part of the island to its southern inhabitants. The lynx-eyed Dickson, so long and faithfully attached to his constant patron, has steadily traced, through all its windings, the obscure path of cryptogamic botany, with peculiar success. No more striking instance can be pointed out, to prove how totally the most consummate practical skill, even in the most difficult part of botany, is independent of theoretical learning. Even those who profit by the certain aids supplied by the discoveries of Hedwig, can with difficulty keep pace with this veteran in their pursuits, who, with conscious independence, neglects all those aids.

Just at the time when the school, whose history we are endeavouring to trace, had most firmly established its credit and its utility, a great additional weight was given to England, in the scale of natural science, by the acquisition of the entire museum, library, and manuscripts of the great Linnæus and his son, which came amongst us, by private purchase, in 1784, after the death of the latter. Hence our nomenclature has been corrected, and our knowledge greatly augmented. These collections have necessarily been consulted by most persons, about to publish on the subject of natural history; and a reference to them, in doubtful cases, secures a general conformity of sentiment and nomenclature, among the botanists of Europe, Asia, and America. We are seldom obliged to waste time in conjecturing what Linnæus, or the botanists with whom he corresponded, meant, for we have before us their original specimens, named by their own hands. An entire London winter was devoted to the almost daily labour, of comparing the Banksian herbarium throughout, with that of Linnæus, and to a copious interchange of specimens between their respective possessors, who, with the aid of Mr. Dryander alone, accomplished this interesting and instructive comparison. Hence the *Hortus Kewensis* of the lamented Aiton, which was at that period preparing for publication, became much more correct in its names, than it, or any other similar performance, could have been, without this advantage. It could scarcely be imagined that Sweden would, unmoved, thus let the botanical sceptre pass from her; but it is much to the honour of the nation, that all her naturalists have ever preserved the most friendly intercourse with us, particularly with the person who deprived them of this treasure. They have not merely pardoned, but publicly sanctioned, the scientific zeal which prompted him to this acquisition, by associating him with all their learned establishments, without any solicitation on his part.

The institution of the Linnæan Society at London in 1788, especially under that name, must be considered as a triumph for Sweden in her turn. By this establishment the intercourse of science is facilitated; essays, which might otherwise have never seen the light, are given to the world; and a general taste for the pleasing study of nature is promoted. Learned and worthy people are thus made acquainted with each other, from the remotest corners of the kingdom, and their information enriches the common stock. The state has given its sanction to this rising establishment. Its publications and its members are spread over the Continent, and other similar institutions have borrowed its name, imitated its plan, and paid respect to its authority. Yet it is not in the name alone of Linnæus, that the members of this Society place their confidence; still less do they bow to that name or to any other, at the expense of their own right of private judgement. Their Transactions are open to the pupils of every school, and the observations of every critic, that have any prospect of being useful to the world. The writer of each communication, must, of course, be answerable for the particulars of his own performance, but the Society is responsible for each being, on the whole, worthy to be communicated to the public. The possession of the very materials with which Linnæus worked, his own specimens and notes, enables us very often to correct mistakes, even of that great man, many of which would be unaccountable without the means of thus tracing each to its source. At the same time, the acquisition of materials to which he never had access, tends to improve and augment the history of what he had left imperfect. His language, his definitions and characters were, for some time, held so sacred, that they were implicitly copied, even though manifestly inapplicable, in some points, to the objects to which they were referred. Synonyms were transcribed from his works by Rose, Hudson, Curtis, and even Gærtner, (we assert it on the positive proof of errors of the press, copied in the transcribing,) without reference to the original books, to see whether such synonyms, or their accompanying plates, agreed with the plant under consideration. The example of Dr. Solander first led the writer of this to avoid such a negligent and unfaithful mode of proceeding; yet he has ever considered as sacred the very words of Linnæus, where they require no correction. They are become a kind of public property, the current coin of the botanical realm, which ought not, with impunity, to be falsified or adulterated. To them we hope to be pardoned if we apply the words of the poet,

> " The solid bullion of one sterling line,
> Drawn to French wire, would through whole pages shine."

Of this it is needless to quote examples. We must be every day more and more sensible of the value of the Linnæan style, in proportion as the number of those who can attain it is evidently so very small. By the light of our master alone can the science, which he so greatly advanced and refined, be preserved from barbarism, while long and tedious, loose and feeble, ill-contrasted and barbarously-worded definitions, press upon it from various quarters. New terms are invented to express old ideas; names and characters are changed for the worse, to conceal the want of new discoveries; and students are often deterred from adopting real improvements, because they know not which guide to prefer.

From the combined effects of the various causes which we have endeavoured to trace, the study of botany in England has, for a long period, been almost entirely practical. To determine the particular species intended, in every case, by Linnæus; to distinguish and to describe new ones; to improve scientific characters, and to correct synonyms; these have been the objects of our writers; and hence many publications of great utility, especially a number of critical and descriptive essays, in the Transactions of the Linnæan Society, not unworthy of the school which gave them birth, have enriched the general stock of knowledge. These are the sound fruits of skill and investigation, the solid advantages of real information, applied to practical use. They are independent of theoretical speculation, and will stand unshaken, amidst any possible changes of system. On such principles the *Flora Britannica* has been attempted, and continued as far as the present unsettled state of some of the latter orders, of the last class, will allow. Such impediments, which depend on the difficulties of systematic discrimination, among the Lichens especially, it is hoped will soon be removed. Meanwhile the *English Botany* of the same writer, illustrated by Mr. Sowerby's expressive and scientific figures, has finished its course, and formed so nearly complete a body of local botany, as, we believe, no other country has produced. In this the liberal contributions of numerous skilful observers, from the alpine heights of Scotland to the shores and circumambient ocean of the south, are preserved and recorded; evincing a degree of general inquiry and acuteness, which hardly any nation can rival. The memory of several benefactors to the science, otherwise in danger of passing away, is embalmed in this national work, which serves at once as their botanical testament, and the monument of their fame. Some of our botanists of the present day have thrown great light on several of the most obscure departments of the science; witness Mr. Sowerby's work on *English Fungi;* the labours of the learned Bishop of Carlisle on *Carices*, and, in conjunction with Mr. Woodward, on *Fuci;* of Mr. Dawson Turner on the latter tribe, and on the *Musci* of Ireland; but especially Mr. Hooker's inimitable display of the British *Jungermanniæ.* Nor shall the contributions of a Winch or an Abbot, a Withering, Knapp, Stackhouse or Velley, nor the more splendid labours of the indefatigable Lambert, be forgotten. Each, in one way or other, has enlarged the bounds of science, or rendered it easier of access. We cannot, in the compass of our present undertaking, pay the tribute due to every individual, our aim being a general picture of the whole. From what we have said, the zeal with which this lovely science has been cultivated in England will sufficiently appear. Nor have public lectures, or botanic gardens, been neglected, in order to render the knowledge of botany as accessible as possible, and to diffuse a taste for its pursuit. The popularity of the study has, at least, kept pace with the means of instruction. The garden and green-house, the woods, fields, and even the concealed treasures of the waters, are now the resource of the young and the elegant, who in the enjoyment of a new sense, as it were, in the retirement of the country, imbibe health, as well as knowledge and taste, at the purest of all sources.

France alone now remains to be considered, in order to finish the historical picture which we have undertaken, of the state of botanical science in Europe. To do justice to this part of our subject, we must turn our attention to times long since gone by, or we shall scarcely render intelligible the state of affairs at present.

The great Tournefort, by the force of his character, his general and particular information, the charms of his pen, and the celebrity which his name gave to his country, through the popularity of his botanical system, was so firmly established, in the ideas of the French, as the *Grand*

Monarque of botany, that they would have as soon allowed the greatness of Louis XIV. to be questioned, as that of this distinguished philosopher. So beneficial was this partiality, in some respects, that it gave an unprecedented impulse and popularity to the science; so disadvantageous was it in others, that it placed a formidable barrier in the way of all improvement. Vaillant, the able and worthy pupil of Tournefort, has never been forgiven for speaking, on some occasions, too freely of his master's defects. Hence his own merit has been kept in the background. The doctrine of the sexes of plants was discountenanced as long as possible, because it was proved by Vaillant, after having been rejected by Tournefort. Nevertheless, when the good seed of science is once sown, it can hardly be totally suffocated by the impediments of prejudice and ignorant partiality. Practical zeal sprung up by the side of speculative jealousy, and the tares withered, while the profitable plants flourished. Some botanists followed the steps of Tournefort to the Levant, exploring afresh those countries which he has for ever rendered classic ground. Others visited America, which they traversed in different directions. The indefatigable Plumier performed three separate voyages to the western world, and though his discoveries have, in a great measure, suffered shipwreck from tardy and imperfect patronage, as a great part of his collections did by the accidents of nature, yet something of value remains. His *Filices* are enough to insure his perpetual remembrance, and his *Nova Genera* are the basis of our knowledge of generic differences in West Indian plants. Most of all has been distinguished, among the French botanists who succeeded the times of Tournefort and Vaillant, the family of the Jussieus. One of these investigated the prolific regions of Peru, and discovered some things which no succeeding traveller has gathered; other branches of this family, besides being eminent in medical science and practice, have pursued the study of botany with no ordinary success, on the most philosop ical principles. Of these the most eminent are the celebrated Bernard de Jussieu, the contemporary of the earlier days of Linnæus; and his nephew Antoine Laurent de Jussieu, the pride and the ruler of systematic botany at present in France. The views and the performances of these great men lead us to a new branch of our subject, which indeed we have had in our contemplation from the beginning of this essay, the exposition of the principles of a natural scheme of botanical classification, as hinted, and imperfectly sketched, by Linnæus, and brought to the perfection of a regular system by the Jussieus.

Previous to our entering on this detail, and the remarks to which it will give rise, we must conclude all that belongs to the former part of our undertaking, by giving some account of those botanists who have formed and maintained a Linnæan school in France. We must shelter ourselves under the broad banner of truth when we observe that these have, till very lately, been almo t the only French botanists that have supplied us with any practical information; and their labours have been useful in proportion as they have commendably shaken off the prejudices of their predecessors. Of this last proposition Duhamel is a witness, though we may perhaps excite some surprise in classing him among Linnæan botanists His preface to his *Traité des Arbres* sufficiently shows how fearful he was of being taken for such, and yet how he was held by vulgar prejudice alone, to the nomenclature, or rather the generical opinions, of Tournefort. He tells us, while he adopts these, that his judgement went with Linnæus, whom he follows in all new discoveries. The plan of his book, confined to hardy trees and shrubs, justifies his use of an alphabetical arrangement, in preference to any system, unless he had thought sufficiently well of Tournefort's to prefer that. But he has prefixed to his work, as a practical method of discovering scientifically what it contained, no other than a sexual classification. His practical botany was so limited, being entirely subservient to his great objects, of forest planting and vegetable physiology, that he had no attention to spare for the consideration of methodical systems. According y he tells us, that some such is necessary for the use of botanists, especially of those who explore the productions of foreign countries, but whether the method of Ray, Tournefort, Boerhaave, Van-Royen, Linnæus, or Bernard de Jussieu be adopted, is of no importance. Six years before Duhamel's work came out, Dalibard had published, in 1749, his *Floræ Parisiensis Prodromus*, according to the Linnæan system.

It has always appeared to the writer of this, from the conversation and writings of French botanists, that the judgement of the learned Le Monnier, and the countenance of his patron the Duke D'Ayen, afterwards Marechal de Noailles, first established the reputation of Linnæus in France ; not so much possibly for the sake of his system, as his discoveries, his commodious nomenclature, and his clear principles of discrimination. When Le Monnier botanized in Chili, in the company of the astronomers with whom he was associated, he soon found, like Dr. Garden in South Carolina, that the classification of Tournefort was no key to the treasury of a new world. He however made his remarks and collections, and studied them subsequently under the auspices of a more comprehensive guide. The Marechal de Noailles, a great cultivator of exotic trees and shrubs, corresponded with the Swedish naturalist, and endeavoured to recommend him to the notice of the lovers of plants in France. Meantime Gerard and Gouan in the south, both introduced themselves to the illustrious Swede, and promulgated his principles and discoveries, though only the latter adopted his classification. Villars we have already noticed as the author of a Linnæan *Histoire des Plantes de Dauphiné*. He died lately, Professor of Botany at Strasburgh, where he succeeded the very able and philosophical Hermann, one of the truest

Linnæans who had imbibed all the technical style of the Swedish school, as well as its accuracy of discrimination. We may now safely announce Hermann as the real author in conjunction perhaps with Baron Born, of that ingenious but bitter satire the *Monachologia*, in which the several *species* of monks are affectedly discriminated, and their manners detailed, like the animals in the Linnæan *Systema Naturæ*. This ludicrous performance has long since appeared in a, not very exact, English translation, and was rendered into French by the late M. Broussonet. As we are led again to name this amiable man, too soon lost to his country, after experiencing every vicissitude of revolutionary peril and alarm, we cannot help distinguishing him as one most zealous in the cultivation and diffusion of Linnæan learning, a taste for which he chiefly imbibed in England. He had no indulgence for those prejudices, which cramped the talents of his countrymen, and prevented their deriving knowledge from any quarter where it was to be had. He recommended the younger Linnæus to their personal acquaintance and favour, which service he also rendered, a few years after, to the person who now commemorates his worth, and who will ever remember, with affection and regret, his many virtues, his agreeable converse, and his various and extensive acquirements.

The intimacy which subsisted between this enthusiastic naturalist, and the distinguished botanist L'Heritier, confirmed, if it did not originally implant, in the mind of the latter, that strong bias which he ever showed for the Linnæan principles of botany. According to these his numerous splendid works are composed. He moreover imbibed, if we mistake not, from the same source, a peculiar preference for uncoloured engravings of plants, instead of the coloured ones which had long been in use. It cannot be denied that the merit of these last is very various, and sometimes very small. They do, nevertheless, present to the mind a more ready idea of each species, than a simple engraving can do, nor is the latter less liable to incorrectness. When plates are taken from the delineations of such exquisite artists as L'Heritier employed, they have a good chance of excellence; but the engravings of Cavanilles, done after miserable drawings, though they deceive the eye by their neat finishing, are really less exact than many a rude outline. Coloured plates, if executed with the uniformity and scientific exactness of Mr. Sowerby's, or the characteristic effect of Jacquin's, speak to the eye more readily than most engravings. The art of printing in colours, practised formerly in England with small success, was revived at Paris by Bulliard, and is carried to the highest perfection in the recent publications of Redouté and Ventenat, which leave hardly any thing to be wished for, with respect to beauty or exactness. Many of the works of L'Heritier have remained imperfect, in consequence of the political convulsions of his country, and his own premature death. The learned and worthy Desfontaines, who travelled in Barbary, has been more fortunate in the completion of his labours. His elegant *Flora Atlantica*, in 2 vols. 4to, with finely engraved uncoloured plates, is classed and modelled on the plan of the Linnæan school. Such also is the plan of the works of that distinguished botanist La Billardiere, who, besides his account of New Holland plants, has published five elegant *decades* of new species from Syria. That scientificohorticulturist M. Thouin, likewise a most excellent botanist, though he has scarcely written on the subject, is a correct pupil of the Swedish school. His general spirit of liberal communication, and his personal attachment to the younger Linnæus, led him to enrich the herbarium of the latter with the choicest specimens of Commerson's great collection, destined otherwise to have remained in almost entire oblivion. A singular fate has attended the discoveries of most of the French voyagers, such as Commerson, Sonnerat, and Dombey, that, from one cause or other, they have scarcely seen the light. So also it has happened to those of Tournefort, Sarrazin, Plumier, and others, whose acquisitions have long slept in the Parisian museums. Happily there seems to have arisen of late a commendable desire to render them useful by publication, and thus many fine plants, known merely by the slight and unscientific appellations of Tournefort, and therefore never adopted by Linnæus, have recently been clearly defined, or elegantly delineated. The journeys of Olivier and Michaux towards the east have enriched the Paris gardens, and been the means of restoring several lost Tournefortian plants. We believe however that the English nurseries have proved the most fertile source of augmentation to the French collections, as appears by the pages of all the recent descriptive writers in France.

We dare not presume to arrange the indefatigable and very original botanist Lamarck among the Linnæan botanists of his country; but we beg leave to mention him here, as one who has thought for himself, and whose works are the better for that reason. His severe and often petulant criticisms of the Swedish teacher, made him appear more hostile than he really was, to the principles of that great man. Being engaged in the botanical department of the *Encyclopédie Méthodique*, he was obliged to conform to an alphabetical arrangement; but he surely might have chosen the scientific generic names for that purpose, instead of barbarous or vernacular ones, which, to foreigners, would have made all the difference between a commodious and an unintelligible disposition of his work. In the detail of his performance, he has great merit, both with respect to clearing up obscure species, or describing new ones, and he had the advantage of access, on many occasions, to Commerson's collection. Lamarck's *Flore Françoise*, is arranged after a new analytical method of his own. This book however is valuable, independent of its system, as an assemblage of practical knowledge and observation. We have only to regret a wanton and inconvenient change of names, which too often occurs, and which is not always for the better; witness *Cheiranthus hortensis,* instead of the long established *incanus* of Linnæus; *Melampyrum violaceum,* which is not correct, for *nemorosum,* which is strictly so, and which preserves an analogy with the rest of the species.

We shall now undertake the consideration of the principles that have been suggested, and the attempts that have been made, respecting a

Natural Classification of Plants.

The sexual system of Linnæus lays no claim to the merit of being a natural arrangement. Its sole aim is to assist us in determining any described plant by analytical examination. The principles on which it is founded are the number, situation, proportion, or connexion, of the stamens and pistils, or organs of impregnation. These principles are taken absolutely, with the sole exception of their not being permitted to divide the genera, that is, to place some species of a genus in one part of the system, and others in another, though such may differ in the number, situation, proportion, or connexion of their stamens or pistils; those characters being possibly artificial, while the genera are supposed, or intended, according to a fundamental law independent of all systems, to be natural assemblages of species. We need not here explain the mode in which Linnæus has provided against any inconvenience in practice, resulting from such anomalies of Nature herself.

But though this popular system of Linnæus does not profess to be a natural method of classification, it is, in many points, incidentally so, several of its classes or orders whose characters are founded in situation, proportion, or connexion, being more or less perfectly natural assemblages; nor can it be denied that, on the whole, it usually brings together as many groups of natural genera, as occur in most systems that have been promulgated. This fact would be more evident, if the various editors of this system, those who have added new genera to the original ones of Linnæus, or, in general, those who have any way applied his method to practice, had properly understood it. They would then have perceived that its author had always natural affinities in view; his aim, however incompletely fulfilled, according to our advanced knowledge, having constantly been, to place genera together in natural affinity or progression, as far as their relationship could be discerned. At the same time he uses an analytical method, at the head of each class in his *Systema Vegetabilium,* in which the genera are disposed according to their technical characters. Murray, in compiling the fourteenth edition of that work, has been inadvertent, respecting this essential part of its plan. Indeed it is probable that he was not competent to judge of the affinities of the new genera, introduced from the *Supplementum,* or from the communications of Jacquin, Thunberg, &c. Yet surely he might have perceived the affinity of *Banksia* to *Protea,* rather than to *Ludwigia* or *Oldenlandia*; and indeed Linnæus himself ought to have discovered the relationship of the latter to *Hedyotis,* if he did not detect their identity, instead of inserting it between two such strict allies of each other as *Ludwigia* and *Ammannia.* To pursue these remarks would be endless. It is hardly necessary to indicate the natural classes or orders of the Linnæan system, such as the *Tetradynamia, Didynamia, Diadelphia, Syngenesia;* the *Triandria Digynia, Gynandria Diandria,* &c. Except the first-mentioned class, which, if *Cleome* be removed, is strictly natural and entire, the others are liable to much criticism We are almost disposed to allow, what we know not that any one has yet observed, that the system in question is the more faulty in theory, for these classes being so natural as they are. Each order of the *Didynamia* presents itself as a natural order, though the character of that class, derived from the proportion of the stamens, serves to exclude several genera of each order, and to send them far back, into the second class. If all ideas of natural affinity be discarded from our minds, there is no harm whatever in this; but if the *Didynamia* claims any credit, as a class founded in nature, the above anomaly is a defect. So, still more, under the same point of view, is the *Diadelphia*, or at least its principal order *Decandria*, liable to exception. This order consists entirely of the very natural family of *Papilionaceæ*. They are characterized as having the ten stamens in two sets. Now it happens that there are many papilionaceous genera, indeed a great number of such have been discovered since Linnæus wrote, whose ten stamens are all perfectly distinct. These therefore are necessarily referred to the class *Decandria*, and they come not altogether amiss there, because they meet, in that class, some concomitant genera, which though, like them, leguminous, are less exactly, or scarcely at all, papilionaceous. But the greatest complaint lies against some genera of the *Diadelphia Decandria*, for having the stamens all really combined into one set, so as in truth to answer to the technical character of the preceding class, *Monadelphia*. There is mostly indeed some indication of a disunion upward, where they, more or less perfectly, form two sets; and some of them are so nearly diadelphous, that their complete union at the bottom may easily be overlooked; others, however, have only a fissure along the upper side of their common tube, without any traces of a separate stamen or stamens. The papilionaceous character of the *corolla* therefore, in such cases, is made to overrule that of the particular mode of union among the stamens, and is in itself so clear, as seldom to be attended with any difficulty; but the incorrectness of principle in the system, in the point before us, as being neither professedly natural, nor exactly artificial, cannot be concealed. Part of the objections, to which the sexual system was originally liable, have been obviated. We mean what concerns the last class but one, *Polygamia.* Dr. Forster observed, in his voyage round the world, that this class was subject to great exception, on account of the trees of tropical climates, so many of which are constantly or occasionally polygamous; that is, each individual frequently bears some imperfect flowers, male or female, along with its perfect or united ones. Such a circumstance reduces any genus to the class *Polygamia;* and on this principle Mr. Hudson, thinking perhaps that he made a great improvement, removed our *Ilex Aquifolium*, or Holly, thither, though *Ilex* is well placed by Linnæus in the fourth class. The author of the present essay has ventured to propose a scheme, which is adopted in his *Flora Britannica*, for getting clear of this difficulty. He considers as polygamous such genera only as, besides having that character in their organs of impregnation, have a difference of structure in the other parts of their two kinds of flowers. Thus *Atriplex* has, in its perfect flowers, a regular spreading calyx, in five equal segments; in the attendant female ones a compressed one, of two leaves, subsequently much enlarged.

The genera thus circumstanced are so very few, as far as we have discovered, that possibly the class might, but for the uniformity of the system, be abolished. We cannot indeed tell what future discoveries may be made; and its character, on the above foundation, is sufficiently clear and permanent; for flowers of an essentially different configuration, can hardly vary into each other. The orders of the last class of the Linnæan system, *Cryptogamia*, are natural, and preserved, all nearly the same, by every systematic projector. The original appendix to this system, the *Palmæ*, would be a great blemish therein, as an artificial arrangement: for such an arrangement ought to be so formed as to admit every thing, on some principle or other. But this stumbling-block is now removed. The palm tribe were placed thus by themselves, merely till their fructification should be sufficiently known.

Now they are found to agree well with some of the established classes and orders, where they meet with several of their natural allies.

Whatever advantages might accrue to the practical study of botany, from the convenience and facility of his artificial system, Linnæus was from the beginning intent on the discovery of a more philosophical arrangement of plants, or, in other words, the classification of nature. This appears from the 77th aphorism of the very first edition of his *Fundamenta Botanica*, published in 1736, where he mentions his design of attempting to trace out fragments of a natural method. In the corresponding section of his *Philosophia Botanica*, he, fifteen years afterwards, performed his promise; and the same *Fragmenta*, as he modestly called them, were subjoined to the 6th edition of his *Genera Plantarum*, the last that ever came from his own hands. The interleaved copies of these works, with his manuscript notes, evince how assiduously and constantly he laboured at this subject, as long as he lived. He was accustomed to deliver a particular course of lectures upon it, from time to time, to a small and select number of pupils, who were for this purpose domesticated under his roof. What this great botanist has himself given to the world, on the subject under consideration, is indeed nothing more than a skeleton of a system, consisting of mere names or titles of natural orders, amounting in his *Philosophia* to 67, besides an appendix of doubtful genera; and that number is, in the *Genera Plantarum*, reduced to 58.

Under the title of each order, the genera which compose it are ranged according to the author's ideas of their relationship to each other, as appears by some of his manuscript corrections; and some of the orders are subdivided into sections, or parcels of genera more akin to each other than to the rest. He ingenuously avowed, at all times, his inability to define his orders by characters. He conceived that they were more or less connected with each other, by several points of affinity, so as to form a map, rather than a series. The experienced botanist, who peruses the above-mentioned *Fragmenta,* will in most cases readily imbibe the ideas of their author, as to the respective affinities of the genera. In some few instances, as the *Dumosæ,* where he avows his own doubts, and the *Holeraceæ,* where he is unusually paradoxical, it is more difficult to trace the chain of his ideas. Such however was all the assistance he thought himself competent to afford. His distinguished pupils Fabricius and Giseke fortunately took notes of his lectures on natural orders; and by the care of the latter, to whom Fabricius communicated what he had likewise preserved, their joint acquisitions have been given to the public, in an octavo volume at Hamburgh, in 1792. Nor was this done without the permission of their venerable teacher, who told Giseke by word of mouth, when they took leave of each other, that "as he loved him, he had laboured with pleasure in his service;" adding, that "Giseke was at liberty to publish, whenever he pleased, any thing that he had retained from his own instructions."

Linnæus, according to a conversation with Giseke, recorded in the preface of the volume edited by the latter, declined to the last any attempt to define in words the characters of his orders. His reason for this appears in his *Classes Plantarum,* where he justly remarks, that no certain principle, or key, for any such definition can be proposed, till all the orders, and consequently all the plants, in the world are known. He has however so far expressed his opinion, in the work last quoted, as to point out the situation of the seed itself, with respect to other parts, and the situation and direction of its vegetating point, or *corculum,* as most likely to lead to a scheme of natural classification. Hence the system of Cæsalpinus stood very high in his estimation. He also, in the conversation above mentioned, divides his own orders into three sections or classes, *Monocotyledones,* comprising the first ten orders, with the 15th; *Dicotyledones* (with two or more cotyledons), the 11th to the 54th order, inclusive, except the 15th; and *Acotyledones*, order 55th to 58th, with a hint that the last, or *Fungi*, ought perhaps to be altogether excluded. This distribution of plants, by the number or the absence of the cotyledons, or lobes of the seed, is the great hinge of all the professedly natural modes of arrangement that have been attempted. We shall for the present not enter on the consideration of this principle, as it will more properly be explained when we examine the system of Jussieu. Linnæus did not consider it as absolute, for he told Giseke that he knowingly admitted into his 11th order some plants that are monocotyledonous, with others that are dicotyledonous. The reason of this was the only secret he kept from his pupil, nor could the latter ever dive into it, though he afterwards endeavoured to learn it from the younger Linnæus, who knew nothing, neither did he, as Giseke says, much care, about the matter. We hope to be able to throw some light upon this mystery, when we come to the order in question.

The want of any avowed principle of distinction, precludes all criticism of these natural orders of Linnæus, as a regular system; we can therefore only take a cursory view of them as they follow each other, with such indications of their characters as Giseke has recorded, or as we may ourselves be able to trace. A great part of the substance of the lectures, published by him, consists of remarks on the genera of each order, as to their mutual distinctions; with numerous botanical and even economical matters, which do not all come within the compass of our present consideration. What we have to lay before the reader, is not, in any manner, forestalled, by what he will find in the fourth volume of the Encyclopædia, above cited, which is taken from a different source.

Order 1. Palmæ. "An entirely natural and very distinct order." This tribe of plants, stationed by nature within the tropics, is considered by Linnæus as the original food of man; still supplying the place of corn to the inhabitants of tropical countries. Palms are the most lofty of plants, and yet it is a matter of doubt whether they ought to be called trees or herbs. They do not form wood in concentric circles, year after year, like our trees, though they are extremely long-lived. The author of the sexual system was, as we have just mentioned in speaking of that system, but little acquainted at first with the structure of the flowers of palms, or the number of their stamens or pistils. His predecessors in the establishment of genera of plants, Tournefort and Plumier, had published little or nothing illustrative of this tribe. He had himself seen no more than three or four species in fructification, nor had he any other resource, in founding genera, than the plates of the *Hortus Malabaricus*, (excellent indeed, but not delineated with any particular view of this kind,) and the less complete representations of Rumphius. The growth of these plants is quite simple. Each terminates in a bud, of a large size, called the heart, or by voyagers in general the cabbage, of the palm. When this is cut off, the tree dies, though the growth of many centuries. This bud has a gradual and nearly continual vegetation, unfolding its leaves, which Linnæus rather incorrectly terms fronds, one after another in succession, not all at any particular season. The bud therefore is perennial, not, as in our trees, annual, nor can it, for this reason, be renewed. Fresh buds, in time becoming trees, are furnished from the generally creeping, perennial, and deeply descending roots. What have commonly been denominated the branches of palms, Linnæus very properly declined calling so, because they never increase by producing lesser branches. He objected to calling them leaves, "because they are each attended by no separate annual bud, neither have they the texture of ordinary leaves, nor do they wither and fall off at any particular season." He adopted the term *frond*, which he always used when he could not decide whether the part in question were a branch, leaf, or stem. We cannot but think these are truly leaves, though it must be confessed they differ from the generality of such, being destitute of any line of separation by which they are capable of falling, or being thrown off, from the stem. In this they agree with the foliage of *Musci* and *Jungermanniæ*; there being a perfect continuity of substance throughout. The hardened torn fibres, or rather vessels, which remain on the stems of palms, where the leaves have once been, are precisely the same as what occur in various mosses; and something similar may be observed in many liliaceous plants and their allies, which approach to the nature of palms.

In describing the fructification of this order, Linnæus considered as belonging thereto, what we should presume to be rather the inflorescence. Hence the great branching flowerstalk retains, in a technical sense, the name of *spadix*, derived from the ancients; and its ample containing sheath is denominated a *spatha*. The latter is reckoned a kind of calyx, as the former a sort of branched common receptacle. Linnæus strengthens his terminology in this case, by tracing an analogy between the *spatha* of palms, and the glume of grasses. We doubt whether any such particular analogy exists. Neither does his other comparison, of the part in question to the sheath of a *Narcissus* and its allies, at all, as far as we can judge, elucidate or confirm his principle. He surely swerves in these instances, as well as in his generic distinctions of the umbelliferous plants, from the correctness of an axiom, on which botany as a philosophical science depends, that generic characters, and much more those of classes and orders, should be exclusively derived from the parts of fructification. Surely a very slight consideration of the flowers and fruits of the *Palmæ*, as we have become acquainted with them since the time of Linnæus, will abundantly satisfy any person, that they afford clear characters, on which to found a sufficient number of distinct and very natural genera. Even that author, in the lectures before us, records that some genera have a three-leaved *calyx*, others none at all; some have a *corolla* of three, others one of six, petals; most have six *stamens*, some three, others nine, while the *Nipa* of Thunberg has only one. The *germens* are three in some, solitary in others, and the *style* and *stigma* are subject to like diversity in different genera. The *fruit* is in some, as the *Phœnix dactylifera*, or Date, a single drupa, in others composed of three; in some, like the Cocoa, a nut with a coriaceous coat. The *seeds* are mostly solitary, but in several instances two or three in each fruit. Hence, while the fructification affords sufficient materials for discriminating genera, Linnæus observes that no common character, exclusively descriptive of the whole order, can be founded upon it. The reader will find the essential characters of his genera in our Vol. IV. 288. His *Zamia*, concerning which he avowed considerable doubts, chiefly because it wanted a *spatha*, is now by common consent among botanists, removed either to the Ferns, or to an intermediate order between them and the Palms, to which also *Cycas* belongs. The technical characters which have induced this alteration, are confirmed by circumstances attending the habit and qualities of these genera.

At the end of his proper *Palmæ*, Linnæus subjoins in a distinct section, three genera, which he was doubtful whether to leave there, or to establish as a distinct order. These are *Stratiotes*, *Hydrocharis*, and *Valisneria*. He remarks in his lectures that "they have a *spatha* extremely like the palms; a calyx of three leaves, and a corolla of three petals; leaves perennial and evergreen, folded when they first come forth. *Hydrocharis* cannot be separated from *Stratiotes*, nor *Valisneria* from *Hydrocharis*. They produce their leaves crowded together at the base, like Ferns. Although their strict affinity with the larger Palms of India cannot be demonstrated, they ought nevertheless to be associated therewith. They are all aquatics, whence we may presume that India may afford some aquatic palms, smaller than the others, which may prove a connecting link between the latter and the plants of which we are speaking." Giseke points out several palms, in various authors, which though but imperfectly ascertained, confirm this conjecture of his preceptor. Linnæus in his own copy of the *Genera Plantarum*, enriched with his manuscript notes, to which we shall often refer, has marked this section, or appendage, of his *Palmæ*, as distinguished by "an inferior fruit, with many seeds." He has moreover added 4 genera to this assemblage, *Pandanus*, *Bromelia*, *Tillandsia*, and *Burmannia*. Giseke has amply illustrated the order of *Palmæ*, by observations of his own, or those of various writers; but the most solid acquisitions to our knowledge, in this interesting tribe, are derived from the labours of Dr. Roxburgh, in his *Plants of Coromandel*.

Order 2. PIPERITÆ. "The plants of this order have an acrid flavour, whence the name." They afford no common character to discriminate the order, except possibly the elongated receptacle and sessile anthers, but some *amentaceæ* have the same. They consist of *Zostera*, *Arum*, and its allies, *Orontium*, *Acorus*, *Piper*, and *Saururus*. The last is removed by Linnæus in his manuscript to his 15th order.

Order 3. CALAMARIÆ. "These are closely related to the true grasses, and have almost the same kind of leaves. Their seed is solitary and naked; stamens three; style one, not unfrequently three-cleft at the summit. Their glume is of one valve (whereas most grasses have two valves), except *Schœnus*, which bears several valves irregularly disposed, though in other respects so near the rest of its order, as scarcely to be distinguished without accurate examination of the parts alluded to. The stem of these plants is a *culm*, mostly triangular, rarely round, often leafless, or nearly so. Leaves rather rigid and rough.

Flowers often disposed in an imbricated manner. Seed in a few instances surrounded with bristles. When these are extended into a kind of wool, hanging out beyond the scales, such a character marks the genus *Eriophorum*." Linnæus asserts that "*Scirpus* differs from *Carex*, in having all the flowers united, whereas in the latter some scales are accompanied with stamens only, others with pistils;" but he forgot the tunic, or *arillus*, of the seed, which makes the essential and clear character of *Carex*. He mistakes also in supposing the stamens are always three in this order; in several instances they are but two, in a few they are solitary. Much has been done respecting the genera and species of this order by Rottboll, Vahl, Brown, Schrader, and others. Linnæus has made a manuscript correction in the *Calamariæ*, excluding from thence *Typha* and *Sparganium*, which he would remove to the preceding order, principally, it seems, because he judged the latter to be very closely allied to *Zostera*; as well as on account of its anthers, but we can trace no resemblance in those to the *Piperitæ*. On the contrary they and their filaments agree with the *Calamariæ*. The stamens of *Typha* indeed are somewhat different, and Mr. Brown, in his *Prodromus Floræ Novæ Hollandiæ*, has anticipated this alteration of Linnæus.

Order 4. Gramina. "The true Grasses compose as peculiar a family as the Palms. They are the most common plants in the world, making about a sixth part of the vegetable kingdom, especially in open situations. There they multiply, and extend themselves by their creeping roots, prodigiously. In confined and woody places they scarcely creep, but stand erect. They are the most important of all vegetables, for this reason, that they are the chief support of such animals as depend on vegetable food. They make the verdure of our summers, and the riches of rustic life. Their leaves are not easily hurt by being trampled on, and though the severity of winter may wither and fade them, so that in the early spring no appearance of life remains, yet they revive. The solicitude of the Author of Nature, for the preservation of this important tribe of vegetables, appears from their flowering stems being rendered unfit for the food of cattle, that nothing may hinder the perfecting of their seeds. Besides, the more they are cut and ill-treated, the more vigorously they grow, propagating themselves proportionably under ground; and in order that they may be enabled to thrive any where, their narrow leaves are so contrived, as to insinuate themselves between the divisions or branches of other herbs, without any mutual impediment. There are very few grasses agreeable to our palate. For the most part they are insipid, like pot-herbs; a very small number being fragrant. None are nauseous or poisonous. Grasses are the most simple of all plants; having scarcely any spines, prickles, tendrils, stings, bracteas, or similar appendages to their herbage."

"Their stem is termed a *culm,* being hollow, composed of joints which are separated by impervious knots. In our quarter of the world the culm is usually simple, unless in consequence of cutting away the flowering part; in the Indies most culms are branched. The leaves are mostly alternate, always undivided, and generally flat on both sides, with a rough edge, and either smooth or hairy surface. Each leaf stands on a sheath, which embraces the stem, and is crowned with a membrane, sometimes termed *ligula,* closely embracing the stem, to hinder the admission of water. The sheath springs from a knot, and (with its membrane) answers the purpose of a stipula."

"The fructification of Grasses differs so much from that of other plants, that it was supposed impossible to reduce them to scientific order. They were first distinguished into corn and grasses; but such a distinction is founded merely on the comparatively larger seeds of the former, on which we depend for food, as small birds do on the very minute seeds of the latter. Ray was the first botanist who undertook a regular examination of grasses.

He distributed them according to their outward appearances, but distinctive characters failed him. Neither was Tournefort, however great a botanist, equal to the arrangement of this tribe. Monti followed Ray, but investigated such only as were natives of Italy. John Scheuchzer, first induced by Sherard, paid a most laborious attention to this subject, collecting grasses from all quarters, and describing them with the greatest exactness; but he was deficient in technical terms, and his very long descriptions are nearly all alike, till he arrives at the flowering part. The terms which he uses are *folliculus* for the corolla, *gluma* for the calyx, *locusta* for the spikelet contained in the latter. After him Micheli contrived a new method, dividing grasses according to their spikelets, which he observed to be either compound or simple. He subdivided them by their flowers being united or separated; and subjoined an order of plants "akin to grasses," which really do not belong to them. If their sexes be attended to, the arrangement of grasses becomes less difficult. They are either *monandrous, diandrous, triandrous*, or *hexandrous*. The two latter have either united, monoecious or polygamous flowers."

"The inflorescence in this order of plants is either spiked or panicled. Their spike, properly so called, consists of several flowers, placed on an alternately toothed *rachis*, or stalk. If such a rachis be conceived perfectly contracted, it will become a toothed common receptacle, as in compound flowers, so that grasses may thus be distinguished into simple and compound. Or if we imagine all the flowers to be sessile on one common base, such grasses as are properly spiked will have a scaly receptacle, the rest a naked one, according to the analogy of the syngenesious class; and by this means the corn family may be separated from the rest, for they are scaly.

"The calyx is a husk of two valves, one proceeding from within the base of the other, like the claw of a crab. These husks are concave, and truly the leaves of the plant in miniature. The calyx contains one, two, or more, florets, which are constructed in the same manner, of two leafy husks, called by Linnæus petals, to distinguish them from the former. Within the petals the receptacle bears two very minute, roundish, pellucid, extremely tender, withering scales, often invisible without a magnifier, which Micheli termed petals, Linnæus nectaries. Stamens generally three, in a few one, two, or six with capillary filaments, and oblong incumbent anthers, whose lobes become separated at each end. Micheli erroneously imagined those which have six stamens, to bear, as it were, doubled flowers. The germen is superior, with two styles, sometimes raised on a common stalk or elongated base, and they are usually reflexed to each side, being either longitudinally hairy, or tufted at the summit only. Seed universally solitary, without a capsule, *Lygeum* only having a nut, of two cells, which is very singular. A few have a simple style, as *Zea*, *Nardus*, and *Lygeum*. The seed is occasionally coated by the petals, which closely enfold it, and are almost united with it,—witness *Hordeum* and *Avena*;" (to which examples indicated by Linnæus we may add *Briza*). "Many grasses are furnished with an awn, *arista*, mostly rough, like a prominent bristle, inserted into the back of the outermost petal, either at the bottom, middle, summit, or a little below the latter. This appendage is either straight, or furnished with a joint, and twisted backward, or simply recurved; in some it is woolly: in several it is accompanied by hairs at the base of the corolla. The use of these parts is to attach the ripe seeds to the coats of animals, that they may be the more dispersed."

"Although grasses are destitute of spines properly so called, a few have their leaves longitudinally involute, in such a manner that their rigid permanent points have all the properties of thorns, as in *Spinifex*, and some *Festucæ*. Their foliation is, for the most part, involute, but in some instances, as *Dactylis glomerata*, it is folded. This character has not as yet received sufficient attention, but ought to be noticed in future, as it may throw great light on the distribution of the family of plants in question. Very few indeed are furnished with setaceous leaves."

Order 5. TRIPETALOIDEÆ. "Scheuchzer and other authors have referred *Juncus* and its allies to Grasses, under the title of *Graminibus affines*. In truth, they are so similar to grasses, as scarcely to be distinguishable without fructification. The genera are *Juncus, Aphyllanthes, Triglochin, Scheuchzeria, Elegia* and *Restio* in the first place, then *Flagellaria, Calamus, Butomus, Alisma* and *Sagittaria*." Linnæus, in his manuscript, has hinted, that the three latter may possibly belong to the above-mentioned section at the end of his *Palmæ;* see Ord 1.

Order 6. ENSATÆ. "So called from the form of their leaves, resembling a sword, being perfectly simple, almost linear, alternate, mostly converging by the margins, often cloven longitudinally, so that the edge of one leaf embraces the other, thus constituting what is termed equitant foliage. The root in many cases is oblong and fleshy, lying flat on the ground, or creeping. But some species of *Iris* are truly bulbous, like *Crocus, Ixia, Antholyza*, &c. Stem, in these genera, simple, erect, zig-zag; but in *Commelina*, especially the annual kinds, it is branched, as in *Tradescantia*. *Crocus* and *Bulbocodium* have no stems. Leaves usually sword-shaped; very rarely quadrangular; in the bulbous species of *Iris* involute; in not a few *Commelinæ* ovate; in *Xyris* and various kinds of *Eriocaulon* awl-shaped. *Fulcra*, or appendages, are scarcely to be found in this order. The calyx is a *spatha*, though but of a spurious kind, being mostly a large concave valve, resembling a halved sheath in *Iris*; most beautiful in *Commelina*, where it is heart-shaped. In *Sisyrinchium* however this part is more perfectly bivalve. Corolla generally of six petals; though in *Iris* so united by their claws, as to constitute a monopetalous corolla. In *Commelina* and *Tradescantia* the petals are very distinct, but the three inferior being ruder in texture, and smaller, resemble a calyx. Style with three stigmas, except some *Commelinæ*. Pericarp a capsule of three cells and three valves, with many seeds; generally inferior, but not so in *Commelina*, *Tradescantia*, and *Callisia*. Hence it follows that this order affords no certain mark, on which a distinctive character could be founded."

Order 7. ORCHIDEÆ. "Orchis is a most ancient generic appellation, alluding to the testicular shape of the roots, in many plants of this family, which have, at all times, been believed to possess a stimulating or aphrodisiacal virtue. All the *Orchideæ* might be comprehended in one genus, in which light also the *Umbellatæ*, *Semiflosculosæ*, *Papilionaceæ*, might each likewise be considered. But the science would be overwhelmed in confusion by such extensive genera, which it is therefore found necessary to subdivide.

"Many *Orchideæ* have a tuberous fleshy root; not properly to be termed bulbous, because its fibres are thrown out from the top, or crown, whereas true bulbs produce their fibres from the base. These tubers, or knobs, are mostly in pairs; some of them globose and undivided; others palmate, like the hand. One of these tubers, from whence the plant of the present year has come, being exhausted, will swim in water; the other, destined to blossom next season, is so solid as to sink. In the palmate kinds, the former is vulgarly called the hand of the Devil, the latter the hand of God. *Ophrys corallorrhiza* however has a threadshaped, branched, and jointed root; that of *O. bifolia* is perfectly fibrous. In other genera, particularly *Epidendrum*, the root consists of clusters of fibres."

"The stem is solitary and herbaceous, except in several kinds of *Epidendrum*, quite simple, often leafy. In some however there is merely a leafless, radical flowerstalk, generally round, though not so in *Ophrys Loeselii* and *paludosa*. The leaves are simple, alternate, undivided, sheathing the stem; sometimes wanting, as in *Orchis abortiva.* Appendages none at all, except bracteas. Inflorescence terminal, either spiked or racemose. Fructification irregular, and very singular, for it is impossible to say what is calyx, and what corolla; nor is this point of much importance, nature having placed no limits between them. There are five petals; besides a nectary, which makes, as it were, a sixth. These five seem to constitute an upper lip, the nectary an under one. Or it may be said that the corolla is composed of three outer, often ruder petals; and three inner, the lowermost of which ought rather to be denominated a nectary. This last is various in different genera, having its appropriate figure and dimensions, while the rest of the petals are more uniform. Sometimes the middlemost of the five petals, composing the upper lip, (like that of a ringent or helmet-shaped flower,) is more erect and dilated; but I have received some species from the Cape of Good Hope, in which these petals are united to each other, and elongated at their common base into a spur. Such will constitute a new division or genus, of this family, as it stands in the *Species Plantarum*, many of which have a spur from the base of the lower lip, or nectary. The petals however do not afford sufficient distinctions, for genera or species. The former are determined by the nectary, which is for that purpose principally to be regarded. There is indeed no occasion to advert to any other part than the flower of these plants, for distinguishing either genera or species. Vaillant therefore, and Seguir, have contented themselves with delineating their various flowers alone."

"The stamens consist of two anthers, nearly without filaments, very singular, and peculiar to this order, concealed in a double pouch or hood, but their pollen has not been ascertained. They are 'contracted at the base, naked, or destitute of a skin, divisible like the pulp of an orange, and covered each by a cell open underneath, inserted into the inner margin of the nectary;' as described in the *Genera Plantarum.* It remains therefore for inquiry, whether the anthers burst in these as in other stamens, and whether the pollen explodes upon the female organs? or whether there be any internal communication between the anthers and germen?" This latter opinion Linnæus was inclined to adopt, because, (as he thought,) "the pistil was so obscure, that no one was able to say whether there were any style or stigma." We cannot but remark here that the latter is sufficiently apparent, in the form of a shining glutinous depression or cavity, just below the anthers; nor is there any doubt that the pollen, though different in texture from other plants, and various in the different species of these, performs the office of impregnation by the stigma. It consists of naked elastic or granular masses, being what Linnæus terms the anthers.

"The germen is inferior in the whole order; the style short, inclining, in many hardly manifest, in some American *Orchideæ* very conspicuous. Stigma either obsolete, or funnelshaped, sometimes compressed. A small gland moreover is present, suspected to belong to the female organs of impregnation, but not very decidedly." (Linnæus surely errs in asserting that the sexes of the plants in question are very obscure.) "The fruit is a capsule, of one cell, and three valves, which last are connected by a lateral suture, to which the seeds are attached, as to a receptacle. The capsule does not burst in the usual manner, but the valves separate at their lateral sutures, while their extremities remain united at top and bottom. The seeds are numerous, of a chaffy appearance, like saw-dust."

"Many fine species of this order are found in Europe and America; the Cape of Good Hope is not rich in them;" (Mr. Brown observed a considerable number there;) "both Indies abound with singular ones, especially with *Epidendra.* Their favourite soil is a spongy, moist, friable, rich, but not manured, earth, in rather shady situations. The species of *Epidendrum* are all, perhaps, parasitical, insinuating their roots into the bark of aged trees."

"*Orchideæ* are extremely difficult of culture." We refrain from transcribing the ideas of Linnæus on this subject, as it is now known that some of these plants may be propagated by seed, and that several succeed very well in our stoves, among the rotten bark of trees, accompanied by fresh vegetable mould. Our wild Orchises are best removed when in full bloom, when the mould should be entirely cleared away from their roots, and the latter planted immediately in fresh sifted soil from their native place of growth, with moderate subsequent watering. Thus treated they will come up and flower for many successive years in the same pot.

Order 8. SCITAMINEÆ. "These nearly approach the *Orchideæ* in aspect. The name of the order is an ancient word, synonymous with *aromatic*, and answers to the whole of the tribe, except *Musa*, *Heliconia*, and *Canna*." (The two former certainly do not belong to this order, and the last but imperfectly.)

"The roots of the *Scitamineæ* are fleshy, mostly acrid and aromatic, lying on the surface of the ground, and throwing out fibres from their under side, like some of the 6th order. Stem always quite simple," (to this there are exceptions in *Maranta*,) "in some bearing alternate leaves; in others naked, and separate from the foliage. Leaves lanceolate, quite entire, even, stalked, convoluted contrary to the direction of the sun; their stalks sheathing the stem. Inflorescence either a spike or cluster, the flowers being separated by coriaceous or membranous bracteas. Flower superior. Calyx a perianth of three valves. Corolla always irregular. Pericarp in most instances a capsule of three cells and three valves, with many seeds in each cell." We pass over much of the Linnæan description, recent discoveries having enabled succeeding writers, particularly Mr. Roscoe, in *Trans. of Linn. Soc.* vol. viii., and Mr. Brown in his *Prodr. Nov. Holl.*, to explain the flowers much better. The corolla is monopetalous, with a double limb, and more or less irregular; each limb in three deep segments; the inner most unequal, one of its segments being a dilated, lobed, ornamented lip, like that of the *Orchideæ*, the other two sometimes very small, or obsolete. Stamen one, inserted into the tube, opposite to the lip; its filament mostly dilated, and of a petal-like habit, by the diversity of whose shape Mr. Roscoe has first reduced this order into natural genera, a matter in which preceding botanists had altogether failed. The anther consists of two parallel distinct lobes, united lengthwise with the filament, bursting longitudinally, sometimes spurred at the base. There are usually the rudiments of two abortive stamens, first asserted to be such by Mr. Brown. Germen roundish, with a threadshaped style, lodged between the lobes of the anther, and a dilated, cup-like, often fringed, stigma.

"To this order belong the Ginger, Cardamoms, Grains of Paradise, *Costus*, Galangale and Zedoary of the shops, all aromatic. We have nothing similar to them in Europe, except *Acorus*."

What Professor Giseke has subjoined to the lectures of Linnæus, relative to this order, is, to say the best of it, superfluous.

Order 9. SPATHACEÆ. "These are distinguished by their bulbous root, consisting of a radical bud, formed from the bases of the last-year's leaves, which envelope the rudiments of the future foliage. In a bud the scales are expanded into leaves; in a bulb the permanent base of the leaves becomes fleshy. In this order the leaves are sheathing at the root, so that they exhibit no instance of a scaly bulb, but only a coated one. Their leaves are, with a few exceptions, almost linear, or linear lanceolate. Stem no other than a *scapus*, or radical flowerstalk, either round, two-edged, or triangular. The *spatha*, or sheath, is a terminal membrane, splitting lengthwise, except in *Hæmanthus*, where it divides into six segments, resembling an involucrum, and is permanent. The *spatha* sometimes contains many flowers, and where it naturally bears but one, is liable occasionally to produce more. The flowers are stalked within the *spatha;* in most instances they are superior, but not in *Bulbocodium*, whose corolla is divided to the very base. This plant therefore has erroneously been referred to *Colchicum*. *Tulbaghia* has a perfectly inferior flower, but cannot be referred to *Hyacinthus*, on account of its many-flowered *spatha*." (The nectary, or crown of the tube, abundantly distinguishes it.) "*Allium* has invariably an inferior flower, but its *spatha* shows that it belongs to the order before us. Some of its species bear flowers as big as a *Narcissus*."

"The corolla in most of the genera is monopetalous, inasmuch as the nectariferous tube bears the petals. Otherwise they might all be denominated hexapetalous, except *Colchicum* and *Crinum;* to say nothing of *Gethyllis*, distinguished from all the rest by its very long tube. Stamens six, except in the genus last mentioned, where they are twice that number. Pistil one, except *Colchicum;* but many have a three-cleft stigma, so that in *Colchicum* this part may be considered as only further divided even down to the germen. Capsule in all of three cells, with many seeds." (*Hæmanthus* has a berry.)

"The roots of this tribe grow best if they are dried after the leaves perish, either artificially, or by the arid nature of their place of growth. Many of these roots are nauseous and acrid, therefore poisonous, especially *Colchicum*. The bulb of a *Narcissus* will kill a dog. No analogy holds good between these plants and the Tulip, whose bulb may be eaten with impunity; because they are not of the same natural order. All the species of *Allium* are impregnated with their own peculiar pungent flavour, and nature being disposed to expel them with violence from the stomach, they prove most powerful sudorifics. Much of the substance of these last-mentioned is mucilaginous, which involves and separates their acrid particles. Hence they are not dangerous in substance, but their expressed juice, deprived of viscidity, is fatal."

Order 10. CORONARIÆ. "A coronary or garland flower was anciently such as, on account of its beauty, was used for ornamental wreaths."

"*Ornithogalum* has much in common with *Allium*, but wants the *spatha*. *Scilla* is so nearly related to *Ornithogalum*, that they are scarcely to be distinguished but by the breadth" (some say the proportion) "of their filaments. *Hyacinthus* and *Scylla* are with difficulty distinguishable, though the latter has six petals, the former a monopetalous six-cleft corolla, but this is in some instances so deeply divided as nearly to approach the latter."

"In this order the root is either tuberous, a solid bulb, or, as in *Lilium*, a scaly one. The leaves of *Aloe, Yucca, Agave*, and *Bromelia*, are, as it were, a bulb above ground, whose dilated, fleshy, permanent scales remain year after year; just as the bulb of the Lily consists only of the perennial bases of the foliage. In the *Aloe* tribe, not merely the base, but the whole leaf is perennial. Whoever is ignorant of this, cannot fail to go astray in studying the order in question."

"The stem is simple, often a mere *scapus*, occasionally leafy, in consequence of a partial elevation of the radical leaves."

"The flower, destitute of *spatha* or any sort of calyx, consists of six petals." (Linnæus terms them such, because they fall off when the flowering is over.) "In *Ornithogalum* some species have the under side of the corolla green, which part therefore is permanent here, as consisting of corolla and calyx united. In some kinds of *Anthericum*, and in *Veratrum*, the petals are likewise permanent, but in a faded condition. The stamens are universally six, three of them interior. Germen superior.

In *Aloe* the pistil is solitary, and three-cleft; but the style is divided to the very base into three parts in *Melanthium, Helonias, Veratrum,* and one species of *Ornithogalum.* All the tribe have a capsule of three cells, and three valves, the seeds being placed one above another."

"There is no uniformity in the qualities of the *Coronariæ,* there being among them a great diversity of scent. The nauseous smell of *Fritillaria imperialis* and *Veratrum* indicates a very poisonous quality, of which likewise *Aloe* partakes. *Lilium* is mild; its root inodorous and mucilaginous; its qualities therefore are emollient and lubricating. *Scilla maritima* is in the highest degree acrid and diuretic, dissolving viscid humours. The root of *Ornithogalum umbellatum,* as well as of *O. luteum,* is eatable. The former appears to be the Dove's dung, sold for so high a price during the siege of Samaria, as recorded in the Second Book of Kings, chap. vi. ver. 25; in the first place, because it is very abundant in Palestine, whence the English call it Star of Bethlehem; secondly, because the flower resembles the dung of pigeons and other birds, in its greyish and white partycoloured hue, whence also comes the name *Ornithogalum,* or bird's milk, alluding to the white substance, always accompanying the dung of these animals; and lastly, because the root in question is to this day eaten in Palestine, at least by the poor." (See *English Botany,* t. 130.)

"Wepfer has proved by many experiments, the very poisonous nature of the root of the Crown Imperial, which kills dogs, wolves, and various other animals. The ancients relate that the honey of its flowers caused abortion. No flower, except *Melianthus,* produces more of this fluid, yet the bees do not collect it! We owe this fine plant, now so common, to Clusius, who more than two hundred years ago received it, along with the Horse-Chestnut, from the east. He likewise acquired many other bulbs before unknown, now become the ornaments of our gardens. From his time no one has taken the same pains. Certainly if any person could travel, for this object, into the interior of Persia and the kingdom of the Mogul, he would be likely to obtain many superb plants of this order, as recent travellers to the Cape of Good Hope have made us acquainted with so many novelties among the *Ixiæ*, *Antholyzæ*, &c. of which Hermann, Oldenland, &c., their predecessors, have not mentioned a word. *Tulipa Gesneriana* is so called, because it was procured by Conrad Gesner, from Cappadocia, whence it has become common throughout Europe: its endless varieties are the delight of florists, and some of them fetch a high price."

Linnæus in his own manuscript has, as we have already said, removed *Bromelia*, *Tillandsia*, and *Burmannia*, from this order to the *Palmæ*, or at least an appendix thereto.

Order 11. SARMENTACEÆ. "Sarmenta among the ancients meant unarmed, prostrate, weak branches, unable to support themselves; hence this name is applied to the order before us, many plants belonging to which answer to that character, being of a long, weak, trailing or twining habit. The *Sarmentaceæ* are monocotyledonous. They differ much in fructification, and may be variously arranged; either by their calyx and corolla; the number of their stamens or of their pistils; the nature of their fruit; or the inferior and superior situation of their germen. Hence it appears that no common character, applicable to the whole order, can be deduced from the fructification."

"*Raiania*, *Tamus*, *Dioscorea*, *Smilax*, *Cissampelos*, *Menispermum* and *Ruscus*, form one assemblage, all except the last having the above-mentioned kind of stem, twining to the left, not to the right, except in one species of *Menispermum*. Such a difference is rare between plants of the same natural order. *Smilax* supports itself by two tendrils, springing from near the base of the footstalks; all the rest are spiral, and without examination of the fructification, may easily be confounded. The above are dioecious, except one or two species of *Ruscus*." (*Centella* ranged among these in *Gen. Pl.* is now referred to *Hydrocotyle*.)

"*Dracæna*, *Asparagus*, *Convallaria*, *Uvularia*, *Gloriosa* and *Erythronium*, compose another section. The last is intermediate, as it were, between the present order and the *Coronariæ*. *Gloriosa simplex* is a small plant, not unlike *Erythronium*, with reflexed petals." (What Miller, who is Linnæus's sole authority for this species, intended, nobody has ever been able to make out.)

"*Medeola*, *Paris* and *Trillium* have whorled leaves, except *M. asparagoides*, which scarcely differs from the genus *Asparagus*, except in having three styles instead of one."

Aristolochia, *Asarum* and *Cytinus*, nearly akin to each other, are removed from this order by the author in his manuscript, to the 27th, *Rhoeadeæ*, but not without a query. In the same place we meet with what may perhaps prove a solution of the mystery, which Giseke was so anxious to unriddle, and to which we have already alluded in the beginning of this part of our subject. Linnæus has here mentioned *Nymphæa*, as having in some of its species one cotyledon, in others two. He notes also that *Menispermum* and *Aristolochia* are dicotyledonous. *Nymphæa* however appears to be the great secret which the worthy professor told his pupil that he or some other person might chance to find out in ten, twenty or fifty years, and would then perceive that Linnæus himself had been aware of it. Accordingly, Gærtner and Jussieu have made the same discovery, or rather, fallen into the same mistake; describing *Nymphæa* as monocotyledonous, and *Cyamus* Sm. Exot. Bot. v. i. 59. (their *Nelumbo*, or *Nelumbium*), as in some measure dicotyledonous. The excellent De Candolle, in the *Bulletin des Sciences*, n. 57, published in 1802, has first rightly considered both as dicotyledonous, and akin to the *Papaveraceæ* of Jussieu, the Linnæan *Rhoeadeæ*.

Linnæus, in his lectures, proceeds to observe, that he "wanted to make further inquiry into the cotyledons of his *Sarmentaceæ;* for though he knew that several of these plants were monocotyledonous, he knew two, and did not doubt there were more, perfectly dicotyledonous. Hence he suspected the order might be separated into two, in other respects very closely related."

"The roots of all this family are oblong and fleshy, except *Erythronium*, whose radicles are long and quite simple; those of *Smilax Sarsaparilla* run very deeply into the ground, and are sometimes so thickened at the ends as to become tuberous. The stem at first coming forth is smooth and leafless, mostly branched, except in *Paris* and *Trillium*; in some prostrate. Leaves in every instance simple and undivided, sometimes linear, sometimes lanceolate and acute, or heartshaped, uniform, mostly alternate; except when three or more stand together in a whorl, and in *Dioscorea oppositifolia.* It is rare that alternate and opposite leaves occur in the same natural order. Flowers mostly on simple stalks, *Smilax* excepted, which has umbels; they are drooping, except in *Paris.* Stamens universally six, except in *Menispermum.* Styles three, or three-cleft. All the genera, almost without exception, are deficient in either calyx or corolla. The fruit is generally of three cells. Inflorescence axillary in all except *Erythronium*, which has but one flower, and *Ruscus*, where it springs from the leaf."

"Their qualities are to be judged of by the smell. All of them betray something of malignity, except two insipid ones which are eatable, *Dioscorea* and *Asparagus. Gloriosa* is very poisonous; the dried flowers of Lily of the Valley cause sneezing, like *Veratrum*, that is, they produce convulsions. *Paris* has always been deemed poisonous. One kind of *Cissampelos*, named *Pareira brava*, and *Smilax*, are known by physicians to be highly diuretic, as well as the roots of *Asparagus. Menispermum Cocculus* kills fishes, lice, and men."

"This whole order is entirely without pubescence, even the prickly *Smilaces.*"

Next follow the Dicotyledonous Orders.

Order 12. HOLERACEÆ, pot-herbs, (erroneously printed *holoraceæ* in *Gen. Pl.*, which has misled several writers). "This denomination is given to plants that are tender or brittle in the mouth, and easy of digestion, like many of the order before us." The order is divided into several sections. Of the first *Blitum, Atriplex, Chenopodium, Salsola, Salicornia*, &c. are examples. The second consists of *Petiveria, Calligonum, Ceratocarpus* and *Corispermum. Callitriche* was subsequently removed to the 15th order. In the third section *Axyris* stands alone. Of the fourth *Herniaria, Illecebrum, Amaranthus, Phytolacca,* may serve to give an idea. The fifth begins with *Begonia,* (of whose affinity Linnæus candidly confesses his ignorance, and to which no botanist has yet found an ally). Next follow *Rumex, Rheum, Polygonum,* &c. The sixth section has *Nyssa, Mimusops, Rhizophora, Bucida* and *Anacardium;* and the seventh *Laurus, Winterana* and *Heisteria;* in both which the fleshy receptacle appears, where he could trace it, to have guided Linnæus to an arrangement evidently paradoxical, which he labours, without satisfying us, to justify.

Order 13. SUCCULENTÆ. "Bradley wrote on Succulent Plants, by which he meant such as could not be preserved in a *Hortus Siccus.* When gathered, vegetables of this nature will live, often for a whole year, flowering as they hang up in a house, and throwing out roots afterwards if planted. All such plants, however, do not enter into the present order. *Stapelia, Euphorbia,* and *Aloe* are excluded. The *Succulentæ* grow, and become very turgid, in the driest soil, nor are any found in watery places. If moistened too much they perish, and their roots decay. They afford, in putrefying, a fine vegetable mould, whereas dry plants, like heath and fir, scarcely yield any."

Linnæus has distinguished these into four sections. In the first are *Cactus*, *Mesembryanthemum*, *Tamarix*, and others. *Nymphæa*, placed here in the Linnæan manuscript, as well as in Giseke's publication, was afterwards removed by Linnæus to his *Rhoeadeæ*. *Sarracenia* he conceived to be akin thereto. In his second section are *Sedum* and its numerous allies; in the third *Portulaca*, *Claytonia*, &c.; and in the fourth a very different assemblage, as we should think, composed of *Saxifraga*, *Adoxa*, &c. and even *Hydrangea*. Linnæus however thought all these sections nearly related. "They are," says he, "succulent, insipid, inert, and inodorous, therefore mere pot-herbs, widely different from the other fleshy plants, *Stapelia*, &c. whose fructification is so unlike them, and whose qualities are so poisonous. We find in this order, that opposite or alternate leaves is an indifferent circumstance. These plants have no true spines, no tendrils, nor climbing stems, neither stipulas nor bracteas." (Giseke well remarks, that *Sedum acre* is one exception to their alleged insipidity, though we can scarcely agree with him that *Sempervivum tectorum* is another.)

Order 14. Gruinales. The best-known genera here are *Linum*, *Drosera*, *Oxalis*, *Geranium* and its relations. Linnæus admits also *Quassia*, *Zygophyllum*, *Averrhoa*, &c., and his editor inserts, with well-founded doubt, *Sparmannia*. Their roots and habits are various. Calyx usually of five leaves, and corolla of five petals. Stamens various in number and connexion. Pistils mostly five or ten. Fruits various. Linnæus professed himself unable to define the character of this order. Many of the plants have acid leaves.

Order 15. Inundatæ. "So called because they grow in water, many of them under its surface, except their blossoms." *Potamogeton* is the genus most generally known, to which Linnæus suspected *Orontium* to be related, but not correctly. *Myriophyllum*, *Proserpinaca*, *Hippuris*, &c. are placed here, and even *Elatine*, notwithstanding its numerous seeds. *Chara* and *Najas* form a section at the end. *Callitriche, Lemna,* and even *Pistia,* were proposed to be brought hither; with *Saururus* and *Aponogeton.*

"The qualities of the *Inundatæ* are very obscure. These plants are mostly inodorous, except a fishy scent in some; nor have they any particular taste; hence they are not used medicinally."

This order is out of its place with respect to the arrangement by the cotyledons, of which Linnæus seems aware, from the remarks subjoined to it, in his lectures, concerning that principle. To these we shall hereafter refer.

Order 16. CALYCIFLORÆ. This consists of *Osyris, Trophis, Hippophäe* and *Elæagnus.* No observation relative to it is given in the lectures, except that these genera are removed elsewhere. A manuscript note before us indicates a suspicion of its relationship to the 6th section of the *Holeraceæ.* Linnæus sometimes referred *Memecylon* to one of these orders, sometimes to the other, but finally to his 18th; we should rather presume it belongs to the 19th notwithstanding the definite number of the stamens, which caused Jussieu to range this genus with the Linnæan *Calycanthemæ*; see the next order.

Order 17. CALYCANTHEMÆ. "The title of this order is precisely synonymous with the last, and is applicable in a different manner to the different genera of which the present consists. In those whose germen is inferior, the calyx bears the flower and enfolds the germen; in those where the latter is superior, it is unconnected with the calyx, into which the stamens are, in that case, inserted, like the *Senticosæ* and *Pomaceæ,* not into the receptacle. The germen is inferior in *Epilobium, Oenothera, Gaura, Jussiæa, Ludwigia* and *Isnarda,* as well as in *Mentzelia* and *Loosa*" (or *Loasa*); "in the rest, *Ammannia, Grislæa, Glaux, Peplis, Frankenia, Lythrum, Melastoma, Osbeckia,* and *Rhexia,* it is superior. Some genera have four, others five or six petals. *Glaux* and *Isnarda* have none. *Ammannia* and *Peplis* have occasionally petals, or not, in the same plant. *Melastoma* has a berry; the rest a capsule, usually of four or five cells, in some genera of but two, or one." Linnæus mentions *Melastoma* as the only arboreous genus. The rest are herbaceous, (rarely shrubby,) with opposite or alternate leaves; stamens from four to twelve, pistil always solitary, the stigmas either four or one.

"These plants are mostly inodorous and insipid, except a styptic property in the root of *Lythrum*; none of them are used in the shops. It is remarkable in this order particularly, that some flowers are sessile and axillary, but towards the summit the leaves gradually diminish, and are finally obliterated, so that the inflorescence becomes a spike, as may be seen in *Epilobium*."

Order 18. Bicornes. "So called," by Linnæus, "from the anthers, which in many of this tribe terminate in two beaks. The plants are rigid, hard and evergreen, almost all more or less shrubby; certainly perennial. *Diospyros* is arboreous. The leaves of this order are alternate, simple, undivided, scarcely crenate, permanent. Stipulas and bracteas wanting;" (certainly not always the latter). "Calyx of one leaf, more or less deeply four or five cleft. Corolla usually monopetalous; in *Pyrola*, *Clethra*, and their near allies, pentapetalous. Nectaries none, except in *Kalmia*." (Linnæus can here mean only the pouches which for a while detain the elastic stamens, and those are by no means nectaries.) "Stamens from four to ten, answering to the divisions of the corolla, or twice their number. Pistil 1, except *Royena*, which is digynous. Germen in some superior; in others, as *Vaccinium*, inferior. Some have a capsule, others a berry; the cells of each four or five; but *Diospyros* has a fruit of eight cells. The seeds are either one or many in each cell, mostly small, chaffy." Linnæus remarks that "they can scarcely be raised in a garden, especially as the plants are many of them natives of boggy situations;" but our English gardeners are masters of their treatment, witness the abundance of *Ericæ* from the Cape, now common in every greenhouse, and many other charming shrubs, cultivated in a peat soil. He conceived the whole order to be nearly confined to one meridian, from the North Cape of Lapland, to the Cape of Good Hope; but he is incorrect in saying there are very few in North America, and none in the East or West Indies.

Halesia, Styrax, Spathelia, Citrus and *Garcinia* are subjoined as an appendix to the *Bicornes,* but there is allowed to be a considerable distance between them, and the last is erased in the *Gen. Plant.* as having opposite leaves. Giseke records, p. 345, that when Linnæus said no *Erica* grew in America, he asked him whether *Hudsonia* were not an exception to this? On which he took that genus from his herbarium, and after contemplating and replacing it, wrote something, Giseke knew not what, in his *Genera Plantarum.* We find what he wrote to be as follows: "Videnda *Hudsonia, Empetrum, Ilex, Itea.*" It is interesting to be thus able to trace the thoughts of such a man. He was moreover correct as to the genus *Erica* itself, of which no species has been detected in America.

Order 19. HESPERIDEÆ. Of this nothing is said in the lectures. The original genera are *Eugenia, Psidium, Myrtus,* and *Caryophyllus*; to which Giseke has added *Calyptranthes* and *Legnotis* of Swartz. *Melaleuca* also strictly belongs to this tribe; though, by a strange error, referred in the *Mantissa* to the 40th order, and yet said to be akin in *Ginora,* which belongs either to this or the 17th. *Philadelphus* is subjoined as forming a section by itself, and still with a mark of doubt. The discoveries in New Holland have thrown much light on this fine order of aromatic and elegant shrubs, of which the Myrtle is a familiar type. Linnæus intended to remove *Garcinia* hither.

Order 20. ROTACEÆ. The lectures are also deficient as to this order. It consists of *Trientalis, Centunculus,*

Anagallis, Lysimachia, Phlox, Exacum, Chlora,- Gentiana, Swertia, Chironia and *Sarothra* ; to which *Ascyrum, Hypericum* and *Cistus* stand as an appendix. The wheel-shaped corolla of many of the above plants, has evidently suggested the name.

Order 21. Preciæ. *Primula* and its elegant relatives form the basis of this order. " They are all destitute of stems. Leaves simple. Flowerstalk umbellate, except in *Cyclamen*. Flower regular. Calyx, as well as corolla, five-cleft. Stamens five. Style one. Fruit a simple superior capsule. The umbel is often accompanied by an involucrum. They are vernal-flowering plants, and have, except *Cyclamen*, nothing malignant in their qualities." *Limosella* stands alone in a second section of this order, but rather perhaps belongs to the 40th. *Menyanthes, Hottonia* and *Samolus* form a third section, attended by a mark of doubt. *Sibthorpia* was once inserted in manuscript, but afterwards erased.

Order 22. Caryophyllei. The Pink and Campion tribe. "Root fibrous. Stem herbaceous, scarcely shrubby, jointed ; its branches commonly alternate. Leaves simple, more or less of a lanceolate figure, undivided, hardly crenate in any degree, sessile, with no other appearance of a footstalk than their elongated narrow base, opposite, obvolute. Stipulas none; neither are there any distinct bracteas, nor spines, prickles nor tendrils. The plants are mostly smooth, few are hairy, none prickly or bristly. Flower rarely sessile. Stamens never numerous, but either the same in number as the petals, or twice as many. Pistils from one to five, not more. Fruit a capsule, either of one cell, or of as many as there are styles ; the cells usually with many seeds, *Drypis* only having a solitary seed. A few of these plants with separated flowers occur among the species of *Cucubalus, Silene* and *Lychnis*. The whole order is harmless, without any peculiar taste or smell, except in the flowers. It contains the *flores caryophyllati* of Tournefort, who defined these as having the calyx tubular, and the limb of the corolla flat; but he referred *Statice* and *Linum* hither, which differ widely from this order, while his character excludes the *Alsine*, or Chickweed, tribe." Linnæus thought *Velezia* had been wrongly placed here by Gerard, and was doubtful respecting *Cherleria*; but he was afterwards satisfied that both are *Caryophyllei*. He remarks that "the order consists, as it were, of two leading genera, or rather families, the *Caryophyllus*, or Pink tribe, such as *Dianthus*, *Saponaria*, *Gypsophila*, *Silene*, *Lychnis*, &c.; and the *Alsine*, or Chickweed family, consisting of *Spergula*, *Cerastium*, *Arenaria*, *Stellaria*, and others. In the first division, the calyx is tubular, of one leaf; in the second of five." A third section of this order has *Pharnaceum*, *Glinus*, *Mollugo*, *Polycarpon*, *Minuartia*, *Queria*, *Ortegia*, *Loeflingia*; to which were afterwards added *Gisekia* and *Rotala*. *Holosteum* also, having laciniated stipulaceous membranes, was intended to have been removed to this third section. *Scleranthus*, by itself, makes a fourth, but is erased by Linnæus, and removed to his 31st order. *Polypremum*, with a query, stands at the end.

A most extraordinary remark is subjoined by Professor Giseke at p. 354; that "*Alsine media* and *Holosteum umbellatum* are one and the same plant," and that "Linnæus had no specimen of the former in his herbarium in the year 1771." Swartz is cited in confirmation, who only says in his *Obs. Bot.* 118, that this *Alsine* is a species of *Holosteum*. We trust it is better referred to *Stellaria* in *Flo. Brit.*, and we can affirm that an authentic specimen of this common plant, which Linnæus had when he published the first edition of *Sp. Pl.*, in 1753, still exists in his collection. The real *Holosteum umbellatum*, a rare English plant, is well delineated in *Engl. Bot.* t. 27.

Order 23. Trihilatæ. "So called from its three-celled, and three-grained fruit, for all the cells are distinct. *Melia* however has five cells. The calyx in this order is either of four or five leaves, or of one leaf in five deep segments. Petals four or five. Stamens eight or ten. Pistil one. One part of the fructification is often diminished as to number, for instance the petals; and when they become but four, the stamens are only eight. A nectary is always present; hence the corolla is frequently irregular. The leaves are disposed to be compound, and are both opposite and alternate. The whole order scarcely contains any thing acrid, except *Tropæolum*, nor any thing either fragrant or noxious; on the other hand, the *Tricoccæ*, properly so called, are highly poisonous."

The first section consists of *Melia*, *Trichilia*, *Guarea* and *Turræa*; to which Linnæus has added, from his 54th or miscellaneous order, *Cedrela* and *Swietenia*. The second is composed of *Malpighia*, *Bannisteria*, *Hiræa*, *Triopteris*. *Acer*, and *Aesculus*. Linnæus was inclined to bring hither from his 14th order, the genus *Fagonia*, because of the likeness of its flower to *Malpighia*, but he found a difficulty in the five cells of its fruit. A third section consists of *Staphylea*, *Sapindus*, *Paulinia*, *Cardiospermum* and *Tropæolum*; to which *Hippocratea* is added in manuscript, and a question subjoined, Whether *Staphylea* be not akin to *Celastrus*? Cavanilles has added many new genera to this order, but he is surely complimented to excess by the editor of the *Prælectiones*.

Order 24. Corydales. "The title of this order is synonymous with *Fumaria* amongst ancient writers."

"The genera are *Melianthus*, *Monnieria*, *Epimedium*, *Hypecoum*, *Fumaria*, *Leontice*, *Impatiens*, *Utricularia*, *Pinguicula*, and perhaps *Calceolaria*. The calyx is of two leaves; except in *Pinguicula*, where it is only cloven; and *Melianthus*, where it consists of four leaves. The flower of *Fumaria* is remarkable in its throat, and uniform in that respect throughout the genus; but the various species differ widely in their fruit; which in some, as *officinalis*, &c. contains a solitary seed; in *capnoides*, *claviculata*, &c. it is a genuine pod; in *vesicaria* a large inflated capsule." *Monnieria* and *Melianthus* (two very puzzling genera,) were thought by Linnæus to be nearly related to each other, though differing from the order under consideration, in having several capsules, and a calyx in five deep divisions. But he judged the ringent corolla of *Monnieria* to betray an affinity to *Fumaria;* while the nectary of *Melianthus* is similar to that of *Monnieria,* the combined stamens of the latter being exactly those of *Fumaria.* Hence a relationship is traced between *Melianthus* and this order, which, but for *Monnieria,* could not have been suspected.

"There is a certain fragility and delicacy of texture characteristic of the *Corydales,* with a glaucous hue, which points out their affinity; as well as a bitter flavour. Scarcely any of the order are odoriferous, except *Melianthus,* which is extremely fœtid."

Linnæus professes his inability to point out any exclusive mark of distinction for this order. "The leaves indeed are alternate in all, *Calceolaria* excepted; and many bear stipulas. Their mode of flowering is spiked, racemose, or solitary, their stalk naked or leafy, different in different species. All that we are acquainted with are smooth and unarmed; a very few of them climbing by means of tendrils. *Melianthus* and *Monnieria* only are shrubby. All the tribe prefer shady, moist situations, where the soil is not disturbed." (Some however grow in cultivated ground, as the Fumitories.)

"The *Melianthus,* a Cape plant, produces more honey than any other plant, so that a tea-spoon full may be collected every morning, from each of its numerous flowers. But the offensive odour" (of the bruised plant) "indicates a poisonous quality, as in *Cimicifuga.*"

Order 25. PUTAMINEÆ. On this order, named from the strong rind of the fruit in several instances, there is no commentary in the *Prælectiones,* nor any manuscript note in the *Gen. Pl.* The genera are *Cleome, Cratæva, Morisona, Capparis; Crescentia* and *Marcgravia* being added with hesitation. *Tanæcium* of Swartz, and *Possira* of Aublet, which last is *Rittera* of Schreber, are subjoined by Giseke.

Order 26. Multisiliquæ. This consists of four sections. In the first are *Pæonia, Aquilegia, Aconitum* and *Delphinium,* to which Linnæus, after much diversity of opinion, finally determined to add *Cimicifuga* and *Actæa.* The second contains *Dictamnus, Ruta* and *Peganum:* the third *Nigella, Garidella, Isopyrum, Trollius, Helleborus, Caltha, Ranunculus, Myosurus* and *Adonis:* and the fourth *Anemone, Atragene, Clematis,* and *Thalictrum.* "Most of the order, with a few exceptions, are of European growth; rarely arboreous or shrubby, except such species of *Clematis* as climb trees. The roots are fibrous, sometimes tuberous. Leaves often many-cleft, or compound; but in a few instances simple; all alternate, except in *Clematis integrifolia.* There are no stipulas, spines, nor prickles. One or two kinds of *Clematis* bear tendrils. Flowers in no case monopetalous. Stamens always more than eight, except in the second section. Fruit in some capsular, in some single-seeded. An acrid taste prevails through the whole. Their odour is disagreeable, almost universally, so that none is esculent, and many, if not all, are poisonous, though there is no milky plant among the whole, nor any one with a twining stem." Linnæus remarks, that "a calyx is very rarely present, and when it occurs, manifestly originates from the leaves;" but this is not applicable to *Ranunculus* and its nearest relations, nor to any genus in the second section; that section indeed being a most distinct order of itself, called by Jussieu *Rutaceæ,* but not well defined by him.

Order 27. Rhœadeæ. The poppy tribe. No remark on this order is found in the lectures of Linnæus, but he has made some manuscript notes. He wished to remove it next to the 24th and to place its genera thus, *Argemone, Chelidonium, Papaver, Podophyllum, Sanguinaria* and *Bocconia. Sanguinaria,* he observes, has the flower of

Actæa, which last genus he had once brought hither. He has finally placed here *Aristolochia*, *Asarum* and *Cytinus*, as we have mentioned under the 11th order. *Nymphæa* also is indicated, but afterwards erased, which is unfortunate.

Order 28. Luridæ. The gloomy family of nightshades, henbane and tobacco. "This order is a most distinct and evident one. All the plants have alternate leaves; a five-cleft calyx; monopetalous corolla; stamens four or five; pistil one; germen superior; seed-vessel of two cells, in some a berry, others a capsule. Their corolla folds in a plaited manner."

Digitalis, *Celsia*, *Verbascum*, *Nicotiana*, *Atropa*, *Hyoscyamus*, *Datura*, *Physalis*, *Solanum*, *Capsicum*, are examples of this order. "They are none of them arboreous, though some are shrubby. Colour (of the herbage) mostly dull and lurid; the taste disagreeable, smell nauseous, hurtful to the nerves, hence their generally poisonous qualities." *Ellisia* is properly expunged in the manuscript, and *Nolana* with equal propriety removed hither from the 41st order.

Linnæus observes, that "the poisonous quality of *Verbascum* appears in its power of killing fish, if made up into balls with meal." "*Nicotiana rustica*," he says, "furnishes the Turks with their best tobacco, yet it is not cultivated by us, though it grows readily. *Atropa Mandragora*, a most poisonous and dangerous plant, becomes, under proper management, an excellent and powerful medicine," for instances of which Linnæus referred his hearers to his lectures on the Materia Medica. These, as Giseke notices, were never published. On turning to the manuscripts used by the professor in that course, we find the *Mandragora* mentioned as "virose, acrid, bitterish and nauseous, useful in the gout and colic; the herb boiled in milk, and applied to scirrhous tumours, more active in dispersing them than hemlock or tobacco. Three of the berries boiled in milk, given to a potter, labouring under a dreadful cholic, threw him into a sleep for twenty-four hours, out of which he awoke cured. The ancients gave an infusion of this plant in wine, before they amputated a limb. Its narcotic qualities render it very useful in epilepsy and hysteria, though to be cautiously administered. Nothing can be more dangerous in a state of pregnancy. The editor of Hernandez, *Hist. Nat. Mexic.* Book viii. chapter 28, speaks of this fruit as eatable, without any soporific or injurious effect."

Linnæus himself appears to have been doubtful about *Catesbæa,* which he has marked as akin to his *Dumosæ.* Giseke has subjoined an observation, not well founded, of the *Solandra* of Swartz being hardly distinct from *Datura.*

Order 29. CAMPANACEÆ. These Linnæus has noted as most nearly allied to the 24th order. "They never form trees, rarely shrubs. Leaves in every instance alternate; calyx and corolla five-cleft; stamens five; pistil one, except *Evolvulus,* which has at least a deeply four-cleft style, if not four distinct ones. Fruit a capsule. They are milky plants, at least while young and tender. Their qualities therefore are purgative, and but slightly poisonous."

Convolvulus and *Campanula,* with their respective allies, constitute this order. To the latter *Viola* is supposed to be connected, through the medium of *Lobelia.* *Parnassia,* though in the manuscript rightly said to be not milky, stands at the end, its affinity being indicated by the nature of the flower-stalk, calyx, as well as the seeds and their situation, but especially the nectaries and stigma. The anthers come one after another and impregnate the latter, retiring subsequently in their turns. Their close application to that part, as Linnæus conceived, rendered the access of extraneous pollen impossible, "hence," says he, "no more species of this genus can be produced." This alludes to his hypothesis of new and permanent species, or even genera, having been generated, from time to time, in the vegetable kingdom, by cross impregnation; which we are very unwilling to admit, nor do any of his instances prove satisfactory to us. As to *Parnassia*, we now know several American species, as distinct as those of any other genus.

Order 30. CONTORTÆ. "This order derives its name from the corolla, which," (in all the species known to Linnæus,) "is twisted in the bud, contrary to the course of the sun, its limb being wheel-shaped, when expanded, in such a way that each of its segments, unequally proportioned in their margins, is curved inward under the next segment, the shorter side of the former being beneath the longer one of the latter. Europe is very poor in this tribe, India very abundant. Many of the plants are milky, and, like most other such, poisonous; some indeed so violently, as immediately to destroy animals that eat them. Their medical effects, rightly managed, may be very great. They have all, naturally, an injurious property, even *Asclepias Vincetoxicum,* though this plant, like *Nerium* and *Vinca,* is scarcely milky, but in its very youngest shoots." (It is singularly remarkable that the fruit of one of this family, at Sierra Leone, the size of an orange, yields a copious and wholesome milk, used by the colonists as cream to their tea. See *Sm. Introd. to Botany*, ed. 3. 316.)

"Many of the order of which we are treating are shrubby; the leaves opposite and evergreen, except the species of cold countries. The flower is, in many cases, intricate in structure, because of the peculiar apparatus of the nectaries of various genera."

"The roots are perennial. Leaves all, as far as hitherto known, simple and undivided, and, with very few exceptions, opposite; sometimes ternate, or quaternate; rarely alternate. The inflorescence is often peculiar, in having its flowerstalk not axillary, but proceeding from the side of the stem between the insertion of the leaves. Calyx of one leaf, five-cleft. Corolla of one petal, regular, its segments contorted, as above described, and often notched in the margin. Nectaries, in many instances, singularly formed. Stamens five. Pistils two, or one with a double stigma. Germen superior in all, except *Gardenia, Genipa*, and *Macrocnemum*," (all now certainly not admitted into this order.) "The Fruit, in many genera, as *Vinca, Nerium, Echites, Plumeria, Tabernæmontana, Cameraria, Periploca, Apocynum, Cynanchum, Asclepias, Stapelia, Ceropegia* and *Pergularia*, consists of two distinct follicles, not observable in other plants. This sort of seed-vessel is like a *spatha* amongst the other kinds of calyx, of one valve, splitting longitudinally at the inner edge. But the seeds are not attached to the suture, there being a separate thread-shaped receptacle, extending the whole length of the seed-vessel, over the whole of which the seeds are imbricated, in a downward position. In all the above mentioned the seeds are crowned with a soft hairy tuft, except those of *Vinca*, which have no such appendage. The flowers of the *Contortæ* are usually very handsome, and there is something so singular in the structure of many of them, especially relative to the nectary and stigma, that it is difficult to say, in many instances, whether they have one or two stigmas; especially when two germens seem to bear but a single style. The corolla in all is five-cleft, and the stamens five. Jacquin contends that the latter are really ten. Linnæus from repeated examination of *Asclepias*, was confirmed in the former opinion, and especially from the investigation of *Periploca*, whose flower, evidently constructed on the same principle as *Asclepias*, has, no less evidently, but five stamens."

Giseke very improperly annexes *Embothrium* and *Rhopala* to this order, only because their fruit is a follicle; nor does any other genus which he, or Linnæus, has mentioned, really belong to it, except *Allamanda, Rauwolfia* and *Cerbera* of the latter; *Gynopogon* and *Melodinus* of Forster, with *Willughbeia* of Schreber. The first has a bivalve coriaceous capsule, as if formed of two follicles united, with imbricated seeds; the rest have pulpy fruits. Most of the other genera referred hither, as *Gardenia, Cinchona, Portlandia,* &c. belong to the great order of *Rubiaceæ* in Jussieu, of which the Linnæan *Stellatæ,* No. 47, make a part. It must be allowed, nevertheless, that the corolla of *Gardenia* answers to the character of the *Contortæ.* Mr. R. Brown, in the *Wernerian Transactions,* has thrown much light on the principal genera of this family, under the title of *Asclepiadeæ* and *Apocineæ,* with the addition of numerous new ones.

Order 31. VEPRECULÆ. No explanation of this occurs in the *Prælectiones.* The genera are *Dais, Quisqualis, Dirca, Daphne, Gnidia, Struthiola, Lachnæa, Passerina, Stellera,* with *Thesium,* and in the manuscript *Scleranthus* and *Santalum.* These three last do not properly belong to the others, which constitute a most natural order of generally small shrubs, as the name implies. They are known by their tough branches; silky inner bark; simple entire leaves; acrid and even burning flavour; and sweet-scented flowers, whose calyx and corolla are united into one integument, most coloured within.

Order 32. PAPILIONACEÆ. An extensive and very natural family, "consisting of the *Leguminosæ* of Ray; which Tournefort," (following an idea of Baptista Porta), "called *Papilionaceæ;* Rivinus *flores tetrapetali irregulares;* and Magnol *pentapetali.* They have not all five petals, for in many the claw of their keel is simple; in some the keel is separated towards the base into a double claw; while in a few only, the whole keel is composed of two distinct petals, as in *Spartium.*"

"Their character is as follows. Perianth of one leaf, irregular, inferior, generally withering. Corolla nearly the same in all. Its standard either emarginate or entire, either reflexed or not at the sides, for the most part very large, compared with the other petals. Wings, if present, always two, opposite, frequently large, sometimes, as in *Colutea* and *Hedysarum,* very short. Keel simple, either pointed, obtuse, or abrupt. Stamens ten, nine of which have their filaments united, more than half way up, forming a membranous sheath to the pistil; the tenth sticking closely under the pistil, and being sometimes inserted into the base of the tube composed by the other nine. Hence arise two divisions of the order, without attention to which the genera are with difficulty defined. Pistil generally uniform; the style downy or woolly, either above or below; stigma either acute or capitate. Legume of two valves, which must not be confounded with a *Siliqua*, or Pod, though old writers have so termed it, applying that name equally to the fruit of this order and that of the *Tetradynamia* class. As these fruits differ widely in structure, Linnæus has restricted to the latter the term pod, whose character is to have the seeds attached to each suture of the valves; whereas in the legume, or fruit of the class *Diadelphia*, they are connected with one suture, or margin, only. The name of *legumen* indeed originally belonged to the seed itself of these plants; but for want of a better word, Linnæus has applied it to their seed-vessel. The legume is mostly of one cell, containing many seeds; except *Astragalus* and *Biserrula*, in which one suture is internally dilated, as it were, so as to make a partition, separating the fruit into two cells; whilst *Phaca* has the same part extended only half the breadth of the legume, rendering the separation incomplete. *Geoffræa* has a *drupa*, which still ought to be considered as a single-seeded legume, whose pulp is hardened," (or rather, whose coat is made pulpy.) "The ripe legume bursts along its sutures, and throws out its seeds. There are indeed some which do not open in this manner, but fall off in separate joints, each containing a seed, examples of which are *Hedysarum* and *Ornithopus*."

"The genera of this natural order so nearly approach each other, that it is difficult to detect their discriminative characters. Tournefort, though he distributed other genera by their flowers, divided and determined these by their foliage. But *Hedysarum* forms an objection to such a principle of arrangement, because some of its species have simple leaves, others ternate, conjugate, or pinnate."

" *Lathyrus, Cicer,* and *Vicia* are genera most nearly akin to each other, as are *Phaseolus* and *Dolichos. Coronilla, Ornithopus, Hippocrepis, Scorpiurus, Lotus,* and some species of *Trifolium,* agree in their umbellate inflorescence." (*Sophora,* and its many new-discovered allies, ought to make a section, at least, by themselves.)

" There is no poisonous plant in this whole order, except the seeds of *Lupinus,* with which the *Hippopotamus* is killed, and which fowls will not eat. Indigo becomes poisonous in its preparation, but the plant is originally harmless. On the other hand, none of this tribe is medicinal, except *Glycyrrhiza. Galega,* commended as antipestilential, is not to be trusted. These plants have no remarkable odour," (except in the flowers of a few species.) " Their seeds are flatulent; but afford nourishing food for labouring people."

Order 33. Lomentaceæ. " These are perhaps all shrubby," (or arboreous.) " Leaves alternate, compound, at least in the indubitable plants of this order; pinnate or bipinnate; without a terminal leaflet, *Moringa* excepted. Stipulas always large, particularly to be noticed. Calyx five-cleft. Corolla in some degree irregular, polypetalous, except *Ceratonia,* and several *Mimosæ.* Stamens differing in number; mostly ten. Pistil universally single. Fruit a legume, for the most part having transverse partitions. The leaves fold together at night, except those of *Ceratonia,* and that in a different manner according to the different species. Many of this order possess a purgative quality, while some have a virose or nauseous flavour about them, but this last is not at all the case with *Ceratonia.*"

Of *Polygala,* which stands at the head of this order, nothing is recorded by Giseke from the lectures of Linnæus, nor has he himself made any note. It surely answers but indifferently to the *Lomentaceæ.* Genuine examples of the order are *Bauhinia, Hymenæa, Cæsalpinia, Cassia,* perhaps *Securidaca;* from which *Ceratonia, Mimosa, Gleditsia,* &c. considerably recede in character, though less in habit. *Cercis* ought to be ranged with *Anagyris, Sophora,* &c. either in the preceding order, or rather in a separate one, intermediate between the two.

Order 34. CUCURBITACEÆ. "In this order there are, properly, no trees. Some of the plants indeed have a climbing, woody, perennial stem; others a perennial root only; whilst others again are entirely of annual duration. Leaves in all alternate, simple, always accompanied at their origin by stipulas. There are mostly glands, either on the footstalks, at the base of the leaf, or on its disk. All have tendrils, by which they climb if they have any opportunity; otherwise they are procumbent. These plants seem akin to the *Sarmentaceæ,* order 11th; but the latter have a twining stem, these not; they are monocotyledonous, these dicotyledonous; they are destitute of tendrils, with which these are furnished. The calyx is either of five leaves, or five deep segments. Corolla of one petal, in five deep divisions, but so much cut in many instances, that it is scarcely possible, but from analogy, to say whether it consists of one or five petals. The stamens are inserted, not into the receptacle, but into the interior surface of the calyx, to which also the corolla is attached. Their filaments are often five, but frequently so combined as to appear three only. So also the anthers are often connected, the summit of one to the base of another, making a continued serpentine line. The style is of considerable thickness, with three, frequently cloven, stigmas. Fruit internally of three cells, fleshy, and somewhat juicy. The seeds are, for the most part, capable of being kept for a long time, though they appear of a dry nature; but that they are not really so, is evident from the emulsions prepared from some seeds of this tribe. Gardeners think them better for keeping. The sex of the flowers is, in several cases, distinct, and either monoecious or dioecious.

The whole order is noxious and fœtid, hence it affords some of the most violent medicines, as Colocynth and Elaterium. Even melons themselves, if taken too plentifully, are said to be injurious, though in ripening they part with much of their unwholesome quality." The genera are *Gronovia*, *Anguria*, *Elaterium*, *Sicyos*, *Melothria*, *Bryonia*, *Cucurbita*, *Cucumis*, *Trichosanthes*, *Momordica*, *Feuillea*, *Zannonia*, *Passiflora*. "The last affords some of the most beautiful of all flowers; many of them are fragrant."

Order 35. SENTICOSÆ. The briar and bramble tribe. The genera are *Alchemilla*, *Aphanes*, *Agrimonia*, *Dryas*, *Geum*, *Sibbaldia*, *Tormentilla*, *Potentilla*, *Comarum*, *Fragaria*, *Rubus*, *Rosa*. *Poterium* and *Sanguisorba* are inserted at the head of this list, in the Linnæan manuscript. See the following order.

Order 36. POMACEÆ. The apple and plum kinds, consisting, in the first section, of *Spiræa*, *Ribes*, *Sorbus*, *Cratægus*, *Mespilus*, *Pyrus*; in a second, of *Punica*; and, in a third, of *Chrysobalanus*, *Prunus* and *Amygdalus*. These two orders are treated of together, in the *Prælectiones*, it is not said for what reason, though their strict affinity cannot be overlooked. "Many of these plants," says Linnæus, "are shrubs, most of the whole are perennial, very few annual. They are rarely smooth. The leaves are alternate, mostly compound. Stipulas always two, large. None of the plants properly climb, though some brambles support themselves on their neighbours. Their distinguishing character principally consists in the receptacle of the stamens being equally that of the germen, but raised, at the sides of the calyx, above the germen. Hence, the calyx bearing the stamens, they are *calycanthemi*. The fruit is either superior or inferior, therefore that distinction is not always important. In *Rosa*, for instance, the part in question seems inferior, but is in fact the contrary, for the seeds are really inserted into the inner side of the calyx, exactly as in *Mespilus*, with this difference only, that in the latter they are imbedded in the pulp, which fills the calyx. The segments of the calyx are mostly in a double series, the innermost largest, the outer alternate therewith, and smaller, answering to the petals. Stamens for the most part numerous, but *Sibbaldia, Alchemilla*, and *Aphanes* form an exception, the first having five stamens, the two last only four," (or even fewer). "The pistils vary in number. There is nothing acrid in the whole order, nor much fragrance; there is much of a styptic, little of a mucilaginous quality; nothing poisonous; so that if the fruits are worth tasting, they may certainly be eaten with impunity."

Order 37. Columniferæ. "So called, not because the author meant to express, in the name, the essential character, but in allusion to some distinguished examples of this order, whose stamens are united into a columnar form. Linnæus was really the founder of the order in question, though Tournefort endeavoured to keep together as many of the same plants as possible, under his *Monopetali Campaniformes*. But the corolla has five petals, though they all fall off in one body, being connected with the combined stamens. Some have been denominated *akin to Malvaceæ;* indeed many of the class *Monadelphia* belong to this tribe."

"The root in all these plants is fibrous, in no instance bulbous or tuberous. Stem often herbaceous, but there are many arboreous, and amongst others the kinds of *Bombax*, or Silk Cotton, almost the largest trees in the world. Some of these only bear spines; but some species of *Hibiscus* are prickly. There is scarcely a perfectly smooth plant in the whole order. They have all stipulas, in pairs. The leaves are alternate, never opposite; in numerous instances stalked; plaited in the bud; and, what is remarkable, many of them have glandular pores under the rib. No tendrils are found in the order. The inflorescence is various. Calyx in several simple and five-cleft, but in some genera double, as *Malva, Alcea, Al-*

thæa, Lavatera, Malope, Gossypium and *Hibiscus.* Petals generally five, but as they often adhere to the united filaments, the corolla seems monopetalous. This adhesion contradicts the opinion of Vaillant, who has said that stamens are never inserted but into a monopetalous corolla. Their connected claws often form a nectary between them. The corolla is somewhat abrupt, and twisted contrary to the sun's motion. Pistils usually corresponding in number to the parts of the fruit; as do the stigmas, where the style is simple. *Turnera* has as many styles as there are cells in the capsule. The fruit is always superior, but differs in different genera. *Malva, Alcea, Althæa, Lavatera* and *Malope,* have numerous capsules, ranged like a wheel round the base of the style; nor is the latter placed upon, but in the midst of, them, as in the *Asperifoliæ,* order 41. Each capsule is single-seeded, and falls off with the seed; which is likewise the case in *Urena;* such seed-vessels might perhaps rather be named *arilli,* or tunics, as they burst at their inner side. Many of this order have solitary seeds in their cells, or capsules, like the above, and the genus *Ayenia;* but many others are polyspermous, as *Bombax, Hibiscus, Theobroma,* &c. A few of the genera produce woolly seeds, as *Bombax* and *Gossypium;* in the place of which appendage, *Adansonia* has a mealy powder. Some bear a capsule of five cells, containing many seeds; which in *Hibiscus Malvaviscus,*" (now constituting the genus *Achania,*) " becomes pulpy. It is curious that *Hibiscus Moscheutos* bears its flower-stalk upon the footstalk, like *Turnera;* a rare circumstance in the whole vegetable kingdom."

" *Hermannia* has hooded petals, in a corolla twisted like that of *Malva.* They are auricled and dilated below, forming a nectary by their involution, as the true *Malvaceæ* do by the cohesion, or approximation, of their petals. The calyx is tumid. Capsule of five cells. All the species are shrubby. The flowers are so alike in all, as hardly to be distinguishable from one another; and hence perhaps it may be presumed, that the various species, all natives of the Cape of Good Hope, may in this, as well as other genera, have been produced from the hybrid impregnation of some original one. *H. pinnata* has the only compound leaves in this order." We must protest against this extensive speculation, of the production of permanent mule species, having seen many arise from such a cause, but none continue to propagate itself for any length of time. It is not the least curious particular, in the structure of the genus before us, that the flowers, which commonly grow together in pairs, have the corolla twisted in an opposite direction to each other.

The second section of this order, composed of *Camellia, Thea, Gordonia, Stuartia, Tilia* and *Kiggelaria,* are at least nearly akin to the foregoing genera.

"This whole order contains no disagreeable or hurtful plants, nor are they esculent. None are fœtid, but some agreeably fragrant. Many of the flowers are beautiful. Their quality is generally mucilaginous, particularly *Althæa, Malva* and *Alcea.* The ancients made considerable use of Mallows in their food, but these plants are now out of use in that respect."

Order 38. TRICOCCÆ. "Botanists apply this term to plants whose fruit is, in a manner, composed of three nuts, combined together like that of *Thea.* In the order under consideration, the seed-vessel is generally a roundish three-cornered capsule, rounded on all sides, with single-seeded cells, which bursting elastically, with considerable force, scatter the seeds to a distance. It must be observed, however, that as in this order some genera, like *Mercurialis* and *Cliffortia,* are dicoccous" (having only two cells, or lobes), "so there are tricoccous plants" (as *Thea,* and many more,) "that do not belong to it."

"The plants of this order bear alternate, mostly simple, leaves, often furnished with glands. Many afford a most acrid milk; they are generally offensive, nauseous, purgative, or poisonous. The style is in several highly remarkable, being more or less deeply three-cleft, and each of its branches divided. The calyx, as well as corolla, have always something unusual in their conformation, or in their nectary; and many of the genera are monoecious or dioecious."

"*Euphorbia*, as a familiar and most distinct genus, may serve as a principal example. It is certainly no less singular than extensive. The calyx of one inflated leaf has four or five marginal teeth, and terminates in as many abrupt coloured glands. The latter are remarkably situated on the teeth themselves; but these teeth seem, together with their glands, to be rudiments of petals. In *Euphorbia corollata* the glands are actual petals, as thin, expanded, and delicate, as those of Flax; but scarcely another instance is known, of petals originating in teeth of a calyx."

"*Plukenetia*, a very rare plant, has a four-cleft flower, and four-celled fruit, with a climbing stem."

There are numerous genera besides. *Rumphia* and *Trewia* are added to the list in the Linnæan manuscript.

Order 39. Siliquosæ. "All botanists have acknowledged the common affinity of the genera constituting this order, and have denominated them *Siliquosæ* and *Siliculosæ*. Tournefort called them cruciform flowers; Linnæus, *Tetradynamia*. These plants have mostly inversely-heartshaped cotyledons, except some Cresses, in which those organs are three-cleft; the rest agree with the genus *Convolvulus*; so this character is no proof of affinity."

"The stems are herbaceous, except some species of *Alyssum*, and one *Vella*. There is no real tree among the whole. The roots are all fibrous, none bulbous or tuberous," (except perhaps *Dentariæ*.) "Leaves universally alternate, without stipulas, tendrils, prickles, or venomous stings. Inflorescence usually a *corymbus*, which gradually elongates itself into a *racemus*, so that the flowers are corymbose, and fruit racemose. Calyx always of four leaves, deciduous, except in *Alyssum calycinum* and *Brassica Erucastrum*. Petals four, with claws; some species of *Lepidium* and *Cardamine* only having flat, or straight petals. The receptacle in most, but not in all, is furnished with glands. Stamens six, the two opposite ones shorter, or at least more spreading." (A very few species have only four or two stamens.) "Fruit commonly a pod, with two valves, two cells, and many seeds. A few genera have a solitary seed, either imbedded in pulp, as *Crambe*; or in a lamellated flat seed-vessel, as *Isatis*; or in an angular one, as *Bunias*."

"The plants of this order are distinguishable into *Siliquosæ* and *Siliculosæ*, the former having an oblong, the latter a rounded pod. But it being difficult to define the precise limits of each, Linnæus refers to the *Siliquosæ* such as have a stigma without a style, and to the *Siliculosæ* such as have a style to elevate the stigma, which character is conspicuous in every instance, except in *Draba*, where the style is but short."

"It is of importance to observe whether the calyx in the present order be closed or spreading; that is, whether the leaves composing that part be parallel, so that their sides touch each other, or horizontally distant."

"The nature of a *Siliqua*, or Pod, appears from what has been already mentioned. It differs from a Legume, in having the seeds attached to each suture, or margin."

"All these plants have a more or less acrid watery juice; hence their external application excites redness in the skin, and their internal use irritates the finer fibres. Nature therefore is solicitous to expel them, and, in consequence of their watery nature, by the kidneys, hence they are all diuretic. Salt, being of a corrosive quality, produces scurvy; but salt is secreted from the body by the promotion of urine, though it must first be dissolved in a watery menstruum; consequently the herbs in question rank among the chief antiscorbutics, especially water-cresses and scurvy-grass. They ought never to be used in a dried state, as their acrimony and medical virtues are destroyed by drying. Boiling likewise is destructive of acrimony, especially in these plants; they ought therefore to be taken recent. Their diuretic powers render them eminently serviceable for evacuating water in the dropsy. Yet their use ought not to be too long continued, as their acrimony abrades the minuter fibres, rendering the vessels, and the intestines, in a manner, callous. This appears from the rigidity and torpidity of stomach induced by too much use of mustard."

"There is scarcely any thing odoriferous about these plants, except in their flowers. When they are bruised, indeed, something volatile ascends, of an acrid, rather than odorous nature, irritating the coats of the nerves, and inducing spasms, which do not originate in the medullary substance of the nervous system, but in its coats."

No alteration or addition respecting the genera of the *Tetradynamia* occurs in the Linnæan manuscript.

Order 40. PERSONATÆ. There is no commentary on this order in the lectures of Linnæus. Giseke has given a synoptical arrangement of the genera, according to the shape of the corolla, which is not in every part precisely correct. He justly expresses his doubts respecting *Melaleuca*, of which we have spoken under the 19th order; and he truly observes that there is no order in which so many genera are named after botanists as in the present.

The only manuscript additions or corrections, which occur in the *Genera Plantarum* of Linnæus, are the following: *Martynia, Craniolaria, Torenia* and *Scrophularia* are pointed out as akin to *Pedalium,* in order 28th; *Hyobanche, Lindernia, Pæderota, Manulea, Premna* and *Calceolaria* are inserted, with a question, certainly not well founded, whether the latter should not rather be referred to the 24th order. *Brunfelsia* also is placed among the *Personatæ,* at the suggestion of Van Royen.

Order 41. ASPERIFOLIÆ. "These plants were first

collected into an order by Cæsalpinus, and received the above appellation from Ray, because of their generally harsh or rough habit. Their root is fibrous. Cotyledons two. Stem branched ; the branches alternate and round. Leaves alternate, simple ; neither divided nor compound, for the most part nearly entire, rough with rigid scattered hairs; convolute before they expand. Stipulas none ; nor are there, except very rarely, any other *fulcra*, or appendages. Common flowerstalk having the flowers ranged along one side. Before flowering it is rolled spirally backwards, gradually expanding as the flowers are ready to open, and divided into two parts, each bearing the flowers on its back, in the form of an unilateral spike. Calyx in five divisions. Corolla inferior, of one petal, regular except in *Echium*, five-cleft ; its mouth either furnished with vaulted valves, or crowned with teeth, or naked. Stamens five, equal ; in *Echium* only they are unequal. Fruit superior. Germens four, naked, except in *Cynoglossum, Tournefortia,* and *Nolana* ; inserted in o the receptacle by their base ; hence the lowest part of each seed is of a tapering form, as if artificially rounded. Pistil one. Style not standing upon the germens, but occupying the central space between them ; often divided into two equal parts ; not one longer than the other as in the class *Didynamia.* Seeds four, rarely combined into two ; but it is singular that *Nolana* has five seeds." Linnæus has, as already mentioned, removed this genus to his *Luridæ,* order 28th.

"The *Asperifoliæ* are distributed according to the mouth, or throat, of their corolla, which is naked, or pervious, in *Echium, Pulmonaria, Lithospermum, Heliotropium, Cerinthe,* and *Onosma* ; toothed in *Symphytum* and *Borago* ; closed with vaulted valves in *Cynoglossum, Asperugo, Anchusa, Lycopsis, Myosotis,* and *Tournefortia.*"

In the *Gen. Plant. Messerschmidia, Coldenia, Hydrophyllum,* and *Ellisia* are inserted in manuscript.

"All the *Asperifoliæ* are mucilaginous, and act only as such. The ancients selected their four cordial flowers out of this order, seeming not to have been aware that the motion of the heart depends upon the nerves, which therefore must be strengthened if the force of the heart is to be increased. This end however is not to be attained by either the flowers or the herbs of this tribe, which nevertheless have long been used for the purpose. The leaves may be eaten as food, by which their small medical use may be estimated. The root is perennial and mucilaginous;" (we would rather say, "if perennial, is mucilaginous," which perhaps were the original words of the lecture.) "Among the whole, *Symphytum* abounds most with mucilage, equalling, in quantity as well as quality, the monadelphous plant *Althæa* in this respect. *Symphytum tuberosum* has been recommended in the gout. Possibly its mucilaginous quality may hinder the crystallization of the gouty matter. The root in almost all the *Asperifoliæ* is red, but for the most part externally only. The root of *Lithospermum tinctorium,* now *Anchusa tinctoria,* is used for its colouring properties. Of all plants, the herbs of this order yield the largest proportion of ashes. There is hardly an odoriferous, nor one fragrant, herb in the whole tribe; though *Cynoglossum* has a somewhat fœtid scent. Their taste is nothing, the great quantity of mucilage involving the stimulating particles. These herbs are esculent, especially when young and tender, although their rough surface renders them less agreeable to delicate palates. They generally grow in dry mountainous situations; and it is singular that in proportion as they are found nearer to water, they become smoother."

Order 42. Verticillatæ. "Ray, in constructing his system, founded three classes, which all succeeding botanists hitherto have approved, the *Stellatæ, Asperifoliæ,* and *Verticillatæ;* but he was unable to give proper characters of the genera. Hermann subsequently, establishing a system upon the fruit, called the *Verticillatæ* of Ray *Gymnotetraspermæ,* plants with four naked seeds, but he could not by this means distinguish them from the *Asperifoliæ,* which have the same character. The generality of *Asperifoliæ,* in fact, differ from the *Gymnotetraspermæ,* in their corolla, which in the former is regular, in the latter irregular, though likewise monopetalous. But *Echium,* though it belongs to the *Asperifoliæ,* has still an irregular corolla. The *Asperifoliæ* have alternate leaves, the *Gymnotetraspermæ* opposite ones. These classes might therefore be distinguished from each other, according to Hermann's method, were not *Echium* an obstacle. Linnæus, however, that he might avoid all confusion between the orders in question, has borrowed a character from the stamens, and has referred to his class *Didynamia* such plants as have two stamens longer, and two shorter. He has moreover divided that class into two orders, the first of which comprehends Hermann's *Gymnotetraspermæ,* whose stamens easily distinguish them from the *Asperifoliæ.* But the consideration of the stamens has further obliged the author of the sexual system to refer certain genera, of the natural order under our present consideration, to his class *Diandria.* These are *Verbena, Lycopus, Amethystea, Ziziphora, Cunila, Monarda, Rosmarinus, Salvia,* and *Collinsonia*; of which *Verbena* and *Collinsonia* perhaps ought rather to be placed in the other order of the *Didynamia,* called *Angiospermia.*" This is correct with regard to *Verbena* only.

"The calyx of the *Verticillatæ* is of one leaf, inferior. Corolla of one petal, irregular, in most instances gaping, with two lips, the uppermost of which was called by Rivinus the *galea,* or helmet, the lowermost the *barba,* or beard. Stamens four, except in the several genera just mentioned, where they are only two, inflexed, ascending under the upper lip. Germens four, from between which the style arises, as in the *Asperifoliæ,* which is wavy, solitary, except in *Perilla,* where there are two, and bearing two acute stigmas. Seeds four, naked, *Prasium* excepted, whose seeds have a succulent skin, causing them to resemble berries. A berry, properly speaking, is a seed-vessel; but in *Rosa* it is the calyx, in *Fragaria* the receptacle, and in *Prasium* the skin of the seeds."

"Many of this order are humble shrubs, none are trees, most of them are annual or perennial herbs. The stem is generally square. Leaves in every instance opposite, simple, mostly undivided. None of the plants are furnished with tendrils, nor of a climbing nature. The scent of nearly all of them is highly fragrant, the odoriferous matter being contained in minute cells, which, when the leaves are held against the light, appear like numerous perforations."

"The flowers usually stand in whorls, encircling the stem as with a ring. When these whorls approach very closely together, the stems appear spiked, as in *Ori ganum.*"

"This order is in the highest degree natural; whence arises great difficulty in determining the genera. Linnæus has derived a character from the calyx, according to which the whole order is divided into two sections."

"The first of these comprehends such as have a five-cleft calyx, that is, where all the teeth of this part are nearly of equal size and shape. The second consists of those with a two-lipped calyx, which is indeed five-cleft, but its two upper segments are, in a manner, united into one, which might almost be termed emarginate only; while between these two united segments and the remaining three, there is so deep a fissure, at each side, that the calyx is nearly divided into two parts, or lobes. Linnæus has bestowed great attention in searching for the essential characters of genera in this natural order, and has detected several, which are marked in the *Systema Vegetabilium* with a sign of exclamation."

Order 43. Dumosæ. "*Dumus* and *nemus* are synonymous, meaning a thicket; or wood consisting of shrubs, not of large trees. All the plants of this order are shrubby, but none of them, except in the genera of *Sideroxylum* and *Chrysophyllum*, grow to large trees."

"*Rhamnus* is supposed to be familiar to everybody. Its calyx is tubular, five-cleft at the margin, occasionally coloured, like a corolla, but not perforated at the bottom. A monopetalous corolla falls off, with a perforated tube; which is not the case here. But betwixt every two segments of the calyx is stationed a delicate little scale, which any person might easily take for so many petals. The stamens however, being placed under each scale, are therefore alternate with the divisions of the calyx; whereas if these scales were real petals, the stamens ought, by a general rule, to be alternate with them, and not with the parts into which the calyx is divided. Some species, as the Buckthorn, *R. catharticus*, have four-cleft flowers, but they are mostly five-cleft. This last-mentioned, like *R. alpinus*, is dioecious; *Zizyphus* is polygamous. The stigma in some *Rhamni* is emarginate, in others three- or four-cleft. The fruit of this genus is various; a berry in some with four seeds, in others, as *Paliurus* and *Alaternus*, with three; in others again it has a single seed with two cells, as in *Zizyphus*. *Paliurus* has not, properly, a berry, but a depressed, bordered, shield-like capsule. The stem in some is thorny, in others prickly, in others unarmed."

"French botanists have recommended the dividing of this genus into several, a measure which appears highly proper to those who have not seen the Indian species. If such genera are to be distinguished by their fruit, species most resembling each other will be put asunder, and widely different ones brought together, as any person making the experiment will find. Besides, the structure of the flower, and the habit of the plants, are respectively so alike in all the species in question, and so different from all the rest of the order, that any peasant might perceive their affinity."

"*Phylica* agrees in almost every point with *Rhamnus*, except that its flowers are aggregate, and florets superior. This genus is so nearly akin to *Brunia*, that without seeing the fruit, which very rarely occurs, they can scarcely be distinguished. *Phylica radiata* therefore, universally esteemed a *Phylica*, proves, on the detection of its fruit, to be a *Brunia*."

"*Ceanothus*, with its three-lobed fruit, like that of *Rhamnus Alaternus*, agrees in every character with *Rhamnus*; but the scales of that genus are here drawn out into vaulted petals, supported by long claws."

"*Büttneria* differs in hardly any respect from *Rhamnus*, except its anthers; for the calyx, prickles, and every thing else, answer so well, that at first sight one would decidedly take it for a species thereof."

"*Sideroxylum* has a five-cleft calyx, and at the same time a monopetalous corolla; but between all the segments of the latter stands a little serrated tooth, analogous to the scales of *Rhamnus*. The flowers are likewise sessile on the stem, but the berry has only one seed."

"*Chrysophyllum* is so nearly akin, and so similar, to *Sideroxylum*, as hardly to be distinguishable by its general aspect; but its fruit contains many seeds, though indeed they are disposed in a circle."

"*Achras* differs from *Chrysophyllum* in having a six-cleft flower; and to this genus *Prinos* is very nearly related, differing in the flat form of the corolla, and fewer cells of the fruit."

"*Ilex* so nearly accords with the last-mentioned genus, that the only *Prinos* then known was originally taken for an *Ilex*; but the flower of *Ilex* is four-cleft, not six-cleft."

"*Tomex* and *Callicarpa* only differ from *Ilex* in having a single style, and not four stigmas. The berry of *Callicarpa* is like that of *Ilex*. In *Tomex* the stamens are inserted into the receptacle, whereas in *Callicarpa* they are attached to the tube of the monopetalous corolla." These genera have since been united by the author himself.

"*Euonymus* is so nearly allied to *Tomex*, as scarcely to be distinguishable, except by having a capsule instead of a berry. Its seeds moreover have a pulpy tunic."

"*Celastrus*, though differing from *Euonymus* in having alternate leaves, is so much akin to that genus, as to have been called *Euonymus* by all systematic writers. Yet its fruit differs in number and proportion from *Euonymus*, just as *Peganum* does from *Ruta*. The tunic of the seeds however, though not pulpy, confirms the affinity to which we allude. Some botanists, especially the French, are unwilling to admit plants with opposite leaves and alternate ones into the same natural order, and they are right; yet this character is not absolute, for such a difference often occurs in one and the same genus."

"*Viburnum* and *Cassine* come so near together, that there is rather a question respecting the distinction of the genera themselves, than of their natural order. *Cassine* has three seeds, *Viburnum* one, which seems two combined. The former is akin to *Sambucus*, and, like that genus, emetic in quality. Concerning the affinity of *Viburnum* to *Ilex* and *Callicarpa*, any person, who considers their fructification and habit, can have no doubt. Thus far therefore the matter is clear."

"*Sambucus* may excite some mistrust, because of its inferior fruit; yet this is the case in *Phylica*, about which nobody has ever doubted. The leaves, aspect, and stipulas indeed seem to indicate something extraneous, and leave us in uncertainty."

"So *Rhus* has much the same sort of fructification, and a berry with one seed; as well as the closest affinity to *Sambucus*, insomuch that if *Sambucus* be kept in this order, *Rhus* must accompany it. So also must the sister shrubs *Schinus* and *Fagara*."

"The *Dumosæ* all agree in malignant qualities. They

are either purgative, or altogether poisonous, as *Sideroxylum* is known to be at the Cape. Nor are the species of *Sambucus* clear of this charge, for their qualities are either nauseous or fœtid, and therefore sudorific, especially the berries and flowers. The bark, taken internally, is either emetic, or powerfully purgative, as its vinous infusion proves in the dropsy; externally it is a powerful repellent."

"*Rhus* is the most venomous of trees, particularly its American three-leaved species, called *Toxicodendra*, or Poison-trees. Their fumes in burning are said to have proved mortal, and their effluvia to have blinded an artist who was at work upon some of the wood. Those who, being in a perspiration, hold a branch of one of these shrubs in the hand, are seized with an eruption over the whole body."

"The bark of *Rhamnus Frangula* is our best indigenous purge, and a syrup of *Rhamnus catharticus* is safely used for children."

"In this tribe, therefore, some have opposite, others alternate, leaves, nor is any general character to be derived from the parts of fructification. The corolla affords none, being either of one or five petals, or altogether absent, as appears from a contemplation of the characters of the different genera. No mark is to be obtained from the nature of the fruit, that being either a berry, drupa, or capsule. The seeds in some instances are solitary, in others numerous, though never more than one in each cell; and it is well worthy of observation that they are attached, as in the *Gymnotetraspermæ*, by their base. These plants betray some affinity to the *Tricoccæ*, but can never be referred to the same order."

In the Linnæan manuscript before us, *Diosma* and *Hartogia* are introduced between *Callicarpa* and *Euonymus*;—see our remark on the 26th order. *Staphylea* is also subjoined, near *Celastrus*, but with two marks of doubt, and a note of its having a nectary, as well as opposite leaves.

Order 44. Sepiariæ. 'All these are shrubby or arborescent. Leaves opposite, with scarcely any evident stipulas. Flowers disposed in a more or less dense panicle. Calyx four-cleft. Corolla four-cleft, regular. Stamens two. Pistil one, with a cloven stigma. Fruit either a drupa, with one, two, or many, seeds, or a capsule."

No manuscript remark occurs here, nor is there any observation worth copying in the lectures, except that *Olea* is said by Linnæus to be scarcely a distinct genus from *Phillyrea.*

Order 45. Umbellatæ. "The name of this order is derived from the form of its inflorescence, whose stalks all spread from a centre, like the ribs of an umbrella."

"These plants are either perfectly umbellate or not. The former are required to have a compound umbel, each stalk, or ray, of which ends in a receptacle, producing other stalks bearing flowers, or florets; the latter have a simple umbel, whose stalks are not subdivided. The latter constitute a separate section in Tournefort's system. They are comprehended by Linnæus in one natural order with the former."

"An umbel is properly a receptacle of a compound flower, elongated into stalks; which manifestly appears in *Eryngium,* whose florets are united into a head, just like the proper compound flowers,—see the 49th order; nor are they supported by elongated stalks. Hence an umbel may accurately be considered as a compound flower. Those who controvert the opinion of Linnæus in this point contend, that many umbellate plants, having male and hermaphrodite flowers in the same species, ought to be placed in his class *Polygamia.* But this is a mistake; for no other plants ought to find a place in that class, than such as have distinct male, or female, as well as hermaphrodite, flowers, in the same species. This is not the case with the *Umbellatæ,* in which all the florets of one universal umbel, that is, the whole umbel itself, constitutes but one flower, and this flower is never altogether barren, that is, its florets are never entirely male. On the contrary, these florets are to be considered as the parts of a compound flower; and there being male and hermaphrodite ones intermixed, is exactly a parallel case with the polygamy of the Syngenesious class."

"This order is eminently natural, though all plants which bear umbels do not belong to it, but only those with five stamens, two styles, and two seeds."

"The germen is inferior, simple, solitary, separating, when arrived at maturity, into two equal naked seeds; each of which is furnished with a thread, inserted into its summit. These two threads combine to form a very slender receptacle, at the top of the stalk of the floret. Each floret has a superior perianth, with five teeth, which is often so small as scarcely to be discerned, even with the help of a magnifier. Petals five, caducous, often unequal; hence Rivinus referred these plants to his class of pentapetalous irregular flowers. Stamens five, inserted into an elevated annular or circular receptacle, that surrounds the pistils, deciduous. Styles two, often very short, and hardly visible. Seeds naked, without any seed-vessel."

"The stem is mostly hollow, sometimes filled with spongy pith; rarely shrubby, very rarely arboreous, of which last character *Phyllis* is the only example,—see order 47. Leaves generally alternate, and repeatedly compound. Root mostly quite simple; in *Oenanthe* tuberous, in *Bunium Bulbocastanum* globose."

"Nothing is more arduous than to distinguish the genera of umbelliferous plants by appropriate characters. Tournefort himself, who excelled in the knowledge of this tribe," (perhaps Linnæus meant rather to say, in the discrimination of genera, but his auditors did not take his words accurately,) "has distributed them according to the shape and size of their seeds. But this is a very fallacious mode, as the seeds often differ much in proportion, though not in any other respect. Morison wrote an entire book on umbellate plants; but with little success, their genera not being, as yet, established. Artedi first paid attention to the *involucrum,* which is either universal as well as partial, or only partial, or entirely wanting. This principle has likewise been adopted, as fundamental, by Linnæus, and his three primary divisions are regulated accordingly. The inequality of the petals affords him a principle for his leading subdivisions, some of the umbelliferous family having the outermost petals of their external florets larger than the rest; while in others all the petals are equal. The former are termed radiant flowers. Another subdivision is taken from the sex of the florets. Some of these, having no germen, are furnished with stamens only; and such florets are termed abortive; others, having both germen and stamens, bring their fruit to perfection, and are therefore denominated fertile."

On these principles Linnæus has arranged the umbellate plants, as may be seen in his works. Nothing occurs in his manuscript, except the insertion of *Hermas* next to *Eryngium.*

Order 46. Hederaceæ. The lectures give no new information concerning this order. The six genera stand as in the *Genera Plantarum; Panax, Aralia, Zanthoxylum, Hedera, Vitis,* and *Cissus.*

Order 47. Stellatæ. "This order was founded by Ray, and received its name from the leaves of most of the plants which compose it being placed, four, six, or eight together, in the form of a star, round the stem. It is unusual to see more than two leaves opposite to each other, nor is it the case here. For two of these only are properly leaves, the rest being no other than stipulas, grown to the size of leaves. This appears evident in several Indian plants of the present order, as *Knoxia, Diodia,* &c. which have only two opposite leaves, though between these some small acute stipulas are found, being the same that in the rest of the order attain the magnitude of leaves. Ray believed all the plants of this order to have whorled leaves, which is generally the case, as far as regards those of European growth, but rarely with the Indian ones, of which few were known in his time."

" In this order there is no tree, unless perhaps *Lippia*; there are very few shrubs, most of the tribe being small herbs, growing in barren earth, or coarse sand."

" The roots are in many instances perennial. Leaves opposite, horizontal, mostly rough. Stipulas of the form and aspect of leaves, so that it is impossible to say whether they be truly such or not, hence the leaves appear whorled; but this does not hold good universally. In those however which have no leafy stipulas, there is found, at each side, a sort of toothed membrane, connecting the leaves together, and occupying the place of stipulas."

" The stem is jointed, with mostly tumid knots. Corolla of one petal, either flat, wheel-shaped, or funnel-shaped; in one genus bell-shaped; mostly four-cleft, sometimes almost down to the base; rarely five-cleft. Stamens four, never eight, though sometimes five or six, in which case the corolla has a parallel number of segments. Pistil solitary, divided; in *Richardia* three-cleft, because that genus has a six-cleft corolla, six stamens, and a three-grained fruit, its parts of fructification being all augmented in a similar proportion. Those parts are not augmented with the same regularity in genera furnished with a three-cleft corolla, and five stamens, for their pistil is still bifid, and their fruit two-grained, as is the case with such as have a four-cleft corolla and four stamens.

" The fruit is, for the most part, inferior; though superior in *Houstonia*; and in *Crucianella* superior with respect to the calyx, though inferior to the corolla." This is incorrect, for *Crucianella* has a real superior perianth, like the rest of the order, though so small as to be hardly discernible; what Linnæus here terms calyx, being an involucrum, or perhaps bracteas. " The sexes are rarely separated in this order, though *Valantia*, which is polygamous, can by no means be excluded from it. Many of the genera have a two-grained fruit, of two cells, with a solitary seed in each. But in *Hedyotis* and *Oldenlandia* the cells contain many seeds; while in *Cornus* both cells are united into one seed, which, nevertheless, has two cells. The fruit has a green, fleshy, but not juicy, coat, nor does it usually become coloured in ripening; though in *Rubia* the fruit is a perfect berry."

(Of the remarks on particular genera, we find nothing to extract except the following.)

" *Asperula tinctoria* is used in Gothland instead of Madder, and is preferable."

" *Sherardia* has an oblong fruit, which the permanent calyx renders toothed, or crowned with three points. It was the fate of William Sherard, a man worthy in the highest degree of botanical honour, to have two different genera distinguished by his name, both which were afterwards referred to others. Pontedera, Vaillant, and Dillenius each published, at the same time, a *Sherardia.* Pontedera described his plant so very obscurely, that it was ten years before Linnæus made it out to be his own *Gallenia.* Vaillant called the two-seeded *Verbenæ* by the name of *Sherardia,* but he was to blame in separating them from their proper genus. Dillenius named a *Sherardia,* from among the *Stellatæ,* which Linnæus has retained, though not very certainly distinct. Being unwilling that so meritorious a botanist should remain without a memorial, Linnæus declined referring the plant in question to *Asperula*; especially as the three teeth, at the top of each seed, may serve, if not very satisfactorily, to keep it separate."

" *Valantia* was so named by Tournefort; but Vaillant, perceiving it to be the same with Tournefort's *Cruciata,* thought it a bad genus, which could not support itself. He therefore wished to abolish all generic names, given in honour of botanists, because he supposed his own was untenable. But Tournefort confounded several genera under the appellation of *Cruciata,* so that Linnæus has been enabled to establish a *Valantia* from among them, referring the rest to their proper places."

Order 48. AGGREGATÆ. "These constitute a natural order, first established by Vaillant in the Memoirs of the French Academy of Sciences. They agree so far with the *Compositæ*, that they have generally a common calyx, as well as receptacle, containing many sessile flowers, each of which has always an inferior germen. But there is a total difference with respect to the remaining parts of fructification, nor can these two orders be, by any means, united."

"The calyx, as we have just said, is common to many flowers. Common receptacle either naked, villous, hairy, or scaly. In the place of a partial calyx is the corolla, generally of one petal, regular or irregular, in four or five divisions, rarely polypetalous. Stamens four, with separate anthers. Germen inferior. Fruit single-seeded. The flower is therefore complete in this tribe, except only *Valeriana*, whose calyx is scarcely apparent. The leaves are often opposite, and the stem shrubby."

Order 49. COMPOSITÆ. "A compound flower generally consists of a common calyx, containing several florets. But this definition is not sufficiently discriminative, for there are certain flowers termed *Aggregate*, which though they have numerous florets in one common calyx, are connected by no affinity whatever with these; witness *Cephalanthus, Dipsacus, Scabiosa, Knautia, Allionia*. Hence botanists have tried to discover an appropriate and distinguishing character for a compound flower, but they have scarcely succeeded. There are indeed flowers of this order, furnished with solitary florets in each calyx, as *Seriphium, Corymbium, Strumpfia*. All of them have a monopetalous corolla, but so has *Scabiosa* and others. Most have five stamens, but some have only four. The greater number bear their anthers united into a cylinder, but *Kuhnia*, which belongs to them, has separate anthers; while *Jasione, Viola* and *Impatiens*, which do not, have combined ones. The united anthers burst internally, by which means their pollen is communicated tc the stigma; but the anthers of *Kuhnia* open at the extremity, and resemble the corolla of an *Aristolochia*. All the florets are superior, but this holds good likewise in *Scabiosa*. Hence it appears that no essential character of compound flowers is to be detected, though no order can be more natural than that before us."

"Tournefort first divided the compound flowers into three sections, according to the shape of their partial corollas. These are either ligulate or tubular. Such as consist of ligulate florets only, are called by this writer *semiflosculosi*; such as are formed only of tubular ones, *flosculosi*; while those which have ligulate florets in the radius, and tubular ones in the disk, are denominated *radiati*. This division seems natural enough, and yet is not so. For it refers both the discoid and capitate compound flowers of Linnæus to the *flosculosi*, which nevertheless are too dissimilar to be possibly admitted into the same section. The *discoidei* of Linnæus, Ray's *aggregati*, having aggregate florets, seated on a hemispherical receptacle, are, in fact, more allied to the *radiati*; while the *capitati*, such as Thistles, are widely different, so as necessarily to constitute a division by themselves."

"Vaillant attempted a new botanical system; but it is to be lamented that we are possessed of no more of his labours, than what concerns the compound flowers. In this performance, published in the Memoirs of the Parisian Academy for the years 1718, 1719 and 1720; he has displayed an extensive knowledge of species, and has treated the subject admirably. As the Memoirs of the Academy are not within the reach of every body's purse, a German named Von Steinwehr has collected the anatomical, chemical and botanical papers, into an octavo volume, published in 1754 at Breslaw. In this Vaillant's treatises are preserved entire," (but in the German language.)

"The florets of compound flowers are threefold with respect to sex, being either *hermaphroditi*, perfect, having the organs of both sexes; female, destitute of anthers; or neuter, deprived of both organs, and barren."

"Tournefort, Vaillant, Ray, and almost every botanist who has treated of this tribe, divide it into three or four orders, some of them adding the aggregate flowers to the compound ones, whence arises the fourth order. But they have not fixed limits to their orders, such being scarcely discoverable. The *semiflosculosi* and *capitati*, for instance, though apparently widely different, are proved nearly akin by *Scolymus* and *Elephantopus*. The former of these has all the habit of a *Carduus*, and yet all its florets are ligulate; the latter is intermediate between the *semiflosculosi* and *capitati*, nor are we certain to which of these divisions it belongs. *Perdicium*, a new genus, connects *Inula*, which is radiated, with the semiflosculous genus *Hieracium*, so that accurate limits are hardly to be drawn between them. Most of the *semiflosculosi* are milky, but *Lapsana* and *Cichorium* want this quality."

"Section 1. *Semiflosculosi*; all the florets ligulate."

"These genera are distributed, first by their receptacle, which is either chaffy, villous, or naked. In the next place, they are subdivided by the down of their seeds, *pappus*, which is either absent, or bristle-shaped, or hairy, or feathery. Thirdly, a peculiar distinguishing character is borrowed from the form or nature of their calyx."

"The quality of the *Compositæ* in general is innocent; but some of the present section are milky, which secretion proves, by experience, somewhat of a poisonous nature. So *Lactuca virosa*, in a wild state, is as poisonous as opium; yet by culture it becomes esculent and culinary, though still causing sleep by its debilitating power." Linnæus surely could not mean that this and the garden lettuce are one species. It is possible his hearers mistook him.

" There are no trees, and few shrubs, among the *semi-flosculosi*; no bulbs, scarcely a tuberous root, except in some species of *Hieracium*. Their flowers are mostly yellow; sometimes red underneath, as in *Leontodon*, *Hieracium* and *Crepis*;" (very rarely pink, in *Geropogon* and *Crepis*;) sometimes blue, in *Cichorium* and *Catananche*; never white."

" Section 2. *Capitati*; all the florets tubular, assembled into a head, in one common calyx."

" All these are prickly or spinous, and vulgarly called *Cardui*, Thistles. If however they were all considered as one genus, such a genus would prove too ample; hence it is best to separate them into several, though the task is very difficult. *Centaurea* belongs to them, though necessarily referred, in the sexual system, to the order *Polygamia-frustranea*. Its calyx, always tumid, and often spinous, proves its affinity. The most extensive genera of this section, *Carduus*, and *Serratula*, are the most difficult to distinguish; hence it is best to study the rest, in the first place, that those puzzling ones may prove easier."

" Vaillant divided this capitate tribe by the spines of their calyx, whether simple, spinous, or leafy. But the gradation is so imperceptible, that no accurate principles of discrimination are hence to be obtained. No plant of this section is milky, or poisonous, or arboreous. Some of the *Serratulæ* are shrubby; many of the herbs are destitute of stems, as in *Carlina*, *Atractylis*, *Onopordum*, *Carduus*, and *Centaurea*."

" *Atractylis* has a radiant flower, and the florets of the radius have each both stamens and pistil, a solitary instance among compound flowers, rendering the genus very distinct. The elongated and coloured scales of the calyx in *Carlina* have misled Tournefort to rank it among radiant flowers."

" The *capitati* have a character peculiar to themselves, in the dilatation, or inflation, of the tube of each floret, just below the limb, which causes their florets to project, in a more elongated manner, than in the *discoidei*, or other compound flowers."

"Section 3rd. *Discoidei*. The first subdivision of these, *polygami æqualis*," (consisting of such as have all the florets furnished with stamens and pistils, and all producing seed,) "are distributed according to the receptacle, whether naked, chaffy or hairy, and their seed-down, like the *semiflosculosi*."

"The second subdivision, *polygamia superflua*, have female florets in the circumference, but these are tubular, not ligulate or radiant. So that the flowers, though they have a marginal series of female florets, cannot be called radiated." We have here extracted the ideas of Linnæus from his remarks on *Artemisia*, which seem to refer to the whole of this subdivision, and are certainly correct, though they interfere with the distribution of the order before us in the *Genera Plantarum*, and seem to have been unintelligible to the editor of the *Prælectiones*;—see his note in p. 539 of that work.

"Section 4th. *Radiati*." (Marginal florets radiant.) "The first subdivision is *polygamia superflua*," (all whose florets are capable of producing perfect seed, though the marginal radiant ones have no stamens.)

These are distinguished by the presence or absence of seed-down, or of a membranous border to the seed, and by the nature of their receptacle, whether naked or chaffy.

The second, *polygamia frustranea*, have imperfect or defective female or neuter florets in the circumference, producing no seed. These in *Centaurea* are tubular, and neuter; in the rest ligulate, furnished with rudiments, more or less evident, of a pistil.

The third, *polygamia necessaria*, have effective seed; bearing female florets in the circumference only.

"Section 5th. *Monogamia*." (Such as have but one floret in each partial calyx.)

Seriphium, *Corymbium*, and *Strumpfia*.

"None of the *Compositæ* are poisonous, except *Tagetes*, *Doronicum*, and *Arnica*; the latter is more so than *Doronicum*. They contain much of a bitter flavour; hence many of the order are medicinal and strengthening. Some, less bitter, as *Arctium*, *Cynara*, *Carduus*, are therefore esculent. Many *semiflosculosi* are used as food, though furnished with a milky juice, which in them is not poisonous," (see a remark under Order 30th.) "except *Lactuca virosa*, whose juice as above mentioned, has the quality of opium, and *L. sativa* has a soporific virtue. Boiling entirely destroys the power of this, as well as of the other *semiflosculosi*."

Order 50th. AMENTACEÆ. "An *amentum*, catkin, is a species of calyx, and very like a spike, consisting of a common receptacle, drawn out like a thread, on which the flowers stand in alternate order, subtended by scales or bracteas. Such a calyx is found in the plants of this order, whence Linnæus gave it the above name. They are all either trees or shrubs, with alternate leaves, and separated male and female flowers, being either monoecious or dioecious. Many of them produce but one seed from each flower; but *Salix* and *Populus* bear a seed-vessel of two valves, with many seeds. The styles are usually two or three. The flowers come before the leaves, that the latter may not hinder the access of the pollen of the male to the female blossoms."

"Monoecious genera are *Betula*, *Carpinus*, *Corylus*, *Quercus*, *Juglans*, *Fagus*, and *Platanus*."

"Dioecious ones *Pistacia*, *Myrica*, *Populus*, and *Salix*."

Order 51st. CONIFERÆ. "These are generally evergreen trees of cold climates. In the Indies almost all the trees are evergreen, and have broad leaves; but in our cold regions most trees cast their foliage every year; and such as do not, bear acerose, that is, narrow and acute, leaves. If they were broader, the snow which falls during winter would collect among them, and break the branches by its weight. Their great slenderness prevents any such effect, allowing the snow to pass between them. This precaution is unnecessary in India, where snow is unknown. Nevertheless, *Liquidambar* is to be referred to this order, though it bears no such slender, but rather broad, foliage; nor is it a native of a cold country."

"The plants of the present order are denominated *Coniferæ,* because they bear *Strobili,* which the older botanists called *Coni,* Cones. A cone and a catkin are closely related to each other. The latter bears several imbricated flowers about a common receptacle or axis. Under each flower a membranous scale or bractea is attached, which if it hardens and becomes woody, the catkin becomes a cone. Hence a cone is nothing more than a permanent or hardened catkin."

"All the *Coniferæ* properly bear cones, though in some instances their fruit seems of a totally different nature. For instance, the fruit of *Juniperus* has all the appearance of a berry, and is universally so called. Yet it is no other than a *strobilus,* whose scales are replete with pulp, and do not split asunder; being in fact six fleshy united scales, in each of which is concealed a solitary seed. *Taxus* has a berry, which is merely a fleshy receptacle, dilated so as nearly to cover the seed, so that the apex of the latter only appears. *Liquidambar* has a singular kind of fruit, which nevertheless is a *strobilus,* whose scales are combined, each of them containing several seeds; whereas in other instances one or two seeds only belong to each scale."

"Some have united this order with the last, but they differ essentially. The *Coniferæ* have not only hardened scales, but likewise monadelphous stamens, the filaments of all of them being combined at the base."

"The fruit in this whole order, *Liquidambar* excepted, is biennial. It is produced in the spring, remaining in an unripe state through the summer, and till the following spring, when it gradually ripens, and the gaping scales allow the seeds to escape."

Order 52nd. Coadunatæ. On this order there is no observation in the lectures. *Illicium* is added in manuscript to the genera in *Gen. Pl.*

Order 53rd. Scabridæ. Here also the lectures are silent. *Forskohlea* and *Trophis* are added in the manuscript.

Order 54th. Miscellaneæ. Here, although no remark is preserved in the lectures, great corrections are made in the manuscript. The genera in the second section, *Poterium* and *Sanguisorba*, are referred to the 35th order, immediately before *Agrimonia*. *Pistia* and *Lemna*, constituting the 3rd section, are transferred to the 15th order. The six genera which compose the 5th section, are sent to the 4th section of the *Holeraceæ*, order 12th. *Nymphæa* and *Sarracenia*, the only plants of the 6th section, are referred, as already mentioned, first to the 27th order, but finally, not without a doubt, to the 11th. See the observations under those orders. *Cedrela* and *Swietenia*, which make the 7th section, are removed to the *Trihilatæ*, order 23rd. *Corrigiola*, *Limeum* and *Telephium*, the 8th and last section, are transferred to the 5th section of the 12th order, *Holeraceæ*. No genera therefore remain in this 54th order, but *Reseda*, *Datisca*, *Coriaria*, and *Empetrum*.

Order 55th. Filices.
—— 56th. Musci.
—— 57th. Algæ.
—— 58th. Fungi.

Nothing occurs here, either in the *Prælectiones* or the manuscript, to the purpose of our present inquiry, concerning the ideas of Linnæus on natural classification. These orders are all natural, and acknowledged as such by all systematics. His particular observations on each, although in many points curious, are now superseded by the advanced state of botanical knowledge in the cryptogamic department.

From the foregoing copious exposition of the general principles, and many of the particular opinions of Linnæus, respecting a natural classification of plants, it will appear how far he was from considering his performances, in this line, as complete. His leading ideas may, nevertheless, be traced, and they will often be found to throw great light upon the subject. It must be remembered that he never thought his own, or any other, scheme of natural classification, could or ought to interfere with his artificial system, nor does he ever advert to the one in treating of the other. It is evident, likewise, that he studiously discouraged any attempt at an uniform definition, or technical discrimination, of his several orders. He perceived that plants were not yet sufficiently known to render such a scheme practicable. Possibly he might be aware that the accomplishment of that scheme at present would only turn his natural system into an artificial one.

The authors of most plans of botanical classification have, on the other hand, seldom considered the questions of natural and artificial arrangement, as opposed to each other. The system of every such author seems to have appeared to himself the most consonant to nature, as well as the most convenient in practice; yet nothing betrays a more absolute incompetency to the subject than such an idea, wherever it makes itself manifest. To pretend that the elaborate speculations of a proficient, on a subject of which he can see but a part, and on which his knowledge must necessarily be inferior to that infinite wisdom which planned and perfected the whole, should be an easy and certain mode of initiation for a learner, evinces no more than that the professor wishes his pupil should not be wiser than himself. To teach composition without a grammar, or philology without an alphabet, would be equally judicious. Plants must be known before they can be compared, and the talent of discrimination must precede that of combination. Clearness and facility must smooth the path of the tyro ; difficulties, exceptions, and paradoxes must be combated and unravelled by an adept. The knowledge of natural classification therefore, being the summit of botanical science, cannot be the first step towards the acquirement of that science. No person surely, who has published a natural system, without knowing all the plants in the world, will suppose that he has removed every present obstacle, much less anticipated every future obscurity, so that no insuperable difficulty can occur to the investigator of plants by such a system. Neither can any artificial system claim such perfection. But they may combine their powers, and cooperate in instruction. The one may trace an outline which the other may correct and fill up. The first may propose, and the second elucidate ; the former may educate and improve the memory and observation, for the use of the latter. When they oppose each other, their several defects and weaknesses appear ; by mutual assistance they strengthen themselves.

Whether the leaders of natural system in the French school of botany have thought with us on this subject, it might seem invidious to inquire too nicely. It were too much to expect that every one of their pupils, half learned and half experienced, however commendable their zeal and enthusiasm, should have done so. Nor is science in any danger if they do not. They must improve the system of Jussieu, before they can overturn that of Linnæus; and if this were accomplished, the nomenclature and definitions of the learned Swede would still form an impregnable fortress, before which they must perish, or seek for shelter within. This dilemma has been, long ago, but too clearly perceived by the rivals of the fame of Linnæus, particularly by such of the French school as have been actuated by a truly contemptible national partiality, instead of a disinterested love of science and truth. Hence the so often repeated exclamations against Linnæus, as a mere nomenclator. Of his didactic precision, and philosophical principles of discrimination, such critics were not jealous, for they could not estimate the value nor the consequences of these. But they could all feel that the nomenclature of Tournefort was giving way, and that their efforts to support it were vain. The writer of these remarks has perceived traces of this feeling in almost every publication and conversation, of a certain description of botanists. He has likewise perceived that it would gradually subside, and that the interests of science were secure. The nomenclature of Linnæus has in the end prevailed, and it were unjust now, to the greatest botanists of the French school, to deny them the honour of liberality on this head.

It is time for us to close this article, with a view of the principles, upon which the eminent systematics, to whom we have so often alluded, have planned and executed their schemes of botanical classification.

Here the learned and truly estimable Bernard de Jussieu, the contemporary of Linnæus in the earlier part of his career, first claims our notice. This great practical botanist, too diffident of his own knowledge, extensive as it was, to be over anxious to stand forth as a teacher, did not promulgate any scheme of natural arrangement till the year 1759, when the royal botanic garden at Trianon was submitted to his direction. His system was published by his nephew in 1789, at the head of his own work, of which it makes the basis. It appears in the form of a simple list of genera, under the name of each order, without any definition, just like the *Fragmenta* of Linnæus, at the end of his *Genera Plantarum.*

In 1763 a very active and zealous systematic, M. Adanson, made himself known to the world by the publication of his *Familles des Plantes.* In this learned and ingenious, though whimsical and pedantic, work, the great task of defining natural orders by technical characters is first attempted. His affected orthography and arbitrary nomenclature render it scarcely possible, without disgust, to trace his ideas; which however, when developed, prove less original than they at first appear. His work is written avowedly to supersede the labours of Linnæus, against whom, after courting his correspondence, he took some personal displeasure; and yet many of his leading characters are borrowed from the sexual system. The discriminative marks of his 58 families are taken from the following sources—leaves, sex of the flowers, situation of the flowers with respect to the germen, form and situation of the corolla, stamens, germens, and seeds. Every family is divided into several sections, under each of which the genera are in like manner synoptically arranged, and discriminated by their leaves, inflorescence, calyx, corolla, stamens, pistil, fruit, and seeds. In the detail of his system, Adanson labours to overset the principle, so much insisted on by Linnæus and his school, and to which the great names of Conrad Gesner, and Cæsalpinus are chiefly indebted for their botanical fame, that the genera of plants are to be characterized by the parts of fructification alone. The experienced botanist knows that this is often but a dispute of words; Linnæus having, in arranging the umbelliferous plants, resorted to the inflorescence, under the denomination of a receptacle;—see his 45th natural order. But it appears to us that the characters deduced from thence are in themselves faulty, being often uncertain, and not seldom unnatural; and that the plants in question may be better discriminated by their flowers and seeds. Adanson however prefers the inflorescence, even in the *Verticillatæ* of Linnæus; for no reason, that we can discover, but because Linnæus has so much better defined the genera of those plants by the *calyx* and *corolla*. It were a needless and ungrateful task to carp at the mistakes of this or any writer on natural classification, with regard to the places allotted for difficult genera, because the human intellect must falter in unravelling the intricate mysteries of Nature. But surely, when *Plantago* is placed with *Buddlæa* in one section of the *Jasmineæ,* and *Diapensia* with *Callicarpa* in another; when the most natural genus of *Lavandula* is divided and widely separated; when *Cassytha* is ranged with *Statice, Eriocaulon,* and the *Proteaceæ,* in one place; *Geoffræa* with *Melia, Rhus, Sapindus* and *Ruta* in another, we may be allowed to wonder, and to doubt whether we are contemplating a natural or an artificial system. It does not appear that Adanson made many proselytes. He haunted the botanical societies of Paris in our time, without associating with any; nor was his extensive knowledge turned to much practical account. Linnæus has made but one slight remark, that we can find, in his own copy of the *Familles des Plantes,* nor could he study deeply what was, undoubtedly, very difficult for him to read. He certainly never noticed Adanson's attacks, unless the satirical sketch of the *Botanophili,* at the end of his *Regnum Vegetabile,* (see the beginning of *Syst. Veg. ed.* 14.) be partly aimed at this author. To apply the whole of it to him would be unjust, though much is very characteristic.

The study of botany had never been entirely neglected in France since the days of Tournefort; because one department in the Academy of Sciences was allotted to that and other branches of Natural History, and the seats in the Academy being pensioned places under government, there was something to be got by an apparent attention to such pursuits. Buffon and his pupils engrossed Zoology. Botany was allowed to exist, so far as not to interfere with his honours; but nothing of foreign origin, and above all, nothing Linnæan, dared to lift up its head. Something of true science, and practical knowledge, did nevertheless imperceptibly work its way. Le Monnier, and the Marechal de Noailles, corresponded, as we have already said, with Linnæus, and acquired plants from England, of which they dared to speak, and to write, by his names. A most able and scientific botanist and cultivator, Thouin, was established in the Jardin du Roi, who studied the Linnæan system, and even ventured, though secretly, to communicate new plants to the younger Linnæus when at Paris. Cels, an excellent horticulturist, was unshackled by academic trammels. L'Heritier, Broussonet, and others came forward. An original letter of Rousseau, the idol of the day, in which he paid the most flattering homage to botany and to Linnæus, was published in the Journal de Paris, and had a wonderful effect on the public mind, and on the conversation of literary circles. In short, a Linnæan party had been, for some time, gaining ground; and every thing was done by party at Paris. The old French school was roused from its slumbers. Of the family of the Jussieus one individual remained, who, though he venerated the names and the pursuits of his uncles, had never devoted himself to their studies any further than to sit in their professorial chair. He possessed however an inherent taste for Botany; he had leisure, opulence, and eminent talents; and though his religious principles, and his rather strict devotional habits, might interfere, which they still do, with his credit in certain philosophical circles, and his predilection for animal magnetism might exclude him from the Royal Society of London, yet he has risen above all such obstacles, to the summit of botanical fame and authority in his own country; and his name stands conspicuous, as the leading teacher of a natural classification of plants. The most indefatigable study for about five years, and the constant assistance and encouragement of numerous pupils and correspondents, enabled Professor Antoine Laurent de Jussieu to publish, in 1789, his *Genera Plantarum secundum ordines naturales disposita*. This octavo volume was received by acclamation throughout Europe, and hailed as the most learned botanical work that had appeared since the *Species Plantarum* of Linnæus.

Before we enter into systematic details, we must remark,

that the author of the work before us has judiciously availed himself of the mode of defining genera, by short *essential* characters, as introduced by Linnæus in the 10th edition of his *Systema Naturæ*, and since adopted by Murray, Willdenow, and the generality of botanists, instead of the full or *natural*, characters, of the Linnæan *Genera Plantarum*. These short characters however are not servilely copied by Jussieu, but wherever he had materials they are revised and studied, so as to acquire all the merit of originality. Secondary characters and remarks are subjoined, in a different type, illustrative of the habit, history, or affinities, of the several genera. In his nomenclature Jussieu almost entirely follows Linnæus, retaining only here and there a name of Tournefort's, in preference, and swerving from classical taste and correctness principally with regard to the new genera of Aublet, whose intolerably barbarous names are nearly all preserved. But a note in the preface, p. 24, informs us, that this adoption is only temporary, till the genera themselves shall be perfectly ascertained and defined. Where Jussieu differs from Linnæus, in certain generic appellations, it is principally because the latter fails in respect for his own laws; as in the use of adjectives, like *Gloriosa, Mirabilis, Impatiens*. The inordinate abuse of generic names in honour of botanists, of which Linnæus is, too justly, charged with setting the example, meets with due reprobation from the French teacher; but he has not as yet stemmed the muddy torrent, nor prevented a great additional accumulation of subsequent impurities. His commendation of Linnæus, as the author of a new and commodious system of specific nomenclature, as well as of technical definition, on the best principles, is liberal, manly and just, no less honourable to the writer, than to the illustrious subject of his remarks. The whole preface of Jussieu is a concise and learned review of the physiology and distinctions of plants, more particularly explaining the progress of the author's ideas and principles of botanical classification.

The main end of the whole book, besides defining the characters of all known genera, is to dispose them in a natural series, in various classes and orders, whose technical distinctions are throughout attempted to be fixed and contrasted. With this view, copious explanations and commentaries accompany each other. We learn more from the doubts of Jussieu, than from the assertions of Adanson. The latter has presented us with a finished system, where every genus is referred, at all hazards, to some place or other. Jussieu, on the contrary, has not only a large assemblage of *Plantæ incertæ sedis,* at the conclusion of his system, like Linnæus; but at the end of most of his individual orders we find some genera classed as akin thereto, without answering precisely to the character, or idea, of each. This circumstance, though highly creditable to the candour and good sense of the author, greatly interferes with the practical use of his book, except for the learned. His judicious doubts, critical remarks, and especially the laxity, and consequent feebleness, of his definitions, though eminently instructive to those who want to define, or to class, a new, or obscure genus, could only bewilder a learner of practical botany. A person must already be deeply versed in plants, before he could, by the work of Jussieu, or by any book, that we have seen, classed according to his method, refer any genus to its proper place, or detect any one that may be there described. Nor does the difficulty to which we allude consist so much in the intricacy of the subject, as in the uncertainty, hesitation, and insufficiency of the guide; because that guide, learned as he is, chooses to conduct us by a path, to which neither he nor any other mortal has a perfect clue. His index indeed must be the resource of a young botanist; who, if he knows a *Rosa,* a *Convolvulus,* or an *Erica,* may, by finding their places and their characters, trace out the allies of each, and proceed step by step to acquire more comprehensive ideas. The analytical mode of inquiry, which serves us in the artificial system of Linnæus, is here of no avail but to an adept. This will abundantly appear as we trace the leading principles of this celebrated method, of which we shall now attempt a concise exposition.

THE SYSTEM OF JUSSIEU

Consists of fifteen classes, which are composed, all together, of one hundred orders. The characters of the classes depend first on the number of cotyledons; next the number of petals, and the situation, or place of insertion, of the stamens and corolla.

The author uses the term *stamina hypogyna* for such stamens as are inserted into the receptacle, or below the germen, which therefore we shall call *inferior stamens; stamina perigyna,* (around the germen,) are inserted into either the corolla or calyx, the germen being superior; these we must denominate *perigynous*; *stamina epigyna, superior stamens,* are inserted above the germen, which latter is therefore, in Linnæan language, *inferior*. The same terms apply to the corolla, which when inserted into the calyx is denominated *perigynous*. The following table will show the characters of Jussieu's Classes:

Cotyledons none.

Class 1st.

Cotyledon one.

Class 2nd. Stamens inferior.
—— 3rd. Stamens perigynous.
—— 4th. Stamens superior.

Cotyledons two, (or more).

Class 5th. Petals none. Stamens superior.
—— 6th. —— —— Stamens perigynous.
—— 7th. —— —— Stamens inferior.
—— 8th. Corolla of one petal, inferior.
—— 9th. ———— ————, perigynous.

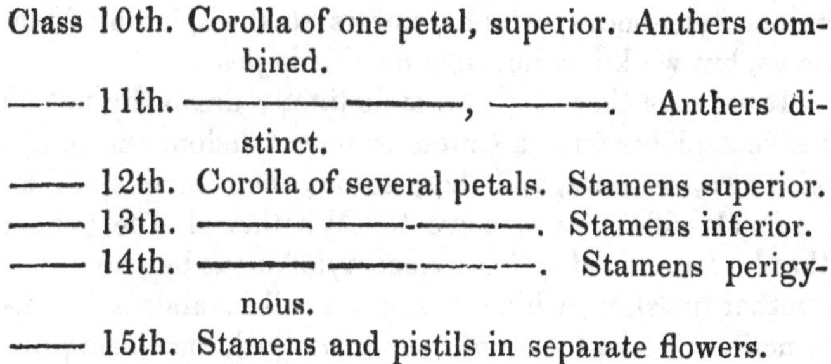
Class 10th. Corolla of one petal, superior. Anthers combined.

—— 11th. ——————, ———. Anthers distinct.

—— 12th. Corolla of several petals. Stamens superior.

—— 13th. ——— —————. Stamens inferior.

—— 14th. ——— ——————. Stamens perigynous.

—— 15th. Stamens and pistils in separate flowers.

In the first place, it is evident that the great hinge, on which this system turns, is the number of the cotyledons. The importance of this character has, from the time of Cæsalpinus and Jungius, been much insisted on. Linnæus, in his *Prælectiones*, p. 329, declares his opinion, that "the monocotyledonous and dicotyledonous plants are totally different in nature, and cannot be combined;" and that "if this distinction falls to the ground, there will never be any certainty. Not that characters should be taken from hence, but sections when formed should be confirmed by the cotyledons." So jealous was this great man of any definition of his natural orders! He subjoins an exception to the above rule, in *Cuscuta* and *Cactus*, which having no leaves, he supposes have no occasion for cotyledons. Linnæus proceeds to observe that "the germination of parasitical plants requires investigation, but that he should greatly wonder if they have any cotyledons." We have already, under the 11th of his natural orders, pointed out other exceptions, made by himself, to the rule just mentioned; but in these he was partly, as we have shown, mistaken; and had he been explicit about the *Sarmentaceæ*, he probably would have proved himself in an error likewise with respect to them. So Adanson asserts the *Juncus* to have two cotyledons, though the rest of its natural order have only one. But Gærtner has demonstrated this genus to be monocotyledonous. Adanson mentions *Orobanche* and *Cuscuta* as monocotyledonous, which answers to the opinion of Linnæus, but we know not how far this is just.

It appears that the line is distinctly drawn by nature between plants with a simple or no cotyledon, and others with two, or more, and that, so far, the principle of Jussieu's classification is correct. Whether all the genera that he has considered as monocotyledonous be truly so, is another question, which does not at all invalidate the distinction. Some have not been examined, and seem principally to be referred to that tribe, because, like others that indubitably belong to it, they are aquatics; or, at least, because of the apparent simplicity of their general structure. Doubts are expressed on this subject by Jussieu himself respecting *Valisneria*, *Cyamus* (his *Nelumbium*), *Trapa*, *Proserpinaca*, and *Pistia*. Some other genera, ranked as acotyledonous, are involved in similar uncertainty.

But with regard to the bulk of the *Acotyledones*, composing the first of Jussieu's classes, there seems to us much greater difficulty. Of his first three orders, *Fungi*, *Algæ*, and *Hepaticæ*, nothing indeed is correctly known, except perhaps what Hedwig has published concerning *Marchantia* and *Anthoceros*, and that is hardly sufficient for our purpose. With the fourth order, *Musci*, this great cryptogamist has made us so well acquainted, that they prove to be any thing else than acotyledonous, or monocotyledonous; at least if his idea of the parts be right. The parts which he takes for cotyledons are peculiarly numerous and complicated; but we are ready to allow with Mr. Brown, at the conclusion of the preface to his *Prodromus Floræ Novæ Hollandiæ*, that these organs are of a most uncertain nature, rather subsequent to germination than its first beginning, like what has been judged the cotyledon of Jussieu's 5th order, the *Filices*. Yet hence a new difficulty arises. The parts in question so complex in *Musci*, are simple in *Filices*, insomuch that no analogy between these orders, otherwise so nearly akin, is to be traced in those parts. On the other hand, it cannot be concealed that the plants termed *monocotyledones* have no cotyledon at all analogous to those of the *dicotyledones*; what Jussieu and others call such, being the *albumen* of the seed, absorbed in the first stage of vegetation. The minute plants assumed to be acotyledonous, must be presumed to be furnished with something analogous, or we cannot conceive how vegetation can take place. By all these observations we mean only to show, that the primary divisions of Jussieu's system are at least totally insufficient to answer that practical purpose, which a student has a right to expect from any methodical arrangement. If the learned be still uncertain, whether the distinctions, on which such divisions are founded do, in a great number of cases, really exist, how can a beginner regulate his first inquiries thereby? We are not the less ready to confess, that the difficulty in question is rather a philosophical speculation, than of any great practical importance. It gives a venerable air of mystery, which may procure respect for other parts of a system, that are more intelligible and more useful, though not free from exception. We allude to the next subdivision of the method of the great French teacher, founded on the petals. This should seem to be obvious and certain, but we soon find ourselves bewildered in an old labyrinth of dispute, concerning the difference between a *calyx* and a *corolla*. We are obliged to submit to a sweeping decision, which allows no *corolla* to monocotyledonous plants; a decision which we cannot safely combat, because of the difficulty of deciding what are such, but which shocks our senses and our judgement, and seems refuted in many instances by Nature herself, as decidedly as any of her laws can be established. Nor do we get clear of this perplexity among the declared dicotyledonous tribes, where the evident *corolla* of the Marvel of Peru is assumed to be an inner *calyx*, there being a real *perianth* besides, subsequently indeed called an *involucrum*. Yet we are at a loss to discern why the terminology here used, should have been different from that applied to the next order, *Plumbagines*. We are ready, most unreservedly, to admit the great difficulty of decision in these cases, as well as in others, occurring in Jussieu's 5th, 6th, and 7th classes; but that very difficulty evinces the precariousness of making any thing connected with this most disputable of all questions, a primary guide in a system of methodical arrangement. When we proceed a step further, and come to the insertion of the stamens, the convenience and clearness of the system indeed improve upon our view; but we must not hope to escape exceptions or inaccuracies, the connection of the *filaments* with the *corolla* being, by no means, uniform or constant, in the orders so characterized, nor even in all the species of particular genera, classed upon that principle. So likewise the insertion of the *stamens* into the *calyx* is attended with such inveterate difficulties, that one of the warmest promulgators and defenders of Jussieu's system, Mr. Salisbury, has thought it easier to deny the existence of any such insertion, than to make it subservient to practical use. We are indeed satisfied that the characters throughout the celebrated method of classification now under our contemplation, are attended with as much difficulty and exception as those of any other system; and we cannot but agree with Mr. Roscoe, *Trans. of the Linn. Soc.* vol. xi. 65, that it forms several as unnatural assemblages as even the professedly artificial system of Linnæus. With regard to practical facility, no person of judgement has ever attempted to invalidate the superiority of the latter.

Having fulfilled the invidious task, which truth has required of us, let us turn to the more pleasing one of pointing out some of the great practical advantages of the labours of Jussieu. We do this with the more readiness, because we conceive that his real merits are better understood in England than any where else. The writer of this cannot disclaim the honour of being the first who announced to his countrymen the performance of his illustrious friend and correspondent, as one of the most learned books ever published. He humbly conceives that few persons, in any country, have studied the work more, or applied it so much to practice. If he has been fortunate in establishing genera, which have not been controverted, he allows his obligations to Jussieu, as much as to Linnæus. The treasures of neither lie on the surface, nor are they to be appreciated by a superficial observer. The foolish contentions of party can neither exalt nor invalidate the reputation of such men; nor is it the counting of stamens and pistils, nor the enunciation of the names of natural orders, implying ideas which do not always exist in the mind of the speaker, that can entitle a pedant or coxcomb to rank as the pupil of either.

We confess ourselves somewhat partial to the Linnæan notion, of conceiving the idea of a natural order in the mind, rather than to the Jussieuan attempt at very precise technical limitation of its characters. If we contemplate the generality of Jussieu's orders in this light, we shall be struck with his profound talents for combination, as well as discrimination; and as we peruse his critical remarks, subjoined to several of these orders, we shall profit more by his queries and difficulties, than by those definitions, at the head of each order, which are, too often, so clogged with exceptions, as to bewilder rather than instruct a student, however intelligible they may be to an adept.

The uninformed reader may, possibly, be surprised to see how great a conformity there is between most of the Natural Orders of Linnæus and those of Jussieu. This will appear by a cursory view of the latter, which, after the detail we have given of the former, will more elucidate the subject than any other explanation that our limits will allow. We shall take the orders of Jussieu in their regular series.

Class 1.

The first five orders, *Fungi, Algæ, Hepaticæ, Musci,* and *Filices,* are the same in both systems, except that Linnæus does not separate the *Hepaticæ* from *Algæ.*

6. *Naïades* are analogous to the *Inundatæ,* ord. 15, of Linnæus.

Class 2.

7. *Aroideæ* answer to the Linnæan *Piperitæ,* ord. 2, though *Piper* itself is removed far away, to the *Urticæ.*

8. *Typhæ* consist of *Typha* and *Sparganium,* two genera first referred by Linnæus to *Calamariæ,* then to *Piperitæ.*

9. *Cyperoideæ* are the Linnæan *Calamariæ,* ord. 3.

10. *Gramineæ* are the *Gramina,* ord. 4, grasses, an order about which there cannot be two opinions, nor do these authors differ, except in the denomination of the integuments of the flower; Jussieu calling the *calyx* a *gluma,* and the *corolla* a *calyx.* This alteration is made, chiefly that he might not allow a *corolla* to monocotyledonous plants.

Class 3.

11. *Palmæ,* palms, necessarily the same in both systems.

12. *Asparagi* answer to the bulk of the *Sarmentaceæ,* ord. 11.

13. *Junci* agree less exactly with *Tripetaloideæ,* ord. 5, both being liable to exceptions, and having undergone subsequent corrections by their respective authors.

14. *Lilia* consist of the latter portion of Linnæus's *Coronariæ,* ord. 10, with the beginning of his next order *Sarmentaceæ.*

15. *Bromeliæ* embrace some others of the *Coronariæ,* about which Linnæus had his doubts to the last, nor is Jussieu satisfied with this order.

16. *Asphodeli* are likewise chiefly *Coronariæ,* except *Allium.*

17. *Narcissi* are Linnæan *Spathaceæ,* ord. 9. We say

nothing of anomalous or doubtful genera, subjoined to this or any other order, and which are sometimes numerous, not unfrequently paradoxical. In the present instance they are *Hypoxis*, *Pontederia*, *Polianthes*, *Alstroemeria*, and *Tacca*, concerning which, the intelligent reader will readily concur with the learned author, that they are "*genera Narcissis non omninò affinia*."

18. *Irides*—Linnæan *Ensatæ*, ord. 6.

CLASS 4.

19. *Musæ*—consist of *Musa*, very mistakenly referred by Linnæus to his *Scitamineæ*; with *Heliconia* and *Ravenala*, Schreber's *Urania*, both nearly akin to *Musa*.

20. *Cannæ* are the *Scitamineæ* of Linnæus, ord. 8.

21. *Orchideæ* are his *Orchideæ*, ord. 7.

22. *Hydrocharides* are an assemblage of water plants, having little else in common. *Valisneria*, *Hydrocharis*, and *Stratiotes*, make a sort of appendix to the Linnæan *Palmæ*. For *Nymphæa* and *Nelumbium* (now called *Cyamus*), see our remarks on the 11th, 54th, and 27th, of the Linnæan orders. *Trapa*, *Proserpinaca* and *Pistia* close the list. Linnæus has the two last in his *Inundatæ*, ord. 15.

CLASS 5.

23. *Aristolochiæ* compose the end of the Linnæan *Sarmentaceæ*, but were afterwards removed to the *Rhoeadeæ*, ord. 27. They are surely best by themselves, and constitute a very natural order, not detected by Linnæus.

CLASS 6.

24. *Elæagni* consist of Linnæan *Calycifloræ*, ord. 16, with various genera besides, referred to almost as many different orders by Linnæus, so that here the two systems exhibit but little analogy, nor is this one of Jussieu's best orders.

25. *Thymelææ*, *Veprecula* of Linnæus, ord. 31, (the *Daphne* tribe,) are very clearly defined.

26. *Proteæ*, an order scarcely known to Linnæus, though an extremely natural one. It makes a part of his

Aggregatæ, ord. 48, in the establishing of which, a sort of artificial character, expressed in the name, has led him into unnatural combinations; a fault which Linnæus, more than any other writer in this department, has generally avoided.

27. *Lauri*, a very good order, not perceived by Linnæus. We cannot say much for the genera of *Myristica* and *Hernandia* annexed to it.

28. *Polygoneæ* make a part of the Linnæan *Holeraceæ*, ord. 12.

29. *Atriplices*, another portion of the same.

Class 7.

30. *Amaranthi*, these, originally a part of the *Miscellaneæ*, ord. 54, were also referred subsequently to the *Holeraceæ*. They are supposed to differ from Jussieu's two preceding orders, in having the stamens inserted into the receptacle, not into the calyx, hence forming a separate class. But there is no instance perhaps in which his system proves more artificial, and at the same time more uncertain in character. Mr. Brown has anticipated the latter part of our remark in his *Prodromus*, 413, nor could it fail to strike any one who ever considered the subject.

31. *Plantagines*.
32. *Nyctagines*.
33. *Plumbagines*.

{ Linnæus has no order analogous to these. Yet he has left manuscript indications of his perceiving the affinity of some of the genera.

Class 8.

34. *Lysimachiæ* embrace many of the *Rotaceæ*, ord. 20, and *Preciæ*, ord. 21. *Globularia*, *Tozzia*, *Samolus*, *Utricularia*, *Pinguicula*, and *Menyanthes*, subjoined as allies, not indeed without many doubts, appear to us greatly misplaced. The first of these is allowed to indicate an order not yet defined.

35. *Pediculares*, an important order, which Jussieu has well selected out of the Linnæan *Personatæ*, ord. 40;

though we are somewhat startled at finding *Polygala* at the head of the list, which Linnæus, not more happily perhaps, ranges with his *Lomentaceæ*, ord. 33.

36. *Acanthi* are a few more of the *Personatæ*.

37. *Jasmineæ* are precisely the Linnæan *Sepiariæ*, ord. 44.

38. *Vitices* consist of more *Personatæ*, separated with judgement from the rest. Linnæus having, in the contemplation of his 40th order, been again seduced by artificial principles, and by the usage perhaps of considering his *Didynamia Angiospermia* as of itself a natural order.

39. *Labiatæ* are precisely the *Verticillatæ*, ord. 42, of Linnæus, a tribe about which no two systematics could differ, and which it is one of the greatest evils of the artificial sexual system to be obliged to disjoin.

40. *Scrophulariæ* are more of the *Personatæ*, ranged here, after the *Labiatæ*, on account of the close affinity of several of them to the next order. But it must be confessed that the *Labiatæ* thus come awkwardly between what are strictly akin, and that this intrusion is a great flaw in the natural character of the system; insomuch that we should gladly remove them to another place, between the *Solaneæ* and *Borragineæ* hereafter mentioned.

41. *Solaneæ* consist principally of *Luridæ*, ord. 28, to which a few more of the *Personatæ* are subjoined as allies. It is remarkable that, in his characters of the seven last-mentioned orders, Jussieu admits those marks, derived from the stamens, on which the classes of the Linnæan artificial system depend. The intelligent reader will easily observe, that the distinctions thence deduced, form a leading principle in the respective positions of these orders and the following. This is the more curious, as the French school is entirely obliged to Linnæus for bringing the organs in question into notice, for the purposes of arrangement, Tournefort and his pupils having never adverted to them.

42. *Borragineæ*, these are the *Asperifoliæ*, ord. 41, of Linnæus, surely better placed by him between his *Perso-*

natæ and *Verticillatæ*. The order is very natural, and Jussieu's criticism upon it excellent.

43. *Convolvuli*.
44. *Polemonia*.
45. *Bignoniæ*.

To these Linnæus has no analogous order, most of the genera in the two first being referred to his *Campanaceæ*, order 29, and of the last to *Personatæ*.

In this instance we cannot but admit the superiority of Jussieu's arrangement.

46. *Gentianæ*—a very natural and distinct order, confounded by Linnæus with his *Rotaceæ*, ord. 20, to which it has but little relationship.

47. *Apocineæ*—precisely the Linnæan *Contortæ*, ord. 30, a most distinct and curious tribe, though both the great authors, of whom we are treating, have been mistaken in referring hither a genus or two, which do not at all belong to it. See our remarks on this 30th order of Linnæus.

48. *Sapotæ*—an order of which Linnæus had no perception. Some of its genera find a place among his *Dumosæ*, ord. 43, an assemblage which, he ingenuously confesses, did not satisfy himself.

Class 9.

49. *Guaiacanæ*. Of this also Linnæus had no distinct ideas. Some of the genera he places with his *Bicornes*, ord. 18. Yet some pupils of Jussieu have refined upon this and the last, and he himself has founded an order of *Ebenaceæ*, upon the first section of his *Guaiacanæ*;—see Brown's *Prodromus*, 524.

50. *Rhododendra*.

51. *Ericæ*.

These two collectively answer to the *Bicornes*, ord. 18, of Linnæus, an error or two, on either part, excepted.

52. *Campanulaceæ* nearly correspond with the genuine *Campanaceæ*, ord. 29, of Linnæus, from whence, as we have before hinted, *Convolvulus* and its allies are well separated in the system of Jussieu.

CLASS 10.

53. *Cichoraceæ,* a most natural order, the *Compositæ semiflosculosæ,* ord. 49. sect. 2. of Linnæus. The essential character of this 10th class is adopted from the artificial system of Linnæus, the united anthers, *antheræ connatæ*; a circumstance never adverted to by any systematic writer before him. Yet it is not absolutely without exception; witness the genera of *Kuhnia, Sigesbeckia,* and *Tussilago.*

54. *Cinarocephalæ* answer nearly, at least in principle, to the *Compositæ capitatæ,* ord. 49. sect. 1.

55. *Corymbiferæ* embrace all the remaining *Compositæ,* including the last section of that order, *nucamentaceæ,* some of which Jussieu terms *Corymbiferæ anomalæ*; such as *Iva, Parthenium, Ambrosia, Xanthium,* and even *Nephelium.*

CLASS 11. Distinguished from the last Class, only by having separate anthers.

56. *Dipsaceæ* consist of some of the Linnæan *Aggregatæ,* ord. 48. See our remark under Jussieu's 26th order. There is ample room for speculation on the affinities and distinctions between these *Dipsaceæ,* the *Proteæ,* ord. 26th, and the whole of Jussieu's 10th class last mentioned. Their contemplation involves questions at any time sufficient to excite a botanical war—such as, what belongs to the inflorescence, and what to the flower? what is a calyx, and what the crown of the seed? what is superior and what inferior insertion? what a simple and what a compound flower?

57. *Rubiaceæ,* a vast and important order, composed, not only of the Linnæan *Stellatæ,* ord. 47, but also of numerous tribes of shrubby plants, very few of which had been referred to the *Stellatæ,* and many of them had not fallen under the notice of Linnæus at all. Jussieu shines in the elucidation of this order, and has well indicated certain characters in the habit, especially that of the intrafoliaceous sheathing stipulas.

58. *Caprifolia* are nearly equivalent to the 4th, or last section of Linnæus's *Aggregatæ*, ord. 48, except *Viburnum* and its allies, with *Cornus* and *Hedera*; the former placed, without much reason, in the Linnæan *Dumosæ*; *Cornus* with the *Stellatæ*; and *Hedera* in *Hederaceæ*, ord. 46, nearly agreeing with Jussieu's 59th next mentioned. *Cornus* and *Hedera*, being both allowed to be polypetalous, really belong to the next class, as the author could not but perceive. Indeed Jussieu's 11th and 12th classes, however distinct in theory, naturally slide into each other.

CLASS 12.

59. *Araliæ* answer to the Linnæan *Hederaceæ*, ord. 46, *Hedera*, *Vitis* and *Cissus* excepted, which Linnæus himself appears to have had some idea of removing from *Panax*, *Aralia*, &c.

60. *Umbelliferæ* of course correspond with the *Umbellatæ*, ord. 45, of Linnæus, one of the most natural of the whole.

CLASS 13.

61. *Ranunculaceæ* answer to the Linnæan *Multisiliquæ*, ord. 26. The authors differ in the denomination of the parts of the flower, Jussieu's *calyx* being sometimes the *corolla*, and his *petals* the *nectaries*, of Linnæus.

62. *Papaveraceæ* are, except *Hypecoum* and *Fumaria*, Linnæan *Rhoeadeæ*, ord. 27.

63. *Cruciferæ* the Linnæan *Siliquosæ*, ord. 39, so natural an order, that we can scarcely say to which it is next akin.

64. *Capparides* mostly Linnæan *Putamineæ*, ord. 25, with some very anomalous genera subjoined as related thereto, *Reseda*, *Drosera* and *Parnassia*, not without great and well-founded doubts of the author.

65. *Sapindi*.
66. *Acera*.
67. *Malpighiæ*.
{ These are comprehended in two of the sections of the *Trihilatæ*, ord. 23. of Linnæus.

68. *Hyperica*. *Ascyrum* and *Hypericum*, the only real

genera of this order, are, with *Cistus*, subjoined to the Linnæan *Rotaceæ*, ord. 20, certainly with no very evident reason.

69. *Guttiferæ* constitute a well-marked order, to which Linnæus has nothing analogous. Most of the genera that compose it, are either left by him unarranged, or considered as of dubious affinity to any others. Indeed they are generally tropical trees, respecting which he had but slight information.

70. *Aurantia*. Of this likewise Linnæus seems to have formed no idea, since he refers *Citrus* to his *Bicornes*, and leaves *Limonia* undetermined. *Camellia* and *Thea*, subjoined by Jussieu, with some other genera, to this order, as connecting it with the next, appear to us of very dubious affinity to the *Aurantia*; nor are they much better annexed by Linnæus to his *Columniferæ*, ord. 37.

71. *Meliæ* constitute a good order, comprehended, not very judiciously, under the Linnæan *Trihilatæ*, ord. 23, above mentioned.

72. *Vites*, consisting only of *Cissus* and *Vitis*, we have already mentioned, ord. 59, as included amongst the *Hederaceæ*, ord. 46, of Linnæus.

73. *Gerania* make a part of the Linnæan *Gruinales*, ord. 14, but *Tropæolum*, a puzzling genus, which Jussieu labours to prove in many respects related to them, is referred by Linnæus, as reasonably perhaps, to his *Trihilatæ*.

74. *Malvaceæ* are almost exactly analogous to the *Columniferæ*, ord. 37.

75. *Magnoliæ* form an order certainly as little connected with the preceding as any two could be in the most artificial system. See the following.

76. *Anonæ*. The leading genera of this and the *Magnoliæ* compose the Linnæan *Coadunatæ*, ord. 52.

77. *Menisperma* are referred by Linnæus to his *Sarmentaceæ*, ord. 11, by their habit more than any just character.

78. *Berberides* constitute a curious order, though liable to some exceptions, of which its author was aware. It entirely escaped the penetration of Linnæus.

79. *Tiliaceæ* a good order, likewise overlooked by him, or partly confounded with his *Columniferæ*, to which it betrays some affinity.

80. *Cisti*. *Cistus* which makes this order, is placed by Linnæus, after *Hypericum*, at the end of his *Rotaceæ*, ord. 20. The reader may wonder to find *Viola* considered as related to *Cistus*, or at least to those species which Jussieu separates therefrom, by an incorrect character, and a faulty name, *Helianthemum*. He attributes to these a capsule of one cell; but one of them at least, *Cistus thymifolius*, has three cells. *Viola*, an anomalous genus, 29, with which it seems to have more points of agree-
is ranged by Linnæus at the end of his *Campanaceæ*, ord.
ment.

81. *Rutaceæ*. This is a very natural, and now become a very extensive order, of which the genuine idea is confined to Jussieu's second section, and likewise to the second section of Linnæus's *Multisiliquæ*, ord. 26. The plants which compose it have alternate leaves, without stipulas; their herbage abounding with aromatic acrid essential oil, lodged in pellucid cells, as in Jussieu's *Aurantia*, ord. 70. Calyx four- or five-cleft. Petals four or five, alternate therewith. Stamens usually twice as many as the petals, distinguished by something elaborate or peculiar in their structure, by which the genera are often well defined. Germen lobed. Capsule mostly of four or five cells, each lined with a bivalve elastic tunic, containing one or two polished seeds. *Diosma* and *Empleurum*, subjoined as akin to *Rutaceæ*, are genuine specimens of the order, though the latter has a capsule deprived of three or four of its lobes or cells, and wants petals. *Melianthus* has no business here. It ranks with the Linnæan *Corydales*, ord. 24, much more properly, though a very puzzling genus. The students at Paris,

in our time, used to amuse themselves with the idea, that the Professor would not allow this fine plant a place in the garden, because he knew not where to class it in his system.

82. *Caryophylleæ* are exactly analogous, except a few rather doubtful genera at the end, to the similarly named 22nd order of Linnæus. . But between this very natural tribe and the last, *Rutaceæ,* there is a *hiatus valdè deflendus,* as to any natural affinity; the present order being much more related, as Jussieu candidly indicates, to the *Amaranthi,* ord. 30, and proving that the presence or absence of a corolla, is no more infallible than any other character, for a general principle of arrangement.

Class 14.

83. *Sempervivæ* are the second section of Linnæus's *Succulentæ,* ord. 13.

84. *Saxifragæ* are chiefly the fourth section of the same.

85. *Cacti* consist merely of *Ribes* and *Cactus,* as artificial a combination as most in the sexual system itself. The former Linnæus ranks with his *Pomaceæ,* ord. 36; the latter is the first genus of his *Succulentæ.*

86. *Portulaceæ* are selected out of the first and third sections of the *Succulentæ.*

87. *Ficoideæ* consist of more of the same.

In this part of their respective systems, we find it more difficult than usual to follow the ideas of the learned authors. Habit seems to have guided Linnæus; but Jussieu tracing, in his last five orders, nearly the same affinities, hás somewhat strained his technical characters to confirm them.

88. *Onagræ* accord, in the main, with the Linnæan *Calycanthemæ,* ord. 17. They well connect the five preceding orders with the following. *Bæckea* belongs to the *Myrti.*

89. *Myrti* are the Linnæan *Hesperideæ,* a very natural family, much amplified by Jussieu from recent discoveries.

90. *Melastomæ* are not distinguished by Linnæus from his *Calycanthemæ*.

91. *Salicariæ* are in the same predicament. Jussieu has considerably the advantage here.

92. *Rosaceæ* embrace the *Senticosæ*, ord. 35, and *Pomaceæ*, ord. 36, of Linnæus, nor can there be a more natural assemblage.

93. *Leguminosæ* comprehend, in like manner, two Linnæan orders, *Papilionaceæ*, the 32nd, and *Lomentaceæ*, the 33rd, which we should be disposed to keep distinct, however nearly they must be considered as akin. The Linnæan characters, though often termed artificial, serve Jussieu for the distinctions of his sections.

94. *Terebintaceæ*, on order learnedly sketched out, rather than completed, by Jussieu, which seems entirely to have escaped the perception of Linnæus. It brings together many things which he either did not pretend to arrange, or which clogged some of his orders.

95. *Rhamni* constitute a very natural order, of which the Linnæan *Dumosæ*, ord. 43, are but a sketch, confessedly imperfect.

Class 15.

96. *Euphorbiæ* are Linnæan *Tricoccæ*, ord. 38.

97. *Cucurbitaceæ* agree, in name as well as idea, with the 34th of the Linnæan orders.

98. *Urticæ* are nearly analogous to *Scabridæ*, ord. 53, except that *Piper* is mentioned as related to them, instead of being referred to a monocotyledonous order with *Arum*, *Pothos*, *Acorus*, &c. Yet its germination is rather hinted at than determined, nor does any thing positive seem to be known on that subject.

99. *Amentaceæ* are mostly what Linnæus has, under the same appellation, in his 50th order.

100. *Coniferæ* are his 51st, bearing the same name.

As Linnæus enumerates, at the end of his Natural Orders, 116 genera, which he could not then satisfactorily refer to any one of them; so Jussieu, at the conclusion of his System, reckons up 137, which, as we have already observed, he denominates *Plantæ incertæ sedis.* These are disposed synoptically, by their petals, germens and styles. It is remarkable how nearly, allowing for new discoveries, Jussieu accords with Linnæus in the number of such genera. These lists have both been greatly diminished by subsequent consideration, or more complete information.

The attention of botanists, first directed by Gærtner, to the minute and curious diversities of structure in the parts of the seed, has greatly assisted Jussieu and his followers in correcting and improving the details of his system. Hence he has been led to favour the world with several essays on particular families, or orders, in the *Annales du Museum d'Hist. Nat.*, some of which have appeared in the very valuable *Annals of Botany*, published by Dr. Sims and Mr. König. In these, several of the difficulties, which originally embarrassed their author, are lessened or removed, but on these it is not our purpose to enter. A new edition of Jussieu's *Genera Plantarum*, which has long been preparing, cannot fail to prove almost a new work; more valuable perhaps for the abundant information which it must afford, concerning the characters and affinities of particular genera, than for any thing concerning a general natural system, to perfect which the scientific world has not, as yet, sufficient materials.

As we cannot here undertake to detail Jussieu's own corrections or improvements of his system, neither can we explain what has been attempted, with the same design, by the late ingenious M. Ventenat, or by those excellent living botanists, M. DeCandolle, or Mr. Brown. We shall only observe, that Ventenat, too servile to Jussieu, explicitly contends for the natural method of classification, as superseding the artificial one, and that he aims at proving this to have been the intention of Linnæus. Yet nothing can be more positive to the contrary than the remarks of the latter, in the preface to his *Ordines Naturales* at the end of his *Genera Plantarum.* He there declares that his "artificial method is alone of use to ascertain plants, it being scarcely possible to find a key to the natural one." "Natural orders," he continues, "serve to teach the nature of plants, artificial ones to distinguish one plant from another." If it be said that Jussieu, having invented a key, or a set of distinctive characters, to his orders, has removed this objection, we would ask, What becomes of his doubtful genera, as numerous as those of Linnæus? or moreover, How is any student, using his system analytically, to make out a single unknown plant? That the pupils of Jussieu have ever been aware of this, the writer of the present essay very well knows. He has always found them, in conversation, aiming compliments at their illustrious master, by contending for the great difficulty and uncertainty of the Linnæan artificial system; by which palpable absurdity they betrayed their secret opinion of Jussieu's. On the other hand, the intelligent and candid DeCandolle, adopting the just opinion of Linnæus, that plants are allied to each other rather in the form of a table, or map, than in a linear series, actually proposes such a series as *necessarily artificial*, in his *Theorie Elementaire de Botanique*, 213. Concerning the precise disposition of the genera in this series, we believe scarcely two botanists would agree; nor might their contentions be unprofitable; but they would never teach, either a tyro or an adept, to ascertain an unknown plant. We will venture to go further, and to declare our opinion, founded on long observation, that botanists who are thus perpetually intent on the abstract theory of classification, scarcely attain any excellence in the technical discrimination, or definition, of what are really founded in nature, the species or genera of the vegetable kingdom. Those err greatly who seek to improve the system of Jussieu, or any other, by refining too much on his distinctions, and subdividing his orders; than which nothing is more easy. Judgement and extensive knowledge are displayed in tracing out the most essential points of agreement in natural objects; not in exalting into unmerited importance the most minute differences. Hence the very conciseness of Linnæus gives perspicuity to his descriptions and definitions. These afford the most instructive study, whatever mode of classification we may think most convenient.

The French school has been much flattered by our able countryman Mr. Brown, having classed his *Prodromus* of the New Holland plants after the method of Jussieu; and many a botanist enjoys this national triumph who is certainly not competent to appreciate the merit of that work. The plants of so novel a country could not, at this time of day, have been presented, with so much advantage, to a philosophical botanist, as in some natural arrangement, however imperfect; nor will many students travel thither, to make them out by methodical investigation. The touchstone of our learned friend's book however will be the *Plantæ incertæ sedis,* nor can it be judged, as to the merit of the system employed, till it arrives at that conclusion. He himself will surely not reckon it complete without a Linnæan index,

"To give the precious metal sterling weight."

To the President of the Linnæan Society of London.

Sir, Surat, January 11, 1810.

ABOUT three years back the exigence of professional duty led me to a sequestered province of Malabar, where the *Cardamomum minus* was indigenous, and engaged a very large proportion of the industry of the natives, and was productive of much revenue. As the period of my visit coincided with the season of fructification, I availed myself of the occasion to attempt its botanical description; and having a good draughtsman, I caused him to delineate as scientifically as I could its various parts, and in the different stages of fructification. This drawing, with an account of the culture, as performed by the natives, was the subject of a tract,—however inadequate in the execution, interesting from the erroneous and discrepant descriptions which had hitherto been published either by botanists or others.

Of the merit of my labours I shall only predicate, but with a confidence founded on the fact, that the delineation of the parts of fructification is most accurate, and that I witnessed every thing I advanced relative to the crops, their culture, and collection. The result of my essay was intended by my much lamented friend Dr. James Anderson, of Madras, whose approbation it obtained, for presentation to the Asiatic Society of Calcutta; and this hint was conveyed to Lord William Bentinck, to whom the tract and drawing was sent; but the commotions which occupied and agitated the Governor's leisure during the latter months of his administration, prevented all future reference to a matter relatively so unimportant.

With this preamble for your information, I consign the essay and its merits to your better judgement and decision.

I am, with much respect, Sir,

Your obedient and humble servant,

David White, M.D.

Superintending Surgeon,
Province of Guzerat.

INDEX.

C.

D.

E.

F.

G.

H.

M.

N.

O.

P.

R.

T.

U.

V.

W.

ERRATA.

Vol. I. page 72, *for* Tinnevelley *read* Norwich.
Vol. II. page 382, last line, *for* This *read* His.

THE END.

RICHARD TAYLOR,
PRINTER TO THE UNIVERSITY OF LONDON,
RED LION COURT, FLEET STREET.

Printed in the United States
By Bookmasters